# THE ENVIRONMENT.

## A REVOLUTION
## IN ATTITUDES

ISSN 1532-270X

# THE ENVIRONMENT
## A REVOLUTION
## IN ATTITUDES

Kim Masters Evans

**INFORMATION PLUS® REFERENCE SERIES**
Formerly published by Information Plus, Wylie, Texas

THOMSON

GALE

Detroit • New York • San Francisco • San Diego • New Haven, Conn. • Waterville, Maine • London • Munich

# THOMSON

# GALE

The Environment: A Revolution in Attitudes

Kim Masters Evans

Paula Kepos, Series Editor

**Project Editor**
John McCoy

**Editorial**
Danielle Behr, Ellice Engdahl, Michael L. LaBlanc, Elizabeth Manar, Daniel Marowski, Charles B. Montney, Heather Price, Timothy J. Sisler

**Permissions**
William Sampson, Sue Rudolph, Edna Hedblad, Jacqueline Key, Lori Hines, Denise Buckley, Sheila Spencer

**Composition and Electronic Prepress**
Evi Seoud

**Manufacturing**
Keith Helmling

LIBRARY OF CONGRESS CATALOGING-IN-PUBLICATION DATA

ISBN 0-7876-5103-6 (set)
ISBN 0-7876-9082-1
ISSN 1532-270X

This title is also available as an e-book.
ISBN 0-7876-9269-7 (set)
Contact your Thomson Gale sales representative for ordering information.

Printed in the United States of America
10 9 8 7 6 5 4 3 2 1

# TABLE OF CONTENTS

Public opinion and knowledge about environmental issues have evolved over time and have led to federal and state laws protecting environmental resources. These laws have been debated on their economic impacts and on how well they protect citizens from environmental hazards and crimes. The international community is also concerned about the environment and has, with varying degrees of success, implemented environmental standards of protection.

Scientists generally agree that the Earth is getting warmer, and many scientists believe that the greenhouse effect accounts for this change. Gases such as carbon dioxide, methane, nitrous oxide, and chlorofluorocarbons are major causes of global warming, which could have severe effects on the Earth. Many nations have committed themselves to reducing greenhouse gas emissions.

Ozone can be either a health hazard or a health protectant, depending on where in the atmosphere it is located. Man-made chemicals, such as chlorofluorocarbons, can deplete the protective layer of ozone found in the upper atmosphere, which protects humans, plants, and animals from excess UV radiation exposure. The international community plays an important role in safeguarding environmental health by protecting the ozone layer.

The growth of industrialized nations and consumer societies has created a greater need for collection and disposal of refuse. Humans get rid of much waste through landfills, incinerators, recycling, and composting, while hazardous and nuclear waste require special consideration. Federal and state governments try to ensure that former disposal sites are safe for humans, animals, and plants.

Emissions of chemicals from automobiles, airplanes, factories, and power plants are mostly to blame for air pollution, which can make people sick and harm the environment. Government regulation, along with the implementation of new technologies, is designed to promote clean air, but many people believe these measures do not go far enough.

Water is an essential resource necessary for sustaining all forms of life, but human alteration of the environment and patterns of water usage can have a devastating affect on the water supply. Legislation, including the Clean Water Act, seeks to improve and protect the integrity of America's surface water, groundwater, oceans, coastal water, and drinking water. How effective such laws have been is a matter of debate.

Burning fossil fuels, either naturally caused or man-made, emit sulfur oxides and nitrogen oxides into the atmosphere, leading to acid rain. Acid rain can harm human health and damage ecosystems. Politicians and environmental groups in the United States and worldwide have sought reductions in the pollution that causes acid rain through legislation and other measures.

Many of the substances released into the environment by modern, industrialized society are harmful to humans and other living creatures. Chemicals in pesticides and fertilizers can be detrimental to the environment. Large concentrations of toxins like asbestos, chlorine, lead, radon, and others can pose dangers to human health and the environment. This chapter also examines the environmental dangers of tobacco, chemical and biological weapons, indoor air pollution, noise pollution, and food toxins.

Forests, wetlands, and mountains, as well as other ecosystems, are endangered by the activities of humans. The diversity of life on Earth (biodiversity) is also at risk, with many species facing extinction. Local, national, and international efforts must be linked to deal with pressures on the environment.

Solar energy, wind energy, hydropower, bioenergy, and geothermal power are all alternative sources of energy. Each type has its own advantages and disadvantages when compared with standard energy sources. Hydrogen is also seen by many as a potential fuel for the future.

# PREFACE

*The Environment: A Revolution in Attitudes* is part of the Information Plus Reference Series. The purpose of each volume of the series is to present the latest facts on a topic of pressing concern in modern American life. These topics include today's most controversial and most studied social issues: abortion, capital punishment, care for the elderly, crime, health care, the environment, immigration, minorities, social welfare, women, youth, and many more. Although written especially for the high school and undergraduate student, this series is an excellent resource for anyone in need of factual information on current affairs.

By presenting the facts, it is Thomson Gale's intention to provide its readers with everything they need to reach an informed opinion on current issues. To that end, there is a particular emphasis in this series on the presentation of scientific studies, surveys, and statistics. These data are generally presented in the form of tables, charts, and other graphics placed within the text of each book. Every graphic is directly referred to and carefully explained in the text. The source of each graphic is presented within the graphic itself. The data used in these graphics are drawn from the most reputable and reliable sources, in particular from the various branches of the U.S. government and from major independent polling organizations. Every effort has been made to secure the most recent information available. The reader should bear in mind that many major studies take years to conduct, and that additional years often pass before the data from these studies are made available to the public. Therefore, in many cases the most recent information available in 2004 dated from 2001 or 2002. Older statistics are sometimes presented as well, if they are of particular interest and no more recent information exists.

Although statistics are a major focus of the Information Plus Reference Series, they are by no means its only content. Each book also presents the widely held positions and important ideas that shape how the book's subject is discussed in the United States. These positions are explained in detail and, where possible, in the words of their proponents. Some of the other material to be found in these books includes: historical background; descriptions of major events related to the subject; relevant laws and court cases; and examples of how these issues play out in American life. Some books also feature primary documents, or have pro and con debate sections giving the words and opinions of prominent Americans on both sides of a controversial topic. All material is presented in an even-handed and unbiased manner; the reader will never be encouraged to accept one view of an issue over another.

## HOW TO USE THIS BOOK

The condition of the world's environment is an issue of great concern to both Americans and people worldwide. Since the late nineteenth century, mankind's ability to alter the natural world, both deliberately and unintentionally, has increased. Many people fear that without proper restraint, mankind's actions could forever alter, or even eliminate, life on Earth. There are those, however, who feel that these fears are exaggerated, and that the substantial cost of environmental protection on business and industry should be minimized. The conflict between these two positions has serious environmental, political, and economic ramifications, both within the United States and worldwide. This book examines the steps that have been taken to protect the Earth's natural environment, and the controversies that surround them.

*The Environment: A Revolution in Attitudes* consists of ten chapters and three appendices. Each of the chapters is devoted to a particular aspect of the environment. For a summary of the information covered in each chapter, please see the synopses provided in the Table of Contents at the front of the book. Chapters generally begin with an overview of the basic facts and background information on the chapter's topic, then proceed to examine sub-topics of particular interest. For example, Chapter 4: Waste Disposal begins with a brief discussion of laws governing waste. It then goes on to discuss industrial waste in the United States, including how much is generated, what it is composed of, and how it is dealt with. This is followed by a section on household waste that examines how much waste Americans generate as part of their daily lives, and why there is so much more of it than in the past. A brief history of household waste disposal is given, leading into a description of how household waste is handled in the United States today. Methods for reducing the amount of waste are also discussed. The chapter concludes with sections on Superfund and on nuclear waste. Throughout the chapter special consideration is given to the problem of hazardous waste and the various methods used to clean up contaminated areas. Readers can find their way through a chapter by looking for the section and sub-section headings, which are clearly set off from the text. Or, they can refer to the book's extensive index if they already know what they are looking for.

**Statistical Information**

The tables and figures featured throughout *The Environment: A Revolution in Attitudes* will be of particular use to the reader in learning about this issue. The tables and figures represent an extensive collection of the most recent and important statistics on the environment, as well as related issues—for example, graphics in the book cover the amounts of different kinds of pollutants found in air across the United States; the major sources of groundwater pollution; the amount of wetland that is destroyed each year; and public opinion on whether the state of the environment has become better or worse in the last five years. Thomson Gale believes that making this information available to the reader is the most important way in which we fulfill the goal of this book: to help readers understand the issues and controversies surrounding the environment and reach their own conclusions about them.

Each table or figure has a unique identifier appearing above it, for ease of identification and reference. Titles for the tables and figures explain their purpose. At the end of each table or figure, the original source of the data is provided.

In order to help readers understand these often complicated statistics, all tables and figures are explained in the text. References in the text direct the reader to the relevant statistics. Furthermore, the contents of all tables and figures are fully indexed. Please see the opening section of the index at the back of this volume for a description of how to find tables and figures within it.

**Appendices**

In addition to the main body text and images, *The Environment: A Revolution in Attitudes* has three appendices. The first is the Important Names and Addresses directory. Here the reader will find contact information for a number of government and private organizations that can provide further information on aspects of the environment. The second appendix is the Resources section, which can also assist the reader in conducting his or her own research. In this section, the author and editors of *The Environment: A Revolution in Attitudes* describe some of the sources that were most useful during the compilation of this book. The final appendix is the index. It has been greatly expanded from previous editions, and should make it even easier to find specific topics in this book.

**ADVISORY BOARD CONTRIBUTIONS**

The staff of Information Plus would like to extend their heartfelt appreciation to the Information Plus Advisory Board. This dedicated group of media professionals provides feedback on the series on an ongoing basis. Their comments allow the editorial staff who work on the project to continually make the series better and more user-friendly. Our top priorities are to produce the highest-quality and most useful books possible, and the Advisory Board's contributions to this process are invaluable.

The members of the Information Plus Advisory Board are:

- Kathleen R. Bonn, Librarian, Newbury Park High School, Newbury Park, California

- Madelyn Garner, Librarian, San Jacinto College—North Campus, Houston, Texas

- Anne Oxenrider, Media Specialist, Dundee High School, Dundee, Michigan

- Charles R. Rodgers, Director of Libraries, Pasco-Hernando Community College, Dade City, Florida

- James N. Zitzelsberger, Library Media Department Chairman, Oshkosh West High School, Oshkosh, Wisconsin

**COMMENTS AND SUGGESTIONS**

The editors of the Information Plus Reference Series welcome your feedback on *The Environment: A Revolution in Attitudes*. Please direct all correspondence to:

Editors
Information Plus Reference Series
27500 Drake Rd.
Farmington Hills, MI 48331-3535

# CHAPTER 1
# THE STATE OF THE ENVIRONMENT—AN OVERVIEW

*We have not inherited the Earth from our fathers. We are borrowing it from our children.*

— Native American saying

Photographs from outer space impress on the world that humankind shares one planet, and a small one at that. (See Figure 1.1.) Earth is one ecosystem. There may be differences in race, nationality, religion, and language, but everyone resides on the same orbiting planet.

General concern about the environment is a relatively recent phenomenon. Two closely related factors explain the rising concern during the second half of the twentieth century: global industrialization following World War II and the worldwide population explosion. It was once thought that the life-sustaining resources of planet Earth were infinite; now it is known that these resources are finite and limited.

## HISTORICAL ATTITUDES TOWARD THE ENVIRONMENT

### The Industrial Revolution

Humankind has always altered the environment around itself. For much of human history, however, these changes were fairly limited. The world was too vast and people too few to have more than a minor effect on the environment, especially as they had only primitive tools and technology to aid them. All of this began to change in the 1800s. First in Europe and then in America, powerful new machines, such as steam engines, were developed and put into use. These new technologies led to great increases in the amount and quality of goods that could be manufactured and the amount of food that could be harvested. As a result, the quality of life rose substantially and the population began to boom. The so-called Industrial Revolution was underway.

While the Industrial Revolution enabled people to live better in many ways, it also increased the rate of pollution. For many years pollution was thought to be an insignifi-

cant side effect of growth and progress. In fact, at one time people looked on the smokestacks belching black soot as a healthy sign of economic growth. The reality was that pollution, along with the increased demands for natural resources and living space that resulted from the Industrial Revolution, was beginning to have a significant effect on the environment.

### Twentieth-Century America

For much of the early twentieth century, Americans accepted pollution as an inevitable cost of economic progress. After World War II, however, more and more incidents involving pollution made people aware of the environmental problems caused by human activities. Los Angeles "smog," a smoky haze of pollution that formed like a fog in the city, contributed a new word to the English language. Swimming holes became so polluted they were poisonous. Still, little action was taken.

In the 1960s environmental awareness began to increase, partly in response to the 1962 publication of Rachel Carson's book *Silent Spring* (Boston: Houghton Mifflin), which exposed the toll of the chemical pesticide DDT on bird populations. Other signs of the drastic effects of pollution on the environment became harder and harder to ignore. For example, in 1969 the Cuyahoga River near Cleveland, Ohio, burst into flames due to pollutants in the water.

Environmental protection rapidly became very popular with the public, particularly with the younger generation. The *New York Times,* on November 30, 1969, carried a lengthy article reporting on the astonishing increase in environmental interest. The article's author, Gladwin Hill, noted that concern about the environmental crisis was especially strong on college campuses, where it was threatening to become even more of an issue than the Vietnam War.

FACTORS CONTRIBUTING TO ENVIRONMENTAL ACTIVISM. In 1969, according to Opinion Research of

## FIGURE 1.1

The Earth, as seen from a U.S. satellite. *(U.S. National Aeronautics and Space Administration.)*

Princeton, New Jersey, only 1 percent of Americans polled expressed concern for the environment. By 1971 fully one-quarter reported that protecting the environment was important. What motivated Americans to this new awareness? The following are among the likely factors:

• An affluent economy and increased leisure time

• The emergence of an "activist" upper middle class that was college educated, affluent, concerned, and youthful

• The rise of television, an increasingly aggressive press, and advocacy journalism (supporting specific causes)

• An advanced scientific community with increasing funding, new technology, and vast communication capabilities

**EARTH DAY AND THE BIRTH OF ENVIRONMENTAL PROTECTION.** The idea for Earth Day began to evolve beginning in the early 1960s. Nationwide "teach-ins" were being held on campuses across the country to protest the Vietnam War. Democratic senator Gaylord Nelson of Wisconsin, troubled by the apathy of American leaders toward the environment, announced that a grassroots demonstration on behalf of the environment would be held in the spring of 1970 and invited everyone to participate. On April 22, 1970, 20 million people participated in massive rallies on American campuses and in large cities. Earth Day would go on to develop into an annual event and was still being celebrated as of 2004.

With public opinion loudly expressed by the Earth Day demonstrations, in 1970 Congress and President Richard Nixon passed a series of unprecedented laws and created the Environmental Protection Agency (EPA), an organiza-

tion devoted to setting limits for water and air pollutants and to investigating the environmental impact of proposed, federally funded projects. In the years that followed, many more environmental laws were passed, setting basic rules for interaction with the environment. Most notable among these laws were the Clean Air Act of 1970 (PL 91-604), the Clean Water Act of 1972 (PL 92-500), the Endangered Species Act of 1973 (PL 93-205), the Safe Drinking Water Act of 1974 (PL 93-523), and the Resource Conservation and Recovery Act of 1976 (PL 95-510).

**THE STATUS OF ENVIRONMENTAL ISSUES IN THE UNITED STATES AT THE BEGINNING OF THE TWENTY-FIRST CENTURY.** Since the 1970s the state of the environment has continued to be a major political issue of interest to many Americans. Many activist organizations, such as Greenpeace and the Natural Resources Defense Council, have been created to watch over and protect the environment. Virtually every state has established one or more agencies charged with protecting the environment. Many universities and colleges offer programs in environmental education. Billions of dollars are spent every year by state and federal governments for environmental protection and enhancement. These efforts have, in turn, led to many improvements in the state of the environment. Many of the most dangerous chemicals that once polluted the air and water have either been banned or their emissions into the environment greatly reduced. Yet, other environmental problems have arisen or worsened since 1970, such as the possibility of global warming and the depletion of Earth's natural resources. International conferences addressing these issues have produced mixed results.

Some studies, such as Gallup polls, have suggested that concern about the environment declined in the 1990s. To explain this, many people point to the fact that obvious dangers, such as rivers on fire and belching smokestacks, have seen substantial improvement. Those dangers that remain, such as global warming and ozone depletion in the atmosphere, are largely invisible, and the public may not as easily accept or be concerned about their existence.

Another factor in the decline in environmental activism is money. Many of the cheapest and easiest environmental problems to fix were resolved in the 1970s, 1980s, and 1990s. Most of the remaining problems at the end of the century were so large or complicated that it was believed that tremendous amounts of money would have to be spent before even modest improvements would be seen. Many Americans, especially those who felt their jobs were threatened by environmental regulations, questioned whether these increased costs were worth the relatively small benefits they would provide. Despite the support of those who wanted to see further environmental improvements, such issues were competing for funds with other pressing issues such as Acquired Immune Deficiency Syndrome (AIDS), homelessness, and starvation in many parts of the world. In addition, since the terrorist attacks of September 11, 2001, environmental issues have been overshadowed by the threat of terrorism.

## THE ROLE OF POPULATION IN THE ENVIRONMENTAL EQUATION

Earth's population is believed to have grown more from 1950 through 2000 than it did during the previous 4 million years. For centuries, deaths largely offset births, resulting in a slow population growth. Beginning around 1950, high birth rates in developing countries, coupled with a reduction in mortality rates and reduced infant mortality (which led to an overall lengthening of the life span), dramatically impacted population growth. According to 2003 figures from the United Nations (UN), between 1950 and 1990 the global population more than doubled from 2.5 billion to more than 5.0 billion people. Another billion people are estimated to have been added between 1990 and 2000.

As shown in Figure 1.2 the vast majority of population growth is expected in less developed countries. This is significant to environmental issues because these countries are, or will be, undergoing their own industrial revolutions—a time when economic development or "progress" generally gets priority over environmental issues.

The growing number of people on Earth increases the demands on natural resources. More people require more food, fuel, clothing, and other necessities for life. All of these must be supplied from the planet's resources and from the Sun's energy. These facts—combined with the realization that the Earth's resources are limited, not infinite—pose serious questions about rapid population growth. Can the world's resources support its population while maintaining the environment, or will human needs overwhelm Earth's capacity to provide?

Most scientists believe that population growth will eventually cease. As a nation's economy develops and its standards of living rise, its population generally stabilizes. This "demographic shift" has already been observed in highly industrialized countries such as Japan and the nations of western Europe where the fertility rate is generally at or below replacement level. The UN projects that this slow decline in growth rates will continue until all regions of Earth have developed high standards of living, experienced a demographic shift, and had their populations stabilize. By the time this occurs, however, Earth's population is expected to have reached at least 9 billion people.

It is not known if Earth's resources can support a population of 9 billion. Some believe there is evidence that population growth is already pressing, or has exceeded, the capacity of natural resources in many areas. Environmentalists and others warn that without conservation, resource protection, and drastic measures to curb popula-

**FIGURE 1.2**

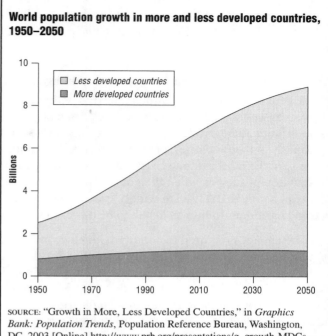

World population growth in more and less developed countries, 1950–2050

SOURCE: "Growth in More, Less Developed Countries," in *Graphics Bank: Population Trends*, Population Reference Bureau, Washington, DC, 2003 [Online] http://www.prb.org/presentations/g_growth-MDCs-LDCs.ppt [accessed March 29, 2004]

tion, humankind and planet Earth might suffer serious harm. Other analysts do not share these opinions, however. Opposing the idea that natural resources and humankind is in danger, some analysts point to the fact that, at many times throughout history, there have been those who claimed in vain that industrial development could not be sustained. For example, at one time it was feared that, as the world ran out of whale oil for lamps, great cities would be left in darkness. This scenario, and other "doomsday" predictions did not come about because new technologies averted the expected disasters.

## THE IMPACT OF ENVIRONMENTAL PROTECTION ON THE U.S. ECONOMY

### How Government Regulations Work

Since federal and state governments began actively protecting the environment in the 1970s, they have acted primarily by creating rules—called regulations—that say how Americans can affect the environment around them. In order to get people and organizations to comply with these regulations, the government fines, imprisons, or otherwise punishes those who violate them.

Most federal regulations are aimed at controlling the environmental practices of businesses and industries, as their behavior is much easier to monitor and control than that of individual citizens. For example, to reduce air pollution the government might regulate the lawn mower industry by not allowing it to make lawn mowers that release more than a certain amount of pollutants. This is much easier for the government than the alternative: checking how much pollution is released when an individual mows his or her lawn, and punishing that person if it is too much.

## Attitudes of Business toward Environmental Regulation

All environmental regulations interfere with the way businesses would otherwise operate. They force businesses to design their products differently, install special machinery in their factories, or even stop certain activities entirely. In the most extreme cases, entire industries might be shut down, such as when the government determined that the chemical DDT, once widely used as a pesticide, was too hazardous to human health to be used at all.

The changes required to comply with environmental regulations almost always result in smaller profits for those companies affected by such restrictions. This is especially true for industries that extract natural resources, such as mining and logging; industries that produce a great deal of pollution, such as electrical power generation; and industries whose products are potentially hazardous, such as the chemical industry. Compliance with governmental regulations is a very significant cost item for some industries. In November 2002 the U.S. Department of Commerce published the results from its latest survey of manufacturing companies on their expenses for environmental compliance. In *Pollution Abatement Costs and Expenditures: 1999* the agency reported that in 1999 manufacturing industries spent $5.8 billion on pollution abatement capital expenditures (such as equipment) and $11.9 billion on pollution abatement operating costs.

It is not surprising that business and industry leaders object to these increased costs. Some claim that environmental regulations will make their businesses unprofitable by driving up prices and production costs, which will in turn force them to close plants and lay off workers, if not shut down entirely. As early as the 1970s these sentiments were finding support among politicians and the general population, and an antiregulatory movement developed.

### DOES ENVIRONMENTAL PROTECTION DESTROY JOBS?
By the end of the twentieth century, the large-scale layoffs that some business-people predicted had not come to pass. Michael Renner, in *State of the World 2000* (New York: Worldwatch Institute and W. W. Norton Company, 2000), states that job loss as a result of environmental regulation has been relatively limited. According to Renner, at least as many people have gained jobs due to the restrictions as have lost them. He points out that environmental regulations have led to the creation of an entirely new industry that earns its profits by assisting other businesses with compliance, mostly by helping them to minimize pollution.

Even for those who accept this positive view of the overall effects of environmental regulation on business, it is unquestionable that some industries and their workers are badly hurt by environmental regulations. Renner contends that policy changes intended to protect the environment must have a clear and predetermined schedule. This way workers will know in advance what is expected of them, what jobs will be in demand, and what training is needed to get them into those positions. As long as environmental regulation results in shrinking profits and loss of jobs, however, attempts to expand such regulation will certainly be met with opposition from those whose livelihood would be affected.

## The Business of Environmental Protection

In order for the United States and other nations to meet their environmental goals, an environmental protection industry has emerged. Its major activities include pollution control, waste management, cleanup of contaminated sites, pollution prevention, and recycling.

Environmental Business International, Inc. (EBI) is a private organization that offers business and market information to the environmental industry. According to EBI the U.S. environmental market was largely driven by major legislation during the 1970s and 1980s. (See Figure 1.3.) During the 1990s overall economic growth and adaptation of ISO standards were important factors driving the market. ISO standards are voluntary standards developed by the International Organization for Standardization (ISO). They were adopted by many industries during the 1990s as a means of showing compliance with certain levels of environmental conduct. The environmental industry is in a transition phase. In the past this industry focused on remedial cleanup; in the future it will be focused more on prevention.

As shown in Figure 1.3 the environmental industry market underwent tremendous growth between 1970 and 2000. The U.S. market for pollution-control technologies and other environmental services topped $200 billion in 2000. Growth is expected to slow a little between 2000 and 2010. Table 1.1 details projected growth rates by sectors within the environmental industry. Wastewater treatment works and solid waste management are the only service sectors expected to increase their market shares. Remediation, hazardous waste management, and related consulting services are expected to continue to decline.

According to EBI, developed nations represent about 90 percent of the global pollution-control industry; the United States alone accounts for approximately 40 percent of the industry. Pollution-control markets are, however, growing rapidly in other countries. Expansion in Asian markets will greatly depend on the region's economic status. If the economy worsens, it may lead governments to give the environment less priority. Also, some govern-

FIGURE 1.3

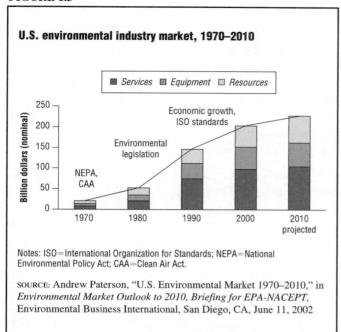

**U.S. environmental industry market, 1970–2010**

Notes: ISO=International Organization for Standards; NEPA=National Environmental Policy Act; CAA=Clean Air Act.

SOURCE: Andrew Paterson, "U.S. Environmental Market 1970–2010," in *Environmental Market Outlook to 2010, Briefing for EPA-NACEPT*, Environmental Business International, San Diego, CA, June 11, 2002

TABLE 1.1

**Growth expected in U.S. environmental industry sectors, 1970–2010**

| Environmental industry segment | 1970 $ | 70–80 growth | 80–90 growth | 90–00 growth | 00–10 growth |
|---|---|---|---|---|---|
| **Services** | | | | | |
| Analytical services | 0.1 | 300% | 314% | −26% | −33% |
| Wastewater treatment works | 4.3 | 116% | 116% | 34% | 36% |
| Solid waste management | 3.2 | 164% | 208% | 45% | 20% |
| Hazardous waste management | 0.1 | 550% | 921% | −15% | −49% |
| Remediation/industrial services | 0.1 | 550% | 1813% | 5% | −36% |
| Consulting & engineering | 0.3 | 367% | 761% | 21% | −21% |
| **Equipment** | | | | | |
| Water equipment and chemicals | 3.2 | 117% | 95% | 57% | 31% |
| Instruments & information systems | 0.1 | 100% | 820% | 84% | 1% |
| Air pollution control equipment | 1.0 | 196% | 258% | 30% | −44% |
| Waste management equipment | 2.0 | 105% | 159% | 20% | 24% |
| Process & prevention technology | 0.0 | 259% | 418% | 192% | 83% |
| **Resources** | | | | | |
| Water utilities | 5.7 | 109% | 67% | 53% | 25% |
| Resource recovery (recycling) | 1.2 | 283% | 197% | 29% | 28% |
| Environmental energy sources | 0.3 | 420% | 15% | 87% | 82% |
| **U.S. totals:** | **$21.4** | **145%** | **178%** | **35%** | **12%** |

SOURCE: Andrew Paterson, "Several Service Sectors Declining; Water, Energy Growing," in *Environmental Market Outlook to 2010, Briefing for EPA-NACEPT*, Environmental Business International, San Diego, CA, June 11, 2002

ments, including many in Eastern Europe and Russia, have a legacy of abusive environmental practices that increases their environmental needs but a lack of funding that limits an otherwise huge market for cleanup and new technology.

## THE ANTIREGULATORY MOVEMENT

Since the mid-1980s dissatisfaction with government regulation has grown. In 1994 the newly elected Republican-controlled Congress attempted to strike down a wide variety of federal regulations, including environmental regulations that they considered overly burdensome. Bills were introduced to relax regulations established under the Clean Water Act, Endangered Species Act, the Superfund toxic-waste cleanup program, the Safe Drinking Water Act, and other environmental statutes. Much of that legislation ultimately failed to pass. However, congressional budget cuts for the agencies responsible for carrying out these acts meant that many of the laws were not strongly enforced.

Several factors contributed to this reaction to federal regulation. During the early days of the environmental era, the United States was experiencing a post-World War II economic boom, leading Americans to regard regulatory costs as sustainable. During the 1970s and 1980s, as economic growth slowed, wages stagnated and Americans became uncertain about the future. An increasing number of Americans started to question the costs of environmental protection. They began to pay more attention to the business leaders and politicians who had claimed since the beginning of the environmental protection movement that regulations would hurt the economy and cost people jobs. Although the national economy improved tremendously in the 1990s, it did not eliminate people's concerns about the potential negative effects of environmental regulations on the economy.

The impact of environmental regulations on private property use has also played a very important role in the change in attitudes. The Endangered Species Act and the wetlands provisions of the Clean Water Act spurred a grassroots "private property rights" movement. Many people became concerned that these acts, as well as other legislation, would allow the government to "take" or devalue properties without compensation. For example, if federal regulations prohibited construction on a plot of land that was protected by law, then the owner of that land often felt that the government was unfairly limiting the use of his or her property. At the very least, the owner wanted government compensation for decreasing the monetary value of the land.

Finally, there are those who feel that environmental regulation by the government, while not necessarily bad, has gone too far. Many people believe that the federal government has overstepped its authority and should allow state and local governments to make their own rules on environmental issues. Similarly, some people feel that existing regulations are too strict and should be relaxed in order to generate economic growth.

## LITIGATION AND ENVIRONMENTAL POLICY

The courts have been an important forum for developing environmental policy because they allow citizens to challenge complex environmental laws and to affect the decision-making process. Both supporters and opponents of environmental protection have successfully used the courts to change environmental policy and law. Successful challenges can force the legislature to change laws or even have the law suspended as unconstitutional. A lawsuit can also be filed to seek compensation for harm to a person, property, or an economic interest. Sometimes, lawsuits have prompted the creation of entirely new laws such as the federal Superfund Law (1980) and the Toxic Substances Control Act (1976). Even the threat of a lawsuit, given the bad publicity it can bring, is sometimes enough to get a business or the government to change its behavior.

There are many different situations under which an individual or organization can go to court over environmental laws and regulations. One common occurrence is for an individual or group to sue the government in order to block a law or regulation from going into effect. For example, when the government halted logging in northwestern forests because of threats to endangered owls, logging companies fought to halt enforcement of those protections because that would decrease the industry's income and cause the loss of jobs.

Some lawsuits are filed not to block an environmental law or regulation from going into effect but because the claimants feel that the government owes them compensation for the negative effects of the law. In 1986 David Lucas bought two residential lots on a South Carolina barrier island. He planned to build houses on these lots, just as had been done on other nearby lots. At the time he bought the land this was entirely legal, but in 1988 South Carolina passed the Beachfront Management Act. Designed to protect the state's beaches from erosion, it prohibited new construction on land in danger of eroding, which included the land Lucas owned.

Lucas went to court claiming that the Beachfront Management Act had violated the Fifth Amendment of the U.S. Constitution by preventing him from building on his property. The Fifth Amendment states, among other things, that no person's "private property shall be taken for public use, without just compensation." Lucas argued that preventing him from building on his property was equivalent to taking it, so the government of South Carolina had to compensate him for it. On June 29, 1992, the U.S. Supreme Court, in *Lucas v. South Carolina Coastal Council* (505 US 1003), agreed with Lucas in a 7–2 decision, and South Carolina was ordered to compensate him.

Supporters of environmental protection also have filed lawsuits. These situations generally occur when people feel the government is not properly enforcing the law. Environmental groups like the Sierra Club and Greenpeace have sued the government on many occasions to compel it to officially recognize certain species as endangered.

## ENVIRONMENTAL JUSTICE—AN EVOLVING ISSUE

The so-called environmental justice issue stems from concerns that racial minorities are disproportionately subject to environmental hazards. The EPA defines environmental justice as "the fair treatment and meaningful involvement of all people regardless of race, color, national origin, or income with respect to the development, implementation, and enforcement of environmental laws, regulations, and policies."

The environmental justice movement gained national attention in 1982 with a demonstration against the construction of a hazardous waste landfill in Warren County, North Carolina, a county with a predominantly African-American population. A resulting 1983 congressional study found that, for three of four landfills surveyed, African-Americans made up the majority of the population living nearby and at least 26 percent of the population in those communities was below the poverty level. In 1987 the United Church of Christ published a nationwide study, *Toxic Waste and Race in the United States,* reporting that race was the most significant factor among the variables tested in determining locations of hazardous waste facilities.

A 1990 EPA report (*Environmental Equity: Reducing Risk for All Communities*) concluded that racial minorities and low-income people bore a disproportionate burden of environmental risk. These groups were exposed to lead, air pollutants, hazardous waste facilities, contaminated fish, and agricultural pesticides in far greater frequencies than the general population.

In 1992 the EPA established the Office of Environmental Justice to address environmental impacts affecting minority and low-income communities. In 1994 President Bill Clinton issued Executive Order 12898 (*Federal Actions to Address Environmental Justice in Minority Populations and Low-Income Populations*), requiring federal agencies to develop a comprehensive strategy for including environmental justice in their decision making.

Examples of environmental injustice include the following claims:

- Low-income Americans, especially minorities, may be more likely than other groups to live near landfills, incinerators, and hazardous waste facilities.

- Low-income and African-American children often have higher than normal levels of lead in their blood.

- Greater proportions of Hispanic Americans and African-Americans than whites live in communities that fail to meet air quality standards.

- Higher percentages of hired farm workers in the United States are minorities. It has been estimated that more than 300,000 farm workers may suffer pesticide-related illnesses each year.

- Low-income and minority fishermen who use fish as their sole source of protein are generally not well informed about the risk of eating contaminated fish from certain lakes, rivers, and streams.

In August 2001 EPA director Christine Todd Whitman reiterated the agency's commitment to environmental justice, defining it as "the fair treatment of people of all races, cultures, and incomes with respect to the development, implementation, and enforcement of environmental laws and policies, and their meaningful involvement in the decision-making processes of the government."

## In Harm's Way

In 1997 residents of Kennedy Heights, a Houston, Texas, neighborhood of about 1,400 people, complained about a variety of illnesses such as cancer, tumors, lupus (an autoimmune disease), and rashes. They eventually discovered that their homes, built 30 years before, were sitting on top of a number of oil pits that had been abandoned in the 1920s. Some tests of the municipal water found traces of crude oil, and residents believed oil sludge from the pits had seeped into the water supply and damaged their health. To make matters worse, their homes lost virtually all their resale value once it became known that the land was contaminated.

The predominantly African-American homeowners had not been told that their property sat over an abandoned oil dump. They blamed Chevron Oil Company, which owned the property for a time after acquiring it from Gulf Oil. Chevron denied contamination could have caused any illness. Accusations of environmental racism mounted, attracting ever-wider attention. Reverend Jesse Jackson visited the site in 1997 to call for a boycott of Chevron products. In March 1999 Chevron agreed to pay $8 million into a fund for distribution to Kennedy Heights claimants.

In 1998 some of the residents of Chester, Pennsylvania, sued the Pennsylvania Department of Environmental Protection. They claimed in their suit (*James M. Seif v. Chester Residents Concerned Citizens et al.*) that the department had violated Title IV of the Civil Rights Act of 1964 (PL 88-352) by issuing a solid waste permit for a facility in a minority community in Chester called Chester Heights. The residents claimed the waste permit violated their civil rights because, while only two permits had been issued in adjoining non-minority communities, five such permits had been approved for their community. Although the case was dismissed when the treatment facility withdrew its application, the suit demonstrated the possibility of claiming environmental discrimination under Title IV. It should be noted, however, that few environmental racism cases have been successful because it is difficult to prove that racial discrimination is the cause of particular environmental policies.

In March 2004 a federal lawsuit against the chemical company Monsanto was settled for $300 million. The suit was spearheaded by a grassroots environmental group called Citizens Against Pollution out of Anniston, Alabama. It was brought on behalf of 18,477 residents living in poor, mostly African-American neighborhoods in the city's west end. The suit alleged that since the 1960s a Monsanto plant had discharged large amounts of polychlorinated biphenyls (PCBs) into a creek running through the area. PCBs are mixtures of synthetic organic chemicals that are now known to be extremely persistent in the environment and toxic to life. Lawyers had evidence linking PCB exposure to a variety of serious illnesses and even deaths suffered by members of the community over decades.

Without admitting fault, Monsanto agreed to pay $300 million to settle the case. According to Ellen Barry in the *Los Angeles Times* ("A Neighborhood of Poisoned Dreams," April 13, 2004), the plaintiffs, who were originally thrilled with the settlement, were shocked and dismayed when they learned the lawyers would receive $120 million of the money. This left $180 million to be split amongst thousands of plaintiffs, resulting in an average payout of only $7,725 per person. A case brought by a different set of plaintiffs in state court resulted in a settlement of $300 million to be split among 2,500 of them. An additional $75 million was earmarked toward cleanup efforts and $25 million was set aside to build a neighborhood health clinic. In total the lawsuits resulted in a settlement of nearly $700 million, the largest payout ever in a tort case (a civil action resulting from a wrongful act) involving toxic chemicals.

## "NEW" CRIME—ENVIRONMENTAL CRIME

Table 1.2 lists the major environmental and wildlife protection acts of the federal government. As recently as the 1980s very few Americans understood that harming the environment could be considered a crime. Since that time, however, a substantial portion of the American public has begun to recognize the seriousness of environmental offenses, believing that damaging the environment is a serious crime and that corporate officials should be held responsible for offenses committed by their firms. Even though the immediate consequences of an offense may not be obvious or severe, environmental crime is a serious problem and does have victims—the cumulative costs in damage to the environment and the toll to humans in illness, injury, and death can be considerable.

Law enforcement agencies generally believe that successful criminal prosecution—even the threat of it—is the best deterrent to environmental crime. Under the dual sov-

**TABLE 1.2**

**Major federal environmental and wildlife protection acts**

**Environmental protection acts**

Clean Air Act (CAA)—Prevent the deterioration of air quality

Clean Water Act (CWA)—Regulate sources of water pollution

Comprehensive Environmental Response, Compensation, and Liability (CERCLA or Superfund)—Address problems of abandoned hazardous waste sites

Emergency Planning & Community Right-To-Know Act (EPCRA)—Help local communities protect public health, safety, and the environment from chemical hazards

Federal Insecticide, Fungicide and Rodenticide Act (FIFRA)—Control pesticide distribution, sale, and use

National Environmental Policy Act (NEPA)—The basic national charter for protection of the environment. It establishes policy, sets goals, and provides means for carrying out the policy

Oil Pollution Act of 1990 (OPA)—Prevent and respond to catastrophic oil spills

Pollution Prevention Act (PPA)—Reduce the amount of pollution produced via recycling, source reduction, and sustainable agriculture

Resource Conservation and Recovery Act (RCRA)—Protect human health and the environment from dangers associated with waste management

Safe Drinking Water Act (SDWA)—Protect the quality of drinking water

Toxic Substances Control Act (TSCA)—Test, regulate, and screen all chemicals produced in or imported into the U.S.

**Wildlife protection acts**

Bald and Golden Eagle Protection Act (BGEPA)—Provide a program for the conservation of bald and golden eagles

Endangered Species Act (ESA)—Conserve the various species of fish, wildlife, and plants facing extinction

Lacey Act—Control the trade of exotic fish, wildlife, and plants

Migratory Bird Treaty Act (MBTA)—Protect migratory birds during their nesting season

SOURCE: Adapted from "Major Environmental Laws," in *Guide to Environmental Issues*, U.S. Environmental Protection Agency, Office of Enforcement and Compliance Assurance, Washington, DC, June 26, 1998 [Online] http://www.epa.gov/epahome/laws.htm [accessed March 31, 2004]

ereignty doctrine, both state and federal governments can independently prosecute environmental crimes without violating the double jeopardy or due process clauses of the U.S. Constitution.

As attitudes toward environmental crimes have changed, the penalties for such offenses have become harsher. Federal criminal enforcement has grown from a misdemeanor penalty for dumping contaminants into waterways without a permit, to a felony for clandestine (secret) dumping. Several federal laws now include criminal penalties. Companies, their officials, and staff can be prosecuted for knowingly violating any one of a number of crimes. Such crimes include transporting hazardous waste to an unlicensed facility, storing and disposing of hazardous waste without a permit, failing to notify of a hazardous substance release, falsifying documents, dumping into a wetland, or violating air quality standards.

Figure 1.4 shows the status of EPA criminal enforcement activities for fiscal years 1999 through 2003. Nearly $400 million in fines were assessed over this period and more than 900 years' worth of sentences were handed out. Figure 1.5 shows both criminal and civil penalties assessed each year from 1998 to 2003. Together, criminal and civil penalties totaled $167 million in 2003.

Since the 1970s, environmental laws have become more complicated. The increasing strictness of those laws may have contributed to the growing incidence of environmental violations. First, many businesses have found compliance increasingly expensive, and many are simply avoiding the costs even if it means violating the law. These companies consider the penalties just another "cost of doing business." Second, businesses and their legal counsel are becoming increasingly savvy in avoiding prosecution through the use of dummy corporations, intermediaries, and procedural techniques.

Smuggling and black-market sales of banned hazardous substances also resulted from environmental legislation. In 1997 federal officials reported that the sale of contraband Freon (a refrigerant used in air conditioning systems) had become more profitable than cocaine at that time. Freon is one of the chlorofluorocarbons (CFCs), a class of chemicals known to cause ozone depletion in the atmosphere. In Mexico, which shares a 2,000-mile border with the United States, Freon is still legal to manufacture and export, but those activities are banned in the United States. However, Freon continues to exist in the cooling systems of many older model cars in the United States, which means a demand also exists.

Another type of environmental crime gaining attention, called "ecoterrorism," occurs when radical environmental groups use economic sabotage to stop what they see as threats to the environment. The Earth Liberation Front is blamed for more than 600 attacks and nearly $43 million in property damage since the late 1990s. The group is accused of setting fire to a genetics laboratory, uprooting experimental crops, and damaging buildings. The Federal Bureau of Investigation (FBI) has named the group one of the most dangerous domestic terrorist groups in America.

## EXPOSING POLLUTERS TO PUBLIC PRESSURE

In 1997 the Clinton administration and the EPA initiated the Sector Facility Indexing Project (SFIP), an effort

FIGURE 1.4

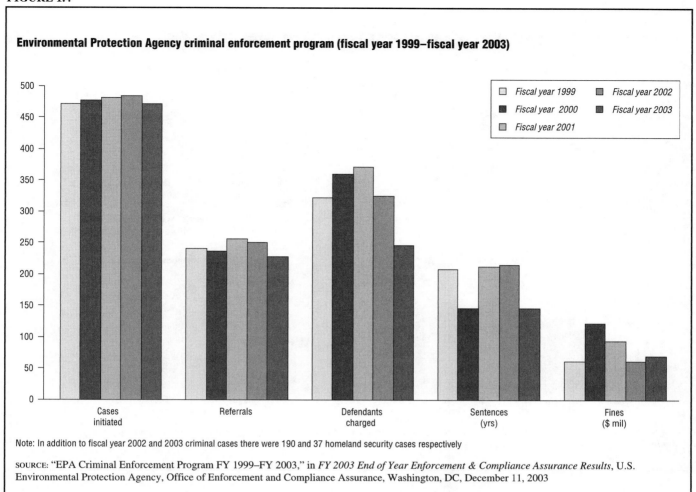

**Environmental Protection Agency criminal enforcement program (fiscal year 1999–fiscal year 2003)**

Note: In addition to fiscal year 2002 and 2003 criminal cases there were 190 and 37 homeland security cases respectively

SOURCE: "EPA Criminal Enforcement Program FY 1999–FY 2003," in *FY 2003 End of Year Enforcement & Compliance Assurance Results*, U.S. Environmental Protection Agency, Office of Enforcement and Compliance Assurance, Washington, DC, December 11, 2003

to keep the public informed about industries that pollute the environment. The latest in an ambitious campaign to expand "right-to-know" initiatives, the SFIP publishes "pollution profiles" online (http://www.epa.gov/sfipmtn1/) on five industrial sectors—oil production, steel, other metals, autos, and paper—in hopes of exposing polluters to possible pressure from an informed public. The profiles include the number of federal inspections, episodes of noncompliance, penalties imposed in recent years, releases of pollutants, pollution spills, injuries and deaths, toxicity of chemicals released, ratios of pollution to production, and demographics (racial and income breakdowns) of the population within a 3-mile radius of a plant. As of 2004, the SFIP had profiled approximately 864 individual facilities.

## THE INTERNATIONAL RESPONSE TO ENVIRONMENTAL PROBLEMS

Environmental issues have never been neatly bound by national borders. Activities taking place in one country often have an impact on the environment of other coun-tries, if not that of the entire Earth. In fact, many of the most important aspects of environmental protection involve areas that are not located within any particular country, such as the oceans, or belong to no one, such as the atmosphere. In an attempt to deal with these issues, the international community has held a number of conferences and concluded numerous treaties.

### A First Step—The Stockholm Conference

In 1972 the United Nations (UN) met in Stockholm, Sweden, for a conference on the environment. Delegates from 113 countries met, with each reporting the state of his or her nation's environment—forests, water, farm-land, and other natural resources. The countries repre-sented essentially fell into two groups. The industrialized countries were primarily concerned about how to protect the environment by preventing pollution and overpopulation and conserving natural resources. The less developed nations were more concerned about problems of widespread hunger, disease, and poverty they all faced. They did consider the environment very important, however, and were willing to protect it as

FIGURE 1.5

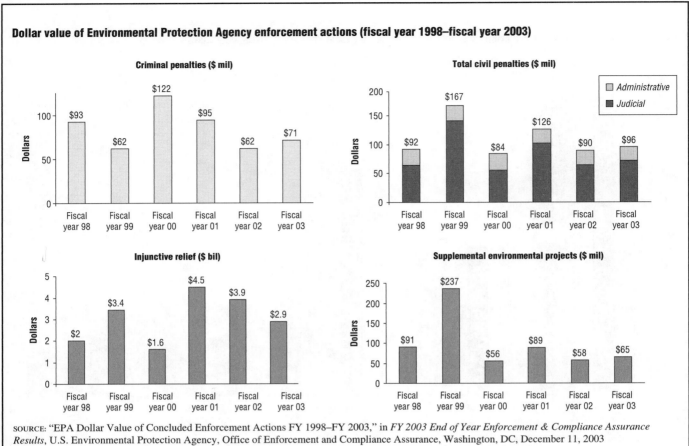

**Dollar value of Environmental Protection Agency enforcement actions (fiscal year 1998–fiscal year 2003)**

SOURCE: "EPA Dollar Value of Concluded Enforcement Actions FY 1998–FY 2003," in *FY 2003 End of Year Enforcement & Compliance Assurance Results*, U.S. Environmental Protection Agency, Office of Enforcement and Compliance Assurance, Washington, DC, December 11, 2003

long as doing so did not have a major negative economic impact on their citizens.

By the end of the two-week meeting, the delegates had agreed that the human environment had to be protected, even as industrialization proceeded in the less developed countries. They established the UN Environment Program (UNEP), which included Earthwatch, a program to monitor changes in the physical and biological resources of the Earth. The most important outcome of the conference was awareness of Earth's ecology as a whole. For the first time in global history, the environmental problems of both rich and poor nations were put in perspective. General agreement emerged to protect natural resources, encourage family planning and population control, and protect against the negative effects of industrialization.

## Some Difficulties Facing International Environmental Protection

Since the 1972 conference, hundreds of environmental treaties have been signed. From this, one might assume that great progress has been made, but this is not truly the case. Most experts believe that international cooperation is not keeping pace with the world's ever-growing interdependence and the rapidly deteriorating condition of much of the environment. Carbon dioxide ($CO_2$) levels are at record highs, water shortages exist around the world, fisheries are becoming depleted, and many scientists are warning that large numbers of species are becoming extinct. The reason for this is that, while nations agree on the fact that the environment must be safeguarded, they disagree sharply on the issue of what role each nation should play in protecting it.

Less developed nations are generally unwilling to alter their laws and economy to end environmentally destructive ways because a shift to environmentally friendly practices would be too expensive, they claim, for their economies to handle. Yet, the richer, industrialized, nations generally refuse to alter their own behavior unless the less developed nations do so as well. Their reason is not so much the cost of change rather than believing it unfair that the less developed nations want them to carry most of the burden of environmental protection.

The less developed nations respond by pointing out that the industrialized nations became rich by using the very same practices they now want the less developed nations to stop using. They claim it is unfair to be expected to limit their economic development in ways that the industrialized nations themselves never would have done.

This is a difficult disagreement but not an impossible one to resolve. When both sides are willing to compromise, agreements can be reached. These compromises usually require the industrialized nations to make bigger changes in their behavior, and to help the less developed nations change without too negative an impact on their economies.

Even agreements like these face many obstacles. Environmental agreements seldom include a means of enforcement but rely instead on each signing country to keep its word. Faced with the actual, immediate costs of implementing environmental agreements, many countries eventually back down from their commitments.

**1992 EARTH SUMMIT.** The 1992 Earth Summit in Rio de Janeiro, Brazil, is an example of a conference where compromises were made and agreements reached but little change actually resulted. Mounting global concern for the environment prompted the UN to convene the summit meeting. Approximately 180 governments participated, making it one of the largest and most important environmental summits ever. As with prior environmental summits, the conference was split between industrialized and developing nations.

The main accomplishments of the Earth Summit were pacts on global warming and biodiversity. President George H. W. Bush signed the global warming treaty for the United States. President Clinton signed the biodiversity treaty in 1994. These agreements came about largely because the industrialized nations also agreed to commit 0.7 percent of their gross national products by 2000 to assist developing countries with compliance.

Problems arose soon after the summit ended. Participating countries submitted annual reports to the 53-nation UN Commission on Sustainable Development (CSD), a standing body set up to implement the Rio agreements. The CSD concluded in 1994 that most countries were failing to provide the money and expertise necessary to implement the plans set at Rio. Chairman of the CSD, Klaus Toepfer of Germany, reported that the world's efforts to finance the goals had fallen "significantly short of expectations and requirements."

By 1996 a number of national governments, including the United States, had prepared plans for environmental protection and submitted them to the CSD. Hundreds of municipalities had also written plans of action. The CSD once again found, however, that other issues had crowded out environmental concerns. As developed and less developed nations alike worried about the potential effects of implementing the Rio agreements, they found reasons to delay implementation and reduce funding for those programs that had been implemented.

**WORLD TRADE ORGANIZATION.** The World Trade Organization (WTO) is an international organization whose purpose is to encourage free trade between its members. Most of the world's nations are members. Although the WTO was officially founded in 1995, it is the result of decades of international cooperation under the General Agreement on Tariffs and Trade (GATT). The WTO continues to administer the free trade system established under GATT.

One of the primary missions of the WTO is to eliminate barriers to free trade. Doing so can have a negative effect on environmental protection, however, because laws designed to protect the environment often have the effect of restricting trade. If the WTO finds that a member nation is restricting trade in violation of GATT, other members are permitted to raise their tariffs (import taxes) on goods from that nation until the barriers to trade are eliminated. Most of the time nations quickly change their laws to eliminate barriers to trade, rather than suffer high taxes.

Since it has forced many environmental laws to be weakened over the years, the WTO is greatly disliked by many environmentalists in the United States. Also, there are groups that think the organization's power over internal U.S. affairs is too great. When the WTO met in Seattle, Washington, in 1999, tens of thousands of activists, including environmental activists, protested in the city. This massive protest succeeded in overshadowing the WTO meeting itself and drew public attention to the problems, environmental and otherwise, with free trade organizations. This was due in no small part to the violent rioting and property damage caused by some protesters.

Americans are not the only ones who take issue with some of the WTO's actions regarding the environment. For example, Europeans opposed to genetically modifying food, a procedure in widespread use in the United States by 2002, wanted restrictions placed on the sale of U.S. food in Europe, but such restrictions would violate GATT and invite retaliation by the United States.

**NORTH AMERICAN FREE TRADE AGREEMENT.** The North American Free Trade Agreement (NAFTA), signed in 1994, is another major free trade agreement with the potential to negatively impact environmental protection in the United States. Members of NAFTA include the United States, Canada, and Mexico, and the purpose of the agreement is to eliminate trade barriers—such as most tariffs, investment restrictions, and import quotas—between these three countries. While its scope is much smaller than the WTO, NAFTA has an even greater impact on the three member countries than GATT.

A significant difference between NAFTA and GATT is that NAFTA is the first treaty of its kind ever to be accompanied by an environmental protection agreement. To discourage countries from weakening environmental standards in the name of increasing foreign trade, the United States, Canada, and Mexico signed the North American Agreement on Environmental Cooperation (NAAEC). Under NAAEC, a member country can be challenged if it or one of its states fails to enforce its environmental laws. A challenge can be brought by one of the member nations, or any interested party (such as an environmental protection group) can petition the NAAEC commission. If the commission finds a member country is showing a "persistent pattern of failure ... to enforce its environmental law," that country may be fined. If the fines are not paid, the other members are permitted to suspend NAFTA benefits in an amount not exceeding the amount of the assessed fine.

Even with NAAEC, some U.S. environmentalists and state officials feared that NAFTA could result in the weakening of numerous humane laws and the reversal of 30 years of advances in animal protection and environmental cleanup. In response, Congress provided more protection for state laws and included more environmental language than in any previous trade agreement. The implementing legislation for NAFTA in the United States allows states much input and requires that they receive notification of actions that may affect them. In addition, during NAFTA discussions, the Border Environment Cooperation Commission and the North American Development Bank were created. Independent of NAFTA itself, these agencies are intended to ensure that policy discussions are open and fairly enforced, to consider allegations that a country is not enforcing environmental laws, to help communities finance environmental infrastructures, and to resolve disputes, particularly those that cross borders.

Despite all these measures designed to make sure that NAFTA does not trample on environmental protection, environmentalists still see the need for concern. They point out that U.S. laws designed to protect certain animals could be challenged as barriers to free trade under NAFTA. They also point to the increased pollution in Mexico and along its border with the United States that has resulted from the increase in trade between these two countries.

## PUBLIC OPINION ON THE ENVIRONMENT

### Quality of the Environment

In March 2004 the Gallup Organization conducted its annual poll dealing with environmental issues. As shown in Figure 1.6, participants were asked to rate the overall quality of the U.S. environment as excellent, good, only fair, or poor. Only 4 percent of those asked gave the environment an excellent rating. Another 39 percent rated the environment in good condition, while 46 percent considered it in fair condition and 11 percent rated it in poor condition. In general the breakdown is similar to that obtained in previous polls. There has been a slight, gradual increase since 2001 in the number of people rating the environment in poor condition.

Another poll question asked if people believed the quality of the environment as a whole was getting better, getting worse, or staying the same. In 2004 a majority of respondents (58 percent) expressed the pessimistic view that the environment is getting worse. Another 34 percent believed that the environment is improving and 6 percent thought it is about the same. This breakdown has remained relatively constant since the question was first asked in 2001.

Figure 1.7 shows the opinions expressed in previous years from Gallup polls about the future of the environment and what actions should be taken to prevent major disruptions in it. In 2003 a majority of those surveyed (approximately 56 percent) felt that some additional actions need to be taken. Nearly a quarter (approximately 23 percent) believed that immediate drastic action was required. Another 20 percent felt that the same actions the U.S. has been taking will be sufficient to protect the Earth from environmental problems. The percentage of people calling for immediate and drastic action has decreased since 1995 when 35 percent of those asked felt it was necessary.

### Specific Environmental Concerns

Gallup also asked poll participants to rate various environmental issues in regard to the amount of concern they feel about them: a great deal, a fair amount, a little, or none. The ten issues assessed were as follows:

- Acid rain
- Air pollution
- Contamination of soil and water by toxic waste
- Damage to the Earth's ozone layer
- Extinction of plant and animal species
- Greenhouse effect (global warming)
- Loss of tropical rain forests
- Maintenance of the nation's supply of fresh water for household needs
- Pollution of drinking water
- Pollution of rivers, lakes, and reservoirs

Figure 1.8 shows the ranking of issues for which people said they felt a great deal of concern. Pollution of drinking water topped the list with 53 percent, followed by pollution of rivers, lakes, and reservoirs with 48 percent, and contamination of soil and water by toxic waste,

FIGURE 1.6

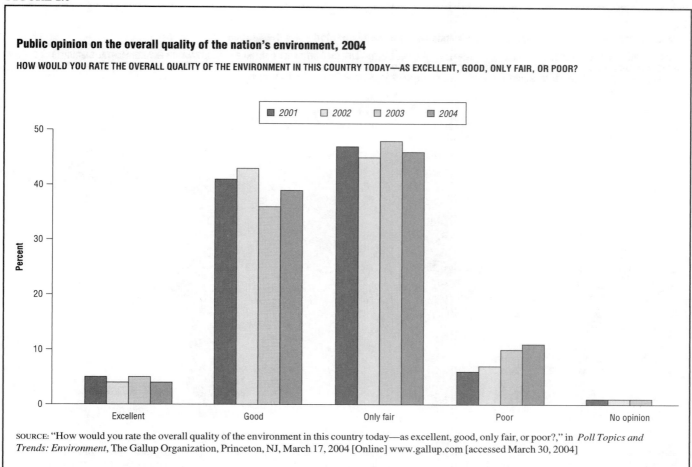

**Public opinion on the overall quality of the nation's environment, 2004**

HOW WOULD YOU RATE THE OVERALL QUALITY OF THE ENVIRONMENT IN THIS COUNTRY TODAY—AS EXCELLENT, GOOD, ONLY FAIR, OR POOR?

SOURCE: "How would you rate the overall quality of the environment in this country today—as excellent, good, only fair, or poor?," in *Poll Topics and Trends: Environment*, The Gallup Organization, Princeton, NJ, March 17, 2004 [Online] www.gallup.com [accessed March 30, 2004]

also with 48 percent. In general, water-related issues garnered the most amount of concern.

### Rating the Government's Role

In general, Gallup poll results indicate dissatisfaction with the government's role in protecting the environment. (See Figure 1.9.) In 2004 a majority of those asked (55 percent) felt that the government was doing too little in this regard. More than a third (37 percent) thought the government was doing about the right amount, and 5 percent believed the government was doing too much. The percentage of people who believe the government is doing too little to protect the environment has decreased since 1992 when 68 percent expressed this view.

In a separate poll conducted during 2003, Gallup asked participants to rank various federal government agencies based on their performance. As shown in Figure 1.10 the Environmental Protection Agency (EPA) was rated dead last behind seven other agencies. Only 39 percent of the people polled felt that the EPA was doing an excellent or good job. This compares with 66 percent who believed that the Centers for Disease Control and Prevention was doing an excellent or good job.

### Personal Participation

Despite an apparent dissatisfaction with government actions regarding the environment there is no indication that Americans feel more compelled to become actively involved in environmental issues. Figure 1.11 shows that in 2004 only 14 percent of the people polled considered themselves active participants in the environmental movement. Far more (47 percent) were sympathetic to the movement, but not active. Another 30 percent expressed neutral feelings about it and 8 percent were unsympathetic.

Overall the percentage of people considering themselves either active or unsympathetic has changed little since 2000. The biggest loss over this period was in the number of people expressing sympathy for the movement, but who are not active participants. This figure fell by 8 percent between 2000 and 2004. The greatest gain is seen in the number of people expressing neutral feelings about the environmental movement. This figure increased by 7 percent, perhaps reflecting an overall sense of apathy among Americans about many sociopolitical affairs.

This assessment appears to be supported by the March 2003 poll results shown in Figure 1.12. A majority of people asked had not personally been active in environ-

FIGURE 1.7

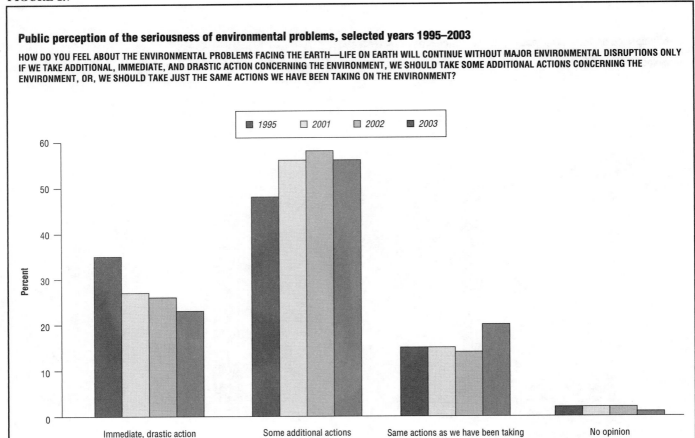

**Public perception of the seriousness of environmental problems, selected years 1995–2003**

HOW DO YOU FEEL ABOUT THE ENVIRONMENTAL PROBLEMS FACING THE EARTH—LIFE ON EARTH WILL CONTINUE WITHOUT MAJOR ENVIRONMENTAL DISRUPTIONS ONLY IF WE TAKE ADDITIONAL, IMMEDIATE, AND DRASTIC ACTION CONCERNING THE ENVIRONMENT, WE SHOULD TAKE SOME ADDITIONAL ACTIONS CONCERNING THE ENVIRONMENT, OR, WE SHOULD TAKE JUST THE SAME ACTIONS WE HAVE BEEN TAKING ON THE ENVIRONMENT?

SOURCE: "All in all, which of the following best describes how you feel about the environmental problems facing the earth–life on earth will continue without major environmental disruptions only if we take additional, immediate, and drastic action concerning the environment, we should take some additional actions concerning the environment, or, we should take just the same actions we have been taking on the environment?," in *Poll Topics and Trends: Environment,* The Gallup Organization, Princeton, NJ, March 17, 2004 [Online] www.gallup.com [accessed March 30, 2004]

mental protection groups or in political activities related to environmental issues during the previous year. More than 40 percent of those asked had contributed money to groups involved in environmental protection, conservation, or wildlife preservation. Far more people (72 percent) had based purchasing decisions on environmental concerns or reduced their household's use of energy (80 percent). The largest percentage of people (89 percent) indicated that they had voluntarily recycled household items within the previous year.

During that same poll participants were also asked whether they had personally changed their shopping or living habits over the previous year to help protect the environment. (See Figure 1.13.) Nearly two-thirds (61 percent) claimed to have made minor changes in their habits. Nearly a quarter (23 percent) said they had made major changes, while 15 percent said they had made no changes.

**Competing Interests: Environment, Energy, and Economy**

For many years the Gallup Organization has polled people about which should take priority: the environment or economic growth. The results are shown in Figure 1.14. The vast majority of polls conducted between 1984 and 2000 showed strong support for the environment even at the risk of curbing economy growth. In all of these polls at least 58 percent of the people asked agreed with this view. The tide began to turn during the early 2000s as economic growth began to gain in priority.

The March 2004 poll showed that 49 percent of those asked believed that environmental protection should be given priority even if it risked curbing economic growth. Nearly as many (44 percent) felt that economic growth should be give priority even if it meant that the environment would suffer to some extent. According to Gallup, a small percentage (4 percent) advocated giving equal priority to environmental protection and economic growth.

According to Wirthlin Worldwide, a Virginia-based marketing research and consulting company, there is a relationship between unemployment rates and public support for the environment. In its report *Environmental Support Softens amid Economic Uncertainty* (McLean, VA,

FIGURE 1.8

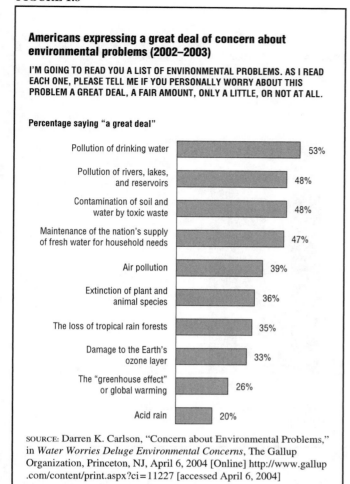

**Americans expressing a great deal of concern about environmental problems (2002–2003)**

I'M GOING TO READ YOU A LIST OF ENVIRONMENTAL PROBLEMS. AS I READ EACH ONE, PLEASE TELL ME IF YOU PERSONALLY WORRY ABOUT THIS PROBLEM A GREAT DEAL, A FAIR AMOUNT, ONLY A LITTLE, OR NOT AT ALL.

Percentage saying "a great deal"

Pollution of drinking water — 53%
Pollution of rivers, lakes, and reservoirs — 48%
Contamination of soil and water by toxic waste — 48%
Maintenance of the nation's supply of fresh water for household needs — 47%
Air pollution — 39%
Extinction of plant and animal species — 36%
The loss of tropical rain forests — 35%
Damage to the Earth's ozone layer — 33%
The "greenhouse effect" or global warming — 26%
Acid rain — 20%

SOURCE: Darren K. Carlson, "Concern about Environmental Problems," in *Water Worries Deluge Environmental Concerns*, The Gallup Organization, Princeton, NJ, April 6, 2004 [Online] http://www.gallup.com/content/print.aspx?ci=11227 [accessed April 6, 2004]

1998), Wirthlin noted that environmental support in the United States generally moves in accord with the economy. The report compared unemployment rates with poll data and found a statistically significant inverse relationship between unemployment rates and support for environmental protection. The report concluded that when the economy is healthy, public concern for the environment runs high; when the economy is bad, the environment becomes a much less important issue in people's minds.

A similar effect is evident in Gallup poll results concerning the competing priorities of environmental protection and developing U.S. energy supplies (such as oil, gas, and coal). As shown in Figure 1.15 support for giving the environment the priority fell by 4 percent between 2001 and 2004 from 52 percent to 48 percent. Support for giving priority to the development of energy supplies grew by 8 percent over the same time period, from 36 percent in 2001 to 44 percent in 2004.

**Young Americans Rank the Country's Problems**

In polls conducted during 2002 and 2003 the Gallup Organization asked young adults aged 18 to 29 to rank a list

of problems facing the United States in terms of the amount of worry they felt about each problem. As shown in Figure 1.16 the quality of the environment ranked fifth on the list of problems for which young adults expressed a great deal of concern. The environment was outranked by terrorism, healthcare, crime and violence, and the economy.

**How Reliable Are Polls on Environmental Issues?**

Some experts suggest that opinion polls are an unreliable guide to how voters actually feel about environmental issues. Although polls of Americans indicate that concern for environmental issues is substantial, this same level of concern does not manifest itself when it comes to actual voting and purchasing decisions. Some observers suggest that people often claim in polls that they are interested in environmental issues because they are trying to give the pollster the answer that he or she wants to hear. In other words, they are giving what they think is the "right" answer. In actuality, respondents may be more interested in other issues and, in the voting booth, may vote other than their poll answers would indicate.

**ENVIRONMENTAL EDUCATION**

**Lacking Basic Knowledge**

In May 2001 the National Environmental Education and Training Foundation, a private nonprofit organization, published the results of its latest annual study of American adults' knowledge of environmental issues. The study, conducted by Roper Starch Worldwide in 2000, found that Americans supported environmental protection but were generally ignorant on the leading causes of pollution, energy sources, and the hazards in some household products.

The study found that only 32 percent of respondents received a passing grade when given 12 multiple choice questions on current environmental topics. Comparison of the answers with those given in 1997 indicated that Americans seemed more knowledgeable about the source of water pollution, but were less knowledgeable about the largest source of carbon monoxide and the function of ozone. However, their attitudes about environmental education were almost unanimous in that 95 percent supported environmental education being taught in grades K–12. Environmental education for adults was also strongly supported with 86 percent of respondents agreeing that government agencies should support programs that educate adults about the environment.

**TEACHING ABOUT THE ENVIRONMENT IN SCHOOLS**

Many states require schools to incorporate environmental concepts, such as ecology, conservation, and environmental law, into many subjects at all grade levels. Some even require special training in environmentalism for teachers. Since 1992 the EPA, with congressional

FIGURE 1.9

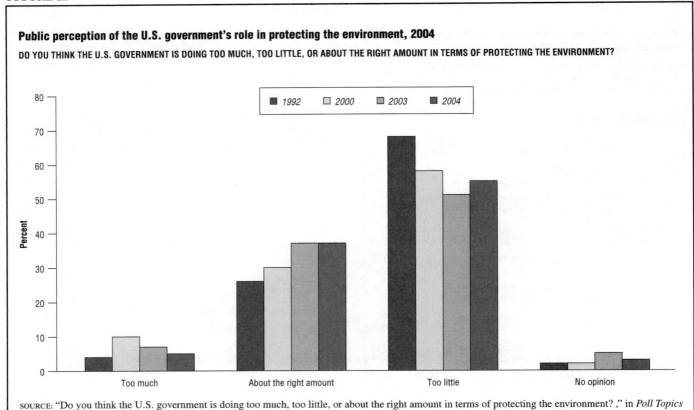

**Public perception of the U.S. government's role in protecting the environment, 2004**

DO YOU THINK THE U.S. GOVERNMENT IS DOING TOO MUCH, TOO LITTLE, OR ABOUT THE RIGHT AMOUNT IN TERMS OF PROTECTING THE ENVIRONMENT?

SOURCE: "Do you think the U.S. government is doing too much, too little, or about the right amount in terms of protecting the environment? ," in *Poll Topics and Trends: Environment*, The Gallup Organization, Princeton, NJ, March 17, 2004 [Online] www.gallup.com [accessed March 30, 2004]

authority to spend a total of $13 million on environmental education, gave grants to about 1,200 such projects.

Although this mandating of environmental education pleases environmentalists, and studies have shown that most Americans support environmental education, some people have protested. Critics claim that most environmental education in the schools is based on flawed information, biased presentations, and questionable objectives. Critics also say it leads to brainwashing and pushing a regulatory mind-set on students. Some critics contend that, at worst, impressionable children are being trained to believe that the environment is in immediate danger of catastrophe because of consumption, economic growth, and free market capitalism.

## NATIONAL AERONAUTICS AND SPACE ADMINISTRATION'S MISSIONS TO PLANET EARTH

In the past three decades humans have changed the way they think about the planet Earth. Missions to other planets have found those planets interesting and diverse but also sterile. Earth is unique; as far as is known, it is the only planet capable of sustaining life.

While environmental change is certain, many things remain unknown. Among those things is how humans

alter Earth. One way to study the effects of humans on Earth as a whole is from space. Since 1991 the National Aeronautics and Space Administration (NASA) has undertaken a comprehensive program to study the planet as an environmental system. The program was originally called Mission to Planet Earth, but is now called Earth Science Enterprise. By using satellites and other tools focused on Earth, scientists hope to expand what is known about how natural processes affect humans and how humans may, in turn, impact the planet.

Earth Science Enterprise has three components: a series of Earth-observing satellites, an advanced data system, and teams of scientists and researchers who study the data collected. Key areas of study include clouds, water and energy cycles, oceans, atmospheric chemistry, land surface, water and ecosystem processes, glaciers and polar ice sheets, and the solid earth.

Phase I of the program comprised satellites, space shuttle missions, and various land- and airborne-based studies. Phase II began April 15, 1999 (delayed from 1998 because of technical problems), with the launch of *Landsat 7* from Vandenberg Air Force Base in California. The first Earth Observing System (EOS) satellite, *Landsat 7* has the goal of taking an estimated 250 photographs of Earth each day. These will be available not only for

**FIGURE 1.10**

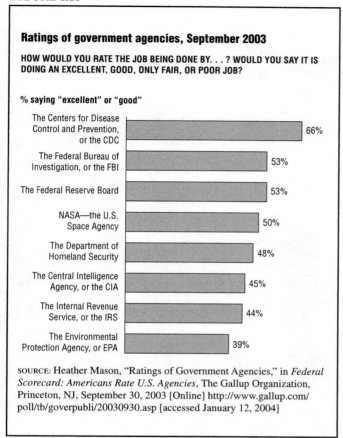

**Ratings of government agencies, September 2003**

HOW WOULD YOU RATE THE JOB BEING DONE BY. . . ? WOULD YOU SAY IT IS DOING AN EXCELLENT, GOOD, ONLY FAIR, OR POOR JOB?

% saying "excellent" or "good"

| Agency | % |
|---|---|
| The Centers for Disease Control and Prevention, or the CDC | 66% |
| The Federal Bureau of Investigation, or the FBI | 53% |
| The Federal Reserve Board | 53% |
| NASA—the U.S. Space Agency | 50% |
| The Department of Homeland Security | 48% |
| The Central Intelligence Agency, or the CIA | 45% |
| The Internal Revenue Service, or the IRS | 44% |
| The Environmental Protection Agency, or EPA | 39% |

SOURCE: Heather Mason, "Ratings of Government Agencies," in *Federal Scorecard: Americans Rate U.S. Agencies*, The Gallup Organization, Princeton, NJ, September 30, 2003 [Online] http://www.gallup.com/poll/tb/goverpubli/20030930.asp [accessed January 12, 2004]

national security purposes but also to private-sector, commercial, and academic segments of society to help answer pressing environmental questions in all the areas of study. The latest satellite launch occurred on June 24, 2002, when the *NOAA-M* satellite was launched to collect meteorological data and to monitor environmental events around the world.

FIGURE 1.11

**Public level of involvement in the environmental movement, 2004**

THINKING SPECIFICALLY ABOUT THE ENVIRONMENTAL MOVEMENT, DO YOU THINK OF YOURSELF AS—AN ACTIVE PARTICIPANT IN THE ENVIRONMENTAL MOVEMENT, SYMPATHETIC TOWARDS THE MOVEMENT, BUT NOT ACTIVE, NEUTRAL, OR UNSYMPATHETIC TOWARDS THE ENVIRONMENTAL MOVEMENT?

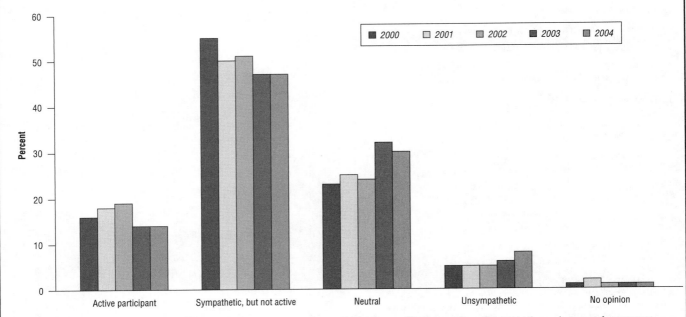

SOURCE: "Thinking specifically about the environmental movement, do you think of yourself as an active participant in the environmental movement, sympathetic towards the movement, but not active, neutral, or unsympathetic towards the environmental movement?," in *Poll Topics and Trends: Environment*, The Gallup Organization, Princeton, NJ, March 17, 2004 [Online] www.gallup.com [accessed March 30, 2004]

**FIGURE 1.12**

**Public participation within the past year in various activities, March 2003**

WHICH OF THESE, IF ANY, HAVE YOU, YOURSELF, DONE IN THE PAST YEAR?

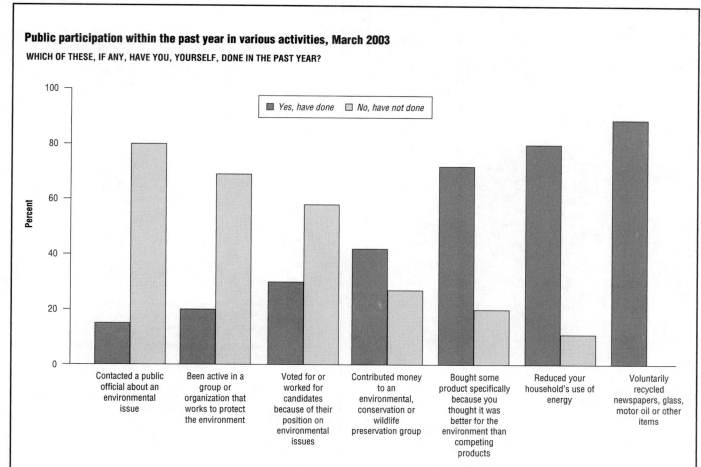

SOURCE: "Which of these, if any, have you, yourself, done in the past year?," in *Poll Topics and Trends: Environment*, The Gallup Organization, Princeton, NJ, March 17, 2004 [Online] www.gallup.com [accessed March 30, 2004]

FIGURE 1.13

**Public participation in activities that affect the environment**

THINKING ABOUT YOUR OWN SHOPPING AND LIVING HABITS OVER THE LAST FIVE YEARS, WOULD YOU SAY YOU HAVE MADE MAJOR CHANGES, MINOR CHANGES OR NO CHANGES TO HELP PROTECT THE ENVIRONMENT?

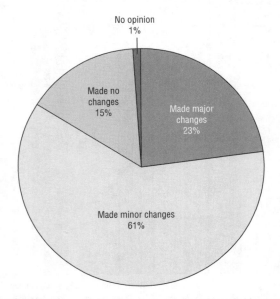

SOURCE: "Thinking about your own shopping and living habits over the last five years, would you say you have made major changes, minor changes or no changes to help protect the environment?," in *Poll Topics and Trends: Environment*, The Gallup Organization, Princeton, NJ, March 17, 2004 [Online] www.gallup.com [accessed March 30, 2004]

FIGURE 1.14

**Public opinion on which should take priority—the environment or the economy, 1984–2004**

WITH WHICH ONE OF THESE STATEMENTS ABOUT THE ENVIRONMENT AND THE ECONOMY DO YOU MOST AGREE—PROTECTION OF THE ENVIRONMENT SHOULD BE GIVEN PRIORITY, EVEN AT THE RISK OF CURBING ECONOMIC GROWTH (OR) ECONOMIC GROWTH SHOULD BE GIVEN PRIORITY, EVEN IF THE ENVIRONMENT SUFFERS TO SOME EXTENT?

Numbers shown in percentages based on yearly averages from 1984–2004

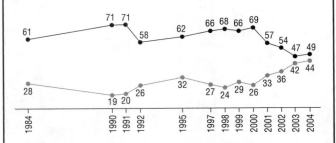

SOURCE: "With which one of these statements about the environment and the economy do you most agree—protection of the environment should be given priority even at the risk of curbing economic growth (or) economic growth should be given priority, even if the environment suffers to some extent?," in *Poll Topics and Trends: Environment*, The Gallup Organization, Princeton, NJ, March 17, 2004 [Online] www .gallup.com [accessed March 30, 2004]

**FIGURE 1.15**

**Public opinion on which should take priority—the environment or the economy, 2004**

WHICH SHOULD BE GIVEN PRIORITY?

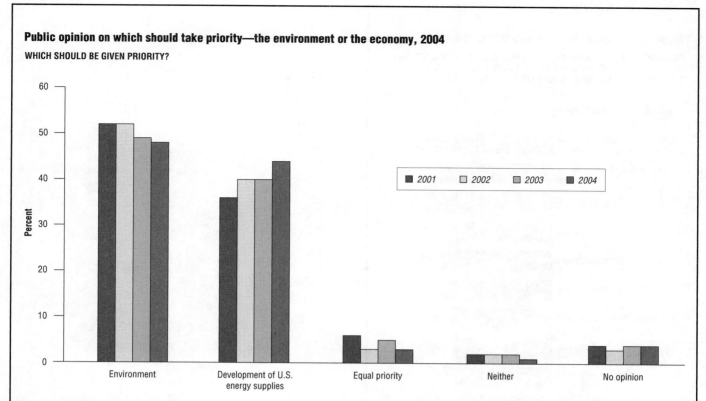

SOURCE: "With which one of these statements about the environment and the economy do you most agree – [ROTATED: protection of the environment should be given priority, even at the risk of curbing economic growth (or) economic growth should be given priority, even if the environment suffers to some extent]?," in *Poll Topics and Trends: Environment*, The Gallup Organization, Princeton, NJ, March 17, 2004 [Online] www.gallup.com [accessed March 30, 2004]

FIGURE 1.16

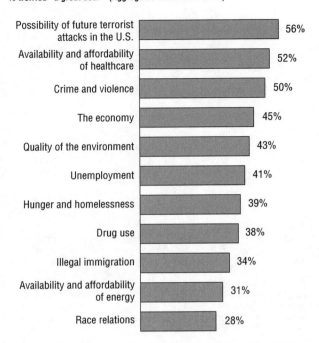

**Worries of young adults based on 2002–2003 data**

I'M GOING TO READ A LIST OF PROBLEMS FACING THE COUNTRY. FOR EACH ONE, PLEASE TELL ME IF YOU PERSONALLY WORRY ABOUT THIS PROBLEM A GREAT DEAL, A FAIR AMOUNT, ONLY A LITTLE, OR NOT AT ALL.

% worried "a great deal"  (Aggregated data. 2002–2003)

| | |
|---|---|
| Possibility of future terrorist attacks in the U.S. | 56% |
| Availability and affordability of healthcare | 52% |
| Crime and violence | 50% |
| The economy | 45% |
| Quality of the environment | 43% |
| Unemployment | 41% |
| Hunger and homelessness | 39% |
| Drug use | 38% |
| Illegal immigration | 34% |
| Availability and affordability of energy | 31% |
| Race relations | 28% |

SOURCE: Linda Lyons, "Worries of Young Adults, Aged 18 to 29," in *Angst Aplenty: Top Worries of Young Americans*, The Gallup Organization, Princeton, NJ, July 15, 2003 [Online] http://www.gallup .com/content/print.aspx?ci=8845 [accessed April 6, 2004]

**22**     The State of the Environment—An Overview

The Environment

# CHAPTER 2
# THE GREENHOUSE EFFECT AND CLIMATE CHANGE

## CLIMATE AND HUMAN EVOLUTION

Earth scientists are in the midst of a revolution in their understanding of how climate change occurred in the past. Researchers are discovering the geological and astronomical forces that have changed the planet's environment from hot to cold, wet to dry, and back again, over hundreds of millions of years. Dramatic climate change is nothing new for planet Earth. The climate of the past 10,000 years, during which civilization developed, is a mere blip in a much bigger history of climate change. In fact, Earth's climate will most certainly continue to go through dramatic changes—even without the influence of human activity. Most scientists believe that, if not for extreme climate shifts some 65 million years ago, most of the animals on Earth today, including humans, would probably not be here.

Astronomical cycles have caused the climate to fluctuate between long periods of cold lasting 50,000 to 80,000 years and shorter periods of warmth lasting about 10,000 years. Some scientists believe that these geologically rapid shifts between warm and cold were the catalyst behind human evolution, forcing humans to adapt physiologically and socially in order to cope with the changing climate.

## THE WORLD CLIMATE

Earth's climate is a delicate balance of energy inputs, chemical processes, and physical phenomena. Temperatures on Venus are too hot for the human body; on Mars, they are too cold. This difference in temperature is due to the varying composition of each planet's atmosphere. All three planets receive huge quantities of solar energy, but the amount radiated back into space as heat depends on the atmospheric composition of the particular planet. Some gases, such as carbon dioxide ($CO_2$) and methane, absorb and maintain heat in the same way that glass traps heat in a greenhouse. These gases in Earth's atmosphere allow temperatures to build up, keeping our planet warm and habitable. That is why the increased buildup of these and other gases caused by pollution is often called the "greenhouse effect." (See Figure 2.1.)

It is necessary, however, to distinguish between the "natural" and a possible "enhanced" greenhouse effect. The natural greenhouse effect creates a climate in which life can exist, causing the mean temperature of Earth's surface to be about 33 degrees Celsius (91 degrees Fahrenheit) warmer than it would be if natural greenhouse gases were not present. Without this process, Earth would be frigid and uninhabitable. An enhanced greenhouse effect, sometimes called the anthropogenic effect, refers to the possible increase in the temperature of Earth's surface due to human activity.

The scorching heat of Venus is the result of an atmosphere that is composed largely of $CO_2$. The atmosphere of Earth, however, is nitrogen- and oxygen-based and contains only 0.03 percent $CO_2$. This percentage has varied little over the past million years, permitting a stable climate favorable to life. The blanket of air enveloping Earth moderates its temperature and sustains life.

### A Revolutionary Idea

Earth's atmosphere was first compared to a glass vessel in 1827 by the French mathematician Jean-Baptiste Fourier. In the 1850s British physicist John Tyndall tried to measure the heat-trapping properties of various components of the atmosphere. By the 1890s scientists had concluded that the great increase in combustion in the Industrial Revolution had the potential to change the atmosphere's load of $CO_2$. In 1896 the Swedish chemist Svante Arrhenius made the revolutionary suggestion that human activities could actually disrupt this delicate balance. He theorized that the rapid increase in the use of coal that came with the Industrial Revolution could increase $CO_2$ concentrations and cause a gradual rise in

**FIGURE 2.1**

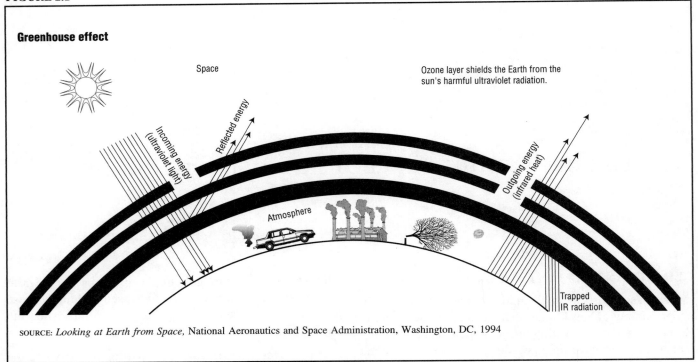

**Greenhouse effect**

Space

Ozone layer shields the Earth from the sun's harmful ultraviolet radiation.

Incoming energy (ultraviolet light)

Reflected energy

Outgoing energy (infrared heat)

Atmosphere

Trapped IR radiation

SOURCE: *Looking at Earth from Space,* National Aeronautics and Space Administration, Washington, DC, 1994

temperatures. For almost six decades his theory stirred little interest.

In 1957 studies at the Scripps Institute of Oceanography in California suggested that, indeed, half the $CO_2$ released by industry was being permanently trapped in the atmosphere. The studies showed that atmospheric concentrations of $CO_2$ in the previous 30 years were greater than in the previous two centuries and that the gas had reached its highest level in 160,000 years.

Findings in the 1980s and 1990s provided more disturbing evidence. Scientists detected increases in other, even more potent gases that contribute to the greenhouse effect, notably chlorofluorocarbons (CFC-11 and -12), methane, nitrous oxide ($N_2O$), and halocarbons (CFCs, methyl chloroform, and hydrochlorofluorocarbons). Atmospheric concentrations of these gases from the 1700s through the 1900s have increased drastically. As shown in Table 2.1, the atmospheric concentration of $CO_2$ increased from 280 parts per million (ppm) in preindustrial times to 370.3 ppm in 2001.

Table 2.2 shows trends for the late 1990s and early 2000s in U.S. greenhouse gas emissions and sinks (repositories, such as forests, that absorb and store carbon). Major sources of greenhouse gas emissions are shown in Figure 2.2. In 2001 electricity generation at power plants continued to account for the largest share (33 percent) of these emissions, followed by transportation (27 percent), industry (19 percent), agriculture (8 percent), commercial sources (7 percent), and residences (5 percent).

Emissions from most sources have increased since 1995. During the late 1990s emissions from industry began to decline and continued that trend into the early 2000s. The Environmental Protection Agency (EPA) attributes the decline to a shift in the overall U.S. economy from a focus on manufacturing industries to service-based businesses. Agricultural emissions are predominantly nitrogen-based, rather than carbon-based. Residential emissions are mainly due to $CO_2$ generated from combustion of fossil fuels (such as oil) for heating purposes.

Scientists know that atmospheric levels of greenhouse gases are increasing. Figure 2.3 shows the dramatic increase in concentrations of $CO_2$ from 1981 through 2002 based on data collected by the National Oceanic and Atmospheric Administration (NOAA) from its Climate Monitoring and Diagnostics Laboratory (CMDL). This buildup of $CO_2$ and other gases could possibly trap energy from the Sun. No one is certain how this accumulation affects Earth's climate.

### Is the Earth Getting Warmer?

As of 2004 even experts are not sure whether the world has already experienced human-induced climate change. Scientists have been unable to provide a definitive answer because they do not know how much the global climate has varied on its own in the relatively recent past (about 1,000 years). Temperature records based on thermometers go back only about 150 years. Investigators have turned, therefore, to "proxy" (indirect) means of

TABLE 2.1

**Global atmospheric concentration, rate of concentration change, and atmospheric lifetime (in years) of selected greenhouse gases**

| Atmospheric variable | $CO_2$ | $CH_4$ | $N_2O$ | $SF_6$[1] | $CF_4$[1] |
|---|---|---|---|---|---|
| Pre-industrial atmospheric concentration | 280 | 0.722 | 0.270 | 0 | 40 |
| Atmospheric concentration[2] | 370.3 | 1.842 | 0.316 | 4.7 | 80 |
| Rate of concentration change[3] | 1.5[4] | 0.007[4] | 0.0008 | 0.24 | 1.0 |
| Atmospheric lifetime | 50–200[5] | 12[6] | 114[6] | 3,200 | >50,000 |

[1]Concentrations in parts per trillion (ppt) and rate of concentration change in ppt/year.
[2]Concentration for $CO_2$ was measured in 2001. Concentrations for all other gases were measured in 2000.
[3]Rate is calculated over the period 1990 to 1999.
[4]Rate has fluctuated between 0.9 and 2.8 ppm per year for $CO_2$ and between 0 and 0.013 ppm per year for $CH_4$ over the period 1990 to 1999.
[5]No single lifetime can be defined for $CO_2$ because of the different rates of uptake by different removal processes.
[6]This lifetime has been defined as an "adjustment time" that takes into account the indirect effect of the gas on its own residence time.
Note: Atmospheric concentrations are in parts per million (ppm) and rate of concentration change is parts per billion (ppb) per year.

SOURCE: "Table 1-1: Global Atmospheric Concentration (ppm Unless Otherwise Specified), Rate of Concentration Change (ppb/year) and Atmospheric Lifetime (Years) of Selected Greenhouse Gases," in *Inventory of U.S. Greenhouse Gas Emissions and Sinks: 1990–2001*, U.S. Environmental Protection Agency, Office of Atmospheric Programs, Washington, DC, April 15, 2003

measuring past temperatures. These methods include chemical evidence of climatic change contained in fossils, corals, ancient ice, and growth rings in trees.

In 1998 Drs. Michael E. Mann and Raymond S. Bradley of the University of Massachusetts at Amherst and Dr. Malcolm K. Hughes of the University of Arizona at Tucson surveyed proxy evidence of temperatures in the Northern Hemisphere since 1400. They discovered that the twentieth century was the warmest century of the past 600 years. They concluded that the warming trend seems to be closely connected to the emission of greenhouse gases by humans. Some experts, however, question whether studies of proxy evidence will ever be reliable enough to yield valuable information on global warming.

Three international agencies have compiled long-term data on surface temperatures—the British Meteorological Office in Bracknell, United Kingdom, the National Climatic Data Center in Asheville, North Carolina, and the National Aeronautics and Space Administration (NASA) Goddard Institute for Space Studies in New York. Temperature measurements from these organizations reported that the 1990s were the warmest decade of the twentieth century and the warmest decade since humans began measuring temperatures in the mid-nineteenth century. The average global surface temperature was approximately 1 degree Fahrenheit warmer than at the turn of the twentieth century, and this rise increased more rapidly since 1980.

According to NASA, the 2002 meteorological year was the second warmest year recorded since the late 1800s. (See Figure 2.4.) A meteorological year runs from the beginning of winter to the end of autumn. Thus, the 2002 meteorological year ran from December 1, 2001, to November 30, 2002. The mean surface temperature for that year was 0.51 degrees Celsius (33 degrees Fahrenheit) warmer than the climatological mean (average for 1951–1980). The warmest temperature occurred in 1998. There has been a strong warming trend over the past three decades.

While some scientists are uncertain whether the greenhouse effect accounts for the change, most believe that it is the most likely explanation. Climate models suggest the potential for a warming from 2 to 6 degrees Fahrenheit over the next 100 years, warmer than Earth has been for millions of years.

In December 2003 the council of the American Geophysical Union (AGU) issued a statement declaring that Earth's climate was warming faster than expected and blamed the increase on anthropogenic (human-related) emissions of greenhouse gases. The statement from the AGU, an international scientific organization with more than 40,000 members, declared that "scientific evidence strongly indicates that natural influences cannot explain the rapid increase in global near-surface temperatures observed in the second half of the 20th century" (David Perlman, "Earth Warming at Faster Pace, Say Top Science Group's Leaders," *San Francisco Chronicle*, December 18, 2003).

However, some scientists observe that major climate events should be viewed in terms of thousands of years, not just a century. A record of only the past century may indicate, but not prove, that a major change has occurred. Is it caused by greenhouse gases, or is it natural variability? While some experts believe it is not possible to conclude that the warming is caused by greenhouse gases emitted by human activity, the rising temperature is roughly on track with that of computer models programmed to predict the course of greenhouse warming.

**FURTHER SIGNS.** In 1990, at the first of several meetings of the Intergovernmental Panel on Climate Change (IPCC), several early signs of actual climate change were noted: the average warm-season temperature in Alaska had risen nearly three degrees Fahrenheit in the previous fifty years; glaciers had generally receded and become thinner on average by about thirty feet in the past forty years; there was about 5 percent less sea ice in the Bering Sea than in the 1950s; and permafrost was thawing, caus-

TABLE 2.2

**Recent trends in greenhouse gas emissions and sinks, in teragrams of carbon dioxide equivalents (Tg $CO_2$ Eq), 1990–2001**

| Gas/source | 1990 | 1995 | 1996 | 1997 | 1998 | 1999 | 2000 | 2001 |
|---|---|---|---|---|---|---|---|---|
| **$CO_2$** | **5,003.7** | **5,334.4** | **5,514.8** | **5,595.4** | **5,614.2** | **5,680.7** | **5,883.1** | **5,794.8** |
| Fossil fuel combustion | 4,814.8 | 5,141.5 | 5,325.8 | 5,400.0 | 5,420.5 | 5,488.8 | 5,692.2 | 5,614.9 |
| Iron and steel production | 85.4 | 74.4 | 68.3 | 71.9 | 67.4 | 64.4 | 65.8 | 59.1 |
| Cement manufacture | 33.3 | 36.8 | 37.1 | 38.3 | 39.2 | 40.0 | 41.2 | 41.4 |
| Waste combustion | 14.1 | 18.5 | 19.4 | 21.2 | 22.5 | 23.9 | 25.4 | 26.9 |
| Ammonia manufacture & urea application | 19.3 | 20.5 | 20.3 | 20.7 | 21.9 | 20.6 | 19.6 | 16.6 |
| Lime manufacture | 11.2 | 12.8 | 13.5 | 13.7 | 13.9 | 13.5 | 13.3 | 12.9 |
| Natural gas flaring | 5.5 | 8.7 | 8.2 | 7.6 | 6.3 | 6.7 | 5.5 | 5.2 |
| Limestone and dolomite use | 5.5 | 7.0 | 7.6 | 7.1 | 7.3 | 7.7 | 5.8 | 5.3 |
| Aluminum production | 6.3 | 5.3 | 5.6 | 5.6 | 5.8 | 5.9 | 5.4 | 4.1 |
| Soda ash manufacture and consumption | 4.1 | 4.3 | 4.2 | 4.4 | 4.3 | 4.2 | 4.2 | 4.1 |
| Titanium dioxide production | 1.3 | 1.7 | 1.7 | 1.8 | 1.8 | 1.9 | 1.9 | 1.9 |
| Carbon dioxide consumption | 0.9 | 1.1 | 1.1 | 1.2 | 1.2 | 1.2 | 1.2 | 1.3 |
| Ferroalloys | 2.0 | 1.9 | 2.0 | 2.0 | 2.0 | 2.0 | 1.7 | 1.3 |
| Land-use change and forestry (Sink) [1] | (1,072.8) | (1,064.2) | (1,061.0) | (840.6) | (830.5) | (841.1) | (834.6) | (838.1) |
| International bunker fuels [2] | 113.9 | 101.0 | 102.3 | 109.9 | 112.9 | 105.3 | 99.3 | 97.3 |
| **$CH_4$** | **644.0** | **650.0** | **636.8** | **629.5** | **622.7** | **615.5** | **613.4** | **605.9** |
| Landfills | 212.1 | 216.1 | 212.1 | 207.5 | 202.4 | 203.7 | 205.8 | 202.9 |
| Natural gas systems | 122.0 | 127.2 | 127.4 | 126.0 | 124.0 | 120.3 | 121.2 | 117.3 |
| Enteric fermentation | 117.9 | 123.0 | 120.5 | 118.3 | 116.7 | 116.6 | 115.7 | 114.8 |
| Coal mining | 87.1 | 73.5 | 68.4 | 68.1 | 67.9 | 63.7 | 60.9 | 60.7 |
| Manure management | 31.3 | 36.2 | 34.9 | 36.6 | 39.0 | 38.9 | 38.2 | 38.9 |
| Wastewater treatment | 24.1 | 26.6 | 26.8 | 27.3 | 27.7 | 28.2 | 28.3 | 28.3 |
| Petroleum systems | 27.5 | 24.2 | 23.9 | 23.6 | 22.9 | 21.6 | 21.2 | 21.2 |
| Rice cultivation | 7.1 | 7.6 | 7.0 | 7.5 | 7.9 | 8.3 | 7.5 | 7.6 |
| Stationary sources | 8.1 | 8.5 | 8.7 | 7.5 | 7.2 | 7.4 | 7.6 | 7.4 |
| Mobile sources | 5.0 | 4.9 | 4.8 | 4.7 | 4.6 | 4.5 | 4.4 | 4.3 |
| Petrochemical production | 1.2 | 1.5 | 1.6 | 1.6 | 1.6 | 1.7 | 1.7 | 1.5 |
| Field burning of agricultural residues | 0.7 | 0.7 | 0.7 | 0.8 | 0.8 | 0.8 | 0.8 | 0.8 |
| Silicon carbide production | * | * | * | * | * | * | * | * |
| International bunker fuels [2] | 0.2 | 0.1 | 0.1 | 0.1 | 0.1 | 0.1 | 0.1 | 0.1 |
| **$N_2O$** | **397.6** | **430.9** | **441.7** | **440.9** | **436.8** | **430.0** | **429.9** | **424.6** |
| Agricultural soil management | 267.5 | 284.1 | 293.2 | 298.2 | 299.2 | 297.0 | 294.6 | 294.3 |
| Mobile sources | 50.6 | 60.9 | 60.7 | 60.3 | 59.7 | 58.8 | 57.5 | 54.8 |
| Manure management | 16.2 | 16.6 | 17.0 | 17.3 | 17.3 | 17.4 | 17.9 | 18.0 |
| Nitric acid | 17.8 | 19.9 | 20.7 | 21.2 | 20.9 | 20.1 | 19.1 | 17.6 |
| Human sewage | 12.7 | 13.9 | 14.1 | 14.4 | 14.6 | 15.1 | 15.1 | 15.3 |
| Stationary combustion | 12.5 | 13.2 | 13.8 | 13.7 | 13.7 | 13.7 | 14.3 | 14.2 |
| Adipic acid | 15.2 | 17.2 | 17.0 | 10.3 | 6.0 | 5.5 | 6.0 | 4.9 |
| $N_2O$ product usage | 4.3 | 4.5 | 4.5 | 4.8 | 4.8 | 4.8 | 4.8 | 4.8 |
| Field burning of agricultural residues | 0.4 | 0.4 | 0.4 | 0.4 | 0.5 | 0.4 | 0.5 | 0.5 |
| Waste combustion | 0.3 | 0.3 | 0.3 | 0.3 | 0.2 | 0.2 | 0.2 | 0.2 |
| International bunker fuels [2] | 1.0 | 0.9 | 0.9 | 1.0 | 1.0 | 0.9 | 0.9 | 0.9 |
| **HFCs, PFCs, and $SF_6$** | **94.4** | **99.5** | **113.6** | **116.8** | **127.6** | **120.3** | **121.0** | **111.0** |
| Substitution of ozone depleting substances | 0.9 | 21.7 | 30.4 | 37.7 | 44.5 | 50.9 | 57.3 | 63.7 |
| HCFC-22 production | 35.0 | 27.0 | 31.1 | 30.0 | 40.2 | 30.4 | 29.8 | 19.8 |
| Electrical transmission and distribution | 32.1 | 27.5 | 27.7 | 25.2 | 20.9 | 16.4 | 15.4 | 15.3 |
| Semiconductor manufacture | 2.9 | 5.9 | 5.4 | 6.5 | 7.3 | 7.7 | 7.4 | 5.5 |
| Aluminum production | 18.1 | 11.8 | 12.5 | 11.0 | 9.0 | 8.9 | 7.9 | 4.1 |
| Magnesium production and processing | 5.4 | 5.6 | 6.5 | 6.3 | 5.8 | 6.0 | 3.2 | 2.5 |
| **Total** | **6,139.6** | **6,514.9** | **6,707.0** | **6,782.6** | **6,801.3** | **6,849.5** | **7,047.4** | **6,936.2** |
| **Net emissions (sources and sinks)** | **5,066.8** | **5,450.7** | **5,646.0** | **5,942.0** | **5,970.9** | **6,008.5** | **6,212.7** | **6,098.1** |

*Does not exceed 0.05 Tg $CO_2$ Eq.
[1] For the most recent years, a portion of the sink estimate is based on historical and projected data. Parentheses indicate negative values (or sequestration).
[2] Emissions from international bunker fuels are not included in totals.
Note: Totals may not sum due to independent rounding.

SOURCE: "Table ES-1: Recent Trends in U.S. Greenhouse Gas Emissions and Sinks (Tg $CO_2$ Eq.)," in *Inventory of U.S. Greenhouse Gas Emissions and Sinks: 1990–2001*, U.S. Environmental Protection Agency, Office of Atmospheric Programs, Washington, DC, April 15, 2003

ing the ground to subside, opening holes in roads, producing landslides and erosion, threatening roads and bridges, and causing local floods. Ice cellars in northern villages have thawed and become useless. More precipitation now falls as rain than snow, and the snow melts faster, causing more running and standing water. While this could be natural variability, it is the kind of change expected of global warming, that is, the Arctic will warm more than the global average.

According to the EPA, global sea levels have risen by six to eight inches over the last century. The increase is attributed to melting mountain glaciers, expansion of ocean water in response to rising temperatures, melting of the polar ice sheets, and surface discharge of groundwater

FIGURE 2.2

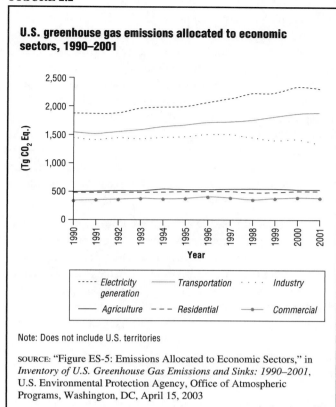

**U.S. greenhouse gas emissions allocated to economic sectors, 1990–2001**

Legend:
- - - - Electricity generation
—— Agriculture
——— Transportation
- - - Residential
· · · · Industry
—●— Commercial

Note: Does not include U.S. territories

SOURCE: "Figure ES-5: Emissions Allocated to Economic Sectors," in *Inventory of U.S. Greenhouse Gas Emissions and Sinks: 1990–2001*, U.S. Environmental Protection Agency, Office of Atmospheric Programs, Washington, DC, April 15, 2003

FIGURE 2.3

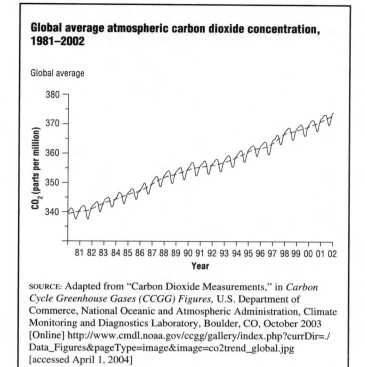

**Global average atmospheric carbon dioxide concentration, 1981–2002**

SOURCE: Adapted from "Carbon Dioxide Measurements," in *Carbon Cycle Greenhouse Gases (CCGG) Figures,* U.S. Department of Commerce, National Oceanic and Atmospheric Administration, Climate Monitoring and Diagnostics Laboratory, Boulder, CO, October 2003 [Online] http://www.cmdl.noaa.gov/ccgg/gallery/index.php?currDir=./ Data_Figures&pageType=image&image=co2trend_global.jpg [accessed April 1, 2004]

that has been pumped out of the ground. The sea level along much of the U.S. coast has risen by ten to twelve inches per century, although the rate varies by location.

Melting glaciers and collapsing ice shelves in Antarctica have been well publicized, particularly the breakup of the Larsen B Ice Shelf in early 2002. Scientists at the U.S. Geological Survey estimate that melting of Antarctica's entire ice sheet will increase the sea level by 73 meters (approximately 240 feet). However, it is unknown whether the ice sheet is actually shrinking or growing.

### The Effects of Volcanic Activity on Climate

Volcanic activity, such as the 1991 eruption of the Mount Pinatubo volcano in the Philippines, can temporarily offset recent global warming trends. Volcanoes spew vast quantities of particles and gases into the atmosphere, including sulfur dioxide ($SO_2$) that combines with water to form tiny supercooled droplets. The droplets create a long-lasting global haze that reflects and scatters sunlight, reducing energy from the Sun and preventing its rays from heating the Earth, thereby causing the planet to cool.

This also occurred in 1982, when the El Chichon volcano in Mexico depressed global temperatures for about four years. NASA reported that satellite sensors measured the $SO_2$ cloud from Mount Pinatubo at 15 million tons, about twice the size of the one emitted by El Chichon.

NASA found that the haze of sulfur from the eruption reflected enough sunlight to cool the Earth by about 1 degree Fahrenheit, as was predicted by computer models.

In 1815 a major eruption of the Tambora volcano in Indonesia produced serious weather-related disruptions, such as crop-killing summer frosts in the United States and Canada. It became known as the "year without a summer." On the other hand, for several years following the Tambora eruption people around the world commented about the beautiful sunsets, which were caused by the suspension of volcano-related particulate matter in the atmosphere.

### The Effects of Clouds on Climate

Clouds may hold a key to understanding climate change. Although we see clouds virtually every day, surprisingly little is known about them—where they occur, their role in energy and water transfer, and their ability to reflect solar heat. Earth's climate maintains a balance between the energy that reaches Earth from the Sun and the energy that radiates back from Earth into space. Scientists refer to this as Earth's "radiation budget." The components of Earth's system are the planet's surface, atmosphere, and clouds.

Different parts of Earth have different capacities to reflect solar energy. Oceans and rain forests reflect only a small portion of the Sun's energy. Deserts and clouds, on the other hand, reflect a large portion of solar energy. A cloud reflects more radiation back into space than the surface would in the absence of clouds. An increase in cloudiness can also act like the panels on a greenhouse roof.

FIGURE 2.4

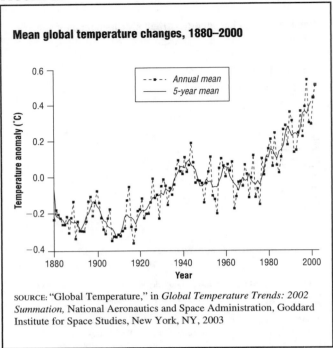

**Mean global temperature changes, 1880–2000**

SOURCE: "Global Temperature," in *Global Temperature Trends: 2002 Summation,* National Aeronautics and Space Administration, Goddard Institute for Space Studies, New York, NY, 2003

NASA's Earth Science Enterprise is a satellite-based program that includes numerous scientific studies of clouds. These studies have revealed that:

- The effect of clouds on climate depends on the balance between the incoming solar radiation and the absorption of Earth's outgoing radiation.

- Low clouds have a cooling effect because they are optically thicker and reflect much of the incoming solar radiation out to space.

- High thin cirrus clouds have a warming effect because they transmit most of the incoming solar radiation while also trapping some of Earth's radiation and radiating it back to the surface.

- Deep convective clouds have neither a warming nor a cooling effect because their reflective and absorptive abilities cancel one another.

### Another Possible Climate Culprit—Solar Cycles

Scientists have known for centuries that the Sun goes through cycles; it has seasons, storms, and rhythms of activity with sunspots and flares appearing in cycles of roughly 11 years. Some scientists contend that these factors play a role in climate change on Earth. Some research, though sketchy and controversial, suggests that the Sun's variability could account for some, if not all, of global warming to date. The biggest correlation occurred centuries ago—between 1640 and 1720—when sunspot activity fell sharply and Earth cooled about 2 degrees Fahrenheit. (The Sun is brighter when sunspots appear and dimmer when they disappear.)

### The Effects of the Oceans on Climate

Oceans have a profound effect on climate, because of their huge capacity to store heat and because they can moderate levels of atmospheric gases that regulate global temperatures. Covering more than 70 percent of Earth and holding 97 percent of the water on the planet's surface, oceans function as huge reservoirs of heat. Ocean currents transport this stored heat and dissolved gases so that different areas of the world serve as either sources or sinks (repositories) for these components. While scientists know a great deal about oceanic and air circulation, they are less certain about the ocean's ability to store additional $CO_2$ or about how much heat it will absorb.

The top eight feet of the oceans hold as much heat as Earth's entire atmosphere. As ocean waters circulate globally, heat is transferred from low altitudes to high altitudes, from north to south, and vertically from surface to deep oceans and back. But how is this heat apportioned? If heat is able to circulate through the entire oceanic depth range, the process could take centuries and the world's oceans could serve to buffer or delay global warming. Researchers are working to determine that possibility, but it remains one of many unanswered questions.

### The Effects of El Niño and La Niña on Climate

For centuries fishermen in the Pacific Ocean off the coast of South America have known about the phenomenon called El Niño. Every three to five years, during December and January, fish in those waters virtually vanish, bringing fishing to a standstill. Fishermen gave this occurrence the name "El Niño," which means "the Child," because it occurs around the celebration of the birth of Jesus, the Christ child. Although originating in the Pacific, the effects of El Niño are felt around the world. Computers, satellites, and improved data gathering have found that the El Niño phenomenon has been responsible for drastic climate change.

An El Niño occurs because of interactions between atmospheric winds and sea surfaces. In normal years, trade winds blow from east to west across the eastern Pacific. They drag the surface waters westward across the ocean, causing deeper, cold waters to rise to the surface. This upwelling of deep ocean waters carries nutrients from the bottom of the ocean that feed fish populations in the upper waters.

In an El Niño the westward movement of waters weakens, causing the upwelling of deep waters to cease. The resulting warming of the ocean waters further weakens trade winds and strengthens El Niño. Without upwelling, the nutrient content of deep waters is diminished, which in turn causes the depletion of fish populations. The warm waters that normally lie in the western waters of the Pacific shift eastward. This turbulence creates eastward weather conditions, in which towering cumulus clouds reach high into the atmosphere with strong vertical forces and the weakening of normal east-to-west trade winds. An El Niño is the warm phase of a phenomenon known as ENSO (El Niño/Southern Oscillation), which can also include a cold phase known as a La Niña.

The worldwide effects of El Niño can include torrential rains, tornadoes, hurricanes, mud slides, flash flooding, beach and cliff erosion, sewage spills, drought, increased snowfall, and disruption in the marine food chain. Such weather events often affect regional energy and economic markets. The El Niño of 1982–83 is estimated to have caused $13.6 billion in damage and killed 2,000 people around the world. In 1997 and 1998 El Niño–related storms in California caused more than 9,000 people to seek federal disaster assistance for property losses. An estimated $3.6 million in aid was distributed to approximately 1,689 people. The American Fisheries Society estimated that 1990s storms were the most devastating for marine life in more than a century.

## POTENTIAL EFFECTS OF A WARMING CLIMATE

### Rising Sea Level

Some observers compare global warming to nuclear war in its potential to disrupt human and environmental systems. While some sources dispute the occurrence of human-induced climate change, if temperature increases of 1.5 to 4.5 degrees Celsius (35 to 40 degrees Fahrenheit) were to take place, substantial changes would occur on Earth's surface. If average temperatures rise this much by 2030, the global sea level could rise 20 to 140 centimeters (8 to 56 inches). The Climate Institute in Washington, D.C., forecasts a further rise of 26 inches by 2100. This rise would be caused by the expansion of seawater as it is warmed, along with melting glaciers and ice caps.

More than half the population of the United States lives within 50 miles of a coastline. A rising sea level would narrow or destroy beaches, flood wetland areas, and either submerge or force costly fortification of shoreline property. Figure 2.5 shows areas along the Southeast coast that could be vulnerable to rising sea levels because they lie at low elevations.

Higher water levels would increase storm damage and many coastal cities worldwide would be flooded. Some islands, such as the Philippines, have already seen encroachment. Residents of the Maldives, islands which lie on average three feet above sea level in the Indian Ocean, are already erecting artificial defenses, such as breakwaters made of concrete, to fend off rising seas.

Some 70 million people in low-lying areas of Bangladesh could be displaced by a one-meter (39.4 inches) rise. Such a rise would also threaten Tokyo, Osaka, and Nagoya in Japan, as well as coastal China and the Atlantic and Gulf Coasts of the United States. The rising waters would also intrude on inland rivers, threatening

fresh water supplies and increasing the salt content of groundwater as the sea encroaches on freshwater aquifers (naturally occurring underground water reservoirs). Much of the increased rainfall would come not in the steady, gentle rains favored by farmers, but from heavy storms and flooding.

### Increased Heat

Warmer temperatures may also increase the evaporation rate, thereby increasing atmospheric water vapor and cloud cover, which in turn may affect regional rainfall patterns. A warmer climate would likely shift the rain belt of the middle latitudes toward the poles as water-laden air of the tropics travels further toward the poles before the moisture condenses as precipitation. This would shift patterns of rainfall around the world. Wetter, more violent weather is projected for some regions. Forests, which are adapted to a narrow temperature and moisture range, would particularly be threatened by climate shifts.

The opposite problem—too little water—could worsen in arid areas such as the Middle East and parts of Africa. Some experts have suggested that the greenhouse effect and global warming are mild terms for a coming era that may be marked by heat waves that may make some regions virtually uninhabitable. Frequent droughts could plague North America and Asia, imperiling food production, as agriculture is particularly vulnerable to the effects of climate change. Most experts believe Africa will be the most vulnerable to climate change because its economy depends largely on rain-fed agriculture, and many farmers are too poor and ill equipped to adapt. Australia and Latin America could also be subject to severe drought.

NASA has begun using computer calculations in an attempt to predict the effects of global warming on some of the major cities in the United States.

### The Effect of Increased Temperatures on Humans

Extra heat alone would be enough to kill some people. Some deaths would occur directly from heat-induced strokes and heart attacks. Air quality also deteriorates as temperatures rise. Hot, stagnant air contributes to the formation of atmospheric ozone, the main component of smog, which damages human lungs. Poor air quality can also aggravate asthma and other respiratory diseases. Increasing ultraviolet rays can increase the incidence of skin cancers, diminish the function of the human immune system, and cause eye problems such as cataracts. Higher temperatures and added rainfall could create ideal conditions for the spread of a host of infectious diseases by insects, including mosquito-borne malaria, dengue fever, and encephalitis.

### Decreasing Biological Diversity

Biological diversity is also predicted to suffer from global warming. Loss of forests, tundra (arctic plains), and

## FIGURE 2.5

**Areas of the Gulf Coast lying at low elevations that are vulnerable to sea level rise**

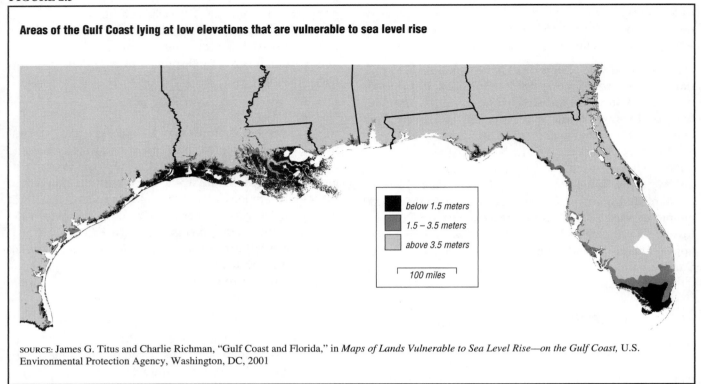

below 1.5 meters
1.5 – 3.5 meters
above 3.5 meters

100 miles

SOURCE: James G. Titus and Charlie Richman, "Gulf Coast and Florida," in *Maps of Lands Vulnerable to Sea Level Rise—on the Gulf Coast,* U.S. Environmental Protection Agency, Washington, DC, 2001

wetlands could irrevocably damage ecosystems. Some species that live in precise, narrow bands of temperature and humidity may find their habitats wiped out altogether. Rising seas would cover coastal mangrove swamps, causing the loss of many species, including the Bengal tiger. Plants and animals of the far north, like the polar bear and the walrus, would die out for lack of an acceptably cold environment. Many species cannot migrate rapidly enough to cope with climate change at the projected rate. Opportunistic species, such as weeds, often out-compete trees and other plants more valued by humans. Pests, such as certain insects, may survive where other species cannot.

The U.S. Forest Service believes that eastern hemlock, yellow birch, beech, and sugar maple forests would gradually shift their ranges northward by 300 to 600 miles but would be severely limited by the warming and largely die out, along with the wildlife they shelter. Studies by World Wildlife Fund International report that more than half the world's parks and reserves could be threatened by climate change. These include the Florida Everglades, Yellowstone National Park, the Great Smoky Mountains, and Redwood National Park in California. The EPA warned in 1988: "If current trends continue, it is likely that climate may change too quickly for many natural systems to adapt."

**SOME POSSIBLE POSITIVE EFFECTS.** There are several possible positive consequences of global warming. Agriculture in the northern United States and southern Canada, on the West Coast, and in interior parts of the West could benefit, as could the evergreen forests of the West Coast. Milder winters could reduce the number of cold-weather deaths, as well as the cost of snow clearance and heating. Northern waters could remain open longer for navigation, and the Arctic Ocean might become ice-free, opening a new trade route between Europe and Asia.

## GENERAL CIRCULATION MODELS

The science of global atmospheric change is still in its infancy. Most of our images of the world's environmental future must come from computer and mathematical models. The most highly developed tools now available to project climatic changes are complex computer models called general circulation models (GCMs). Even these models are crude, imperfect representations of the real world and of the future world. They cannot indicate where changes will be the worst. Nor can they accurately calculate the heat reflected by clouds or absorbed by the ocean. As a result, scientists disagree over whether the forecasts of global warming are reliable.

Although computer models have become more accurate over the years, important uncertainties limit their predictive abilities. Even the most powerful computers are limited in their ability to store and analyze the vast quantity of data required to accurately simulate the global climate. Modelers have tried to overcome these limitations by introducing assumptions into their models that deliberately oversimplify some operations to free the GCMs' capacity for more critical operations. For example, modelers have assumed that the ocean was not warmed by emissions of greenhouse gases before 1985. Although this assumption increases GCM capacity, it increases the uncertainty of the computer's predictions because the ocean will reach its capacity to absorb emissions sooner. Scientists do not know, however, how much or by how long the predictions are distorted because of this error.

In *Global Deception: The Exaggeration of the Global Warming Threat* (Washington, DC: Center for the Study of American Business, Washington University, 1997), Patrick J. Michaels claims that the global warming crisis has been greatly overstated. He notes that newer GCMs, which are more physically realistic than the older models, tend to forecast less warming rather than more.

## GREENHOUSE GASES

The Energy Information Administration of the U.S. Department of Energy (DOE) reported that $CO_2$ accounted for 83 percent of greenhouse gas emissions in the United States in 2001. (See Figure 2.6.) Methane ($CH_4$) was second with 9 percent of the total, followed by nitrous oxide ($N_2O$) with 6 percent and other greenhouse gases, such as hydrofluorocarbons (HFCs), perfluorocarbons (PFCs), and sulfur hexafluoride (SF6), with 2 percent.

### Carbon Dioxide

Carbon dioxide ($CO_2$) is a gas that is a naturally occurring component of Earth's atmosphere. It is also produced by a variety of human and animal sources. These factors are a part of the carbon cycle. (See Figure 2.7.) As shown in Figure 2.6, $CO_2$ was the most prevalent of the greenhouse gases emitted from 1990 to 2001.

The burning of fossil fuels by industry and motor vehicles is, by far, the leading source of $CO_2$ in the atmosphere, accounting for 97 percent of the nation's emission of greenhouse gases in 2001. (See Table 2.2.) As populations and economies expand, they use ever-greater amounts of fossil fuels.

The DOE released its *International Energy Annual 2001* in March 2003. The report presents data collected on $CO_2$ emissions related to fossil fuel use around the world. The United States was responsible for the largest portion (23 percent) of such emissions in 2001, followed by Western Europe (16 percent) and China (13 percent). (See Figure 2.8.) The western European countries responsible for most of that region's $CO_2$ emissions were Germany, the United Kingdom, Italy, and France.

The DOE predicts that $CO_2$ emissions from developing countries could actually surpass those from industrialized countries before 2020. (See Figure 2.9.) This trend is attributed to increased use of coal in developing countries, particularly China and India, while developed countries rely increasingly on natural gas.

FIGURE 2.6

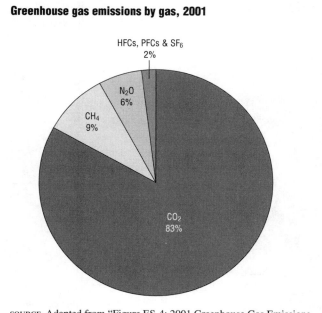

Greenhouse gas emissions by gas, 2001

HFCs, PFCs & $SF_6$
2%

$N_2O$
6%

$CH_4$
9%

$CO_2$
83%

SOURCE: Adapted from "Figure ES-4: 2001 Greenhouse Gas Emissions by Gas," in *Inventory of U.S. Greenhouse Gas Emissions and Sinks: 1990–2001*, U.S. Environmental Protection Agency, Office of Atmospheric Programs, Washington, DC, April 15, 2003

**THE ROLE OF THE FORESTS AS CARBON SINKS.** Forests act as sinks, or repositories, absorbing and storing carbon. Trees naturally absorb and neutralize $CO_2$, although scientists do not agree on the extent to which forests can soak up excess amounts. The increasing levels of $CO_2$ in the atmosphere might conceivably be tolerated in Earth's normal $CO_2$ cycle if not for the additional complicating factor of deforestation. The burning of the Amazon rain forests and other forests has had a twofold effect: the immediate release of large amounts of $CO_2$ into the atmosphere from the fires, and the loss of trees to neutralize the $CO_2$ in the atmosphere. (See Figure 2.10.)

**THE ROLE OF THE OCEANS AS CARBON SINKS.** The oceans are, by far, the largest reservoir of carbon in the carbon cycle. The oceans hold approximately 50 times more carbon than the atmosphere and 20 times more than the terrestrial reservoir (land). Oceanographers and ecologists disagree over the carbon cycle–climate connection and over the ocean's capacity to absorb $CO_2$. Some scientists believe that the oceans can absorb one to two billion tons of $CO_2$ a year, about the amount the world emitted in 1950. Until scientists can more accurately determine how much $CO_2$ can be buffered by ocean processes, the extent and speed of disruption in the carbon supply remains unclear.

### Methane

Methane ($CH_4$) is an important component of greenhouse emissions, second only to $CO_2$. (See Figure 2.6.) While there is less methane in the atmosphere, scientists

**FIGURE 2.7**

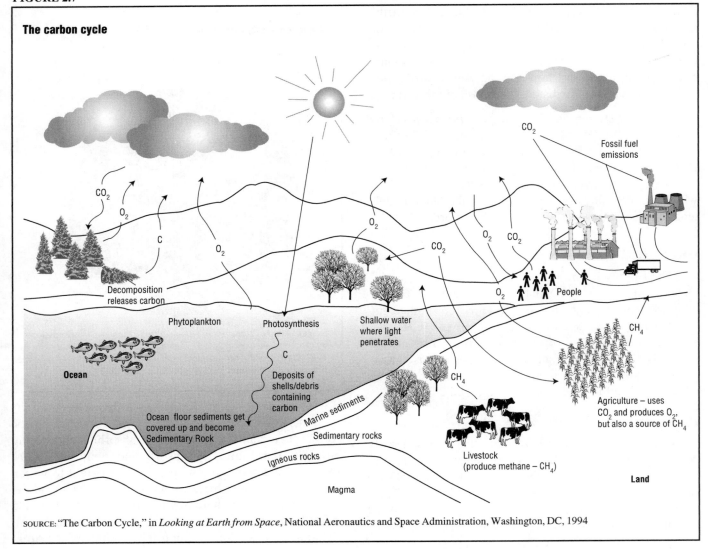

The carbon cycle

SOURCE: "The Carbon Cycle," in *Looking at Earth from Space*, National Aeronautics and Space Administration, Washington, DC, 1994

estimate that it may be 21 times more effective at trapping heat in the atmosphere than $CO_2$. Over the past two centuries, methane's concentration in the atmosphere has more than doubled and scientists generally attribute those increases to human sources, such as landfills, natural gas systems, agricultural activities, coal mining, and wastewater treatment. However, according to the Energy Information Administration, U.S. emissions of methane from energy and waste management sources actually declined between 1990 and 2002, while agricultural emissions increased over the same time period. (See Figure 2.11.) The increase is blamed on domestic animals and the decomposition of their waste, which releases methane. Growing markets for beef and milk products are driving a booming livestock business.

**Nitrous Oxide**

Nitrous oxide ($N_2O$) is formed by natural biological sources and by a number of human activities. Although $N_2O$ makes up a much smaller portion of greenhouse gases than $CO_2$, it is much more (perhaps 310 times more) powerful than $CO_2$ at trapping heat. Figure 2.12 shows that agriculture is and has been the major source of $N_2O$ emissions in the United States, followed by energy and industrial sources. Specific agricultural sources of $N_2O$ emissions are shown in Figure 2.13.

**OTHER GREENHOUSE GASES**

Other primary greenhouse gases are what the DOE calls "engineered gases." These are gases specially designed for modern industrial and commercial purposes. They include hydrofluorocarbons (HFCs), perfluorocarbons (PFCs), and sulfur hexafluoride (SF6).

HFCs are chemicals that contain hydrogen, fluorine, and carbon. They are popular substitutes in industrial applications for chlorofluorocarbons (CFCs). CFCs are commonly used in cooling equipment, fire extinguishers, as propellants, and for other uses. They are blamed for depletion of the ozone layer in the stratosphere, which shields the Earth from deadly ultraviolet radiation.

FIGURE 2.8

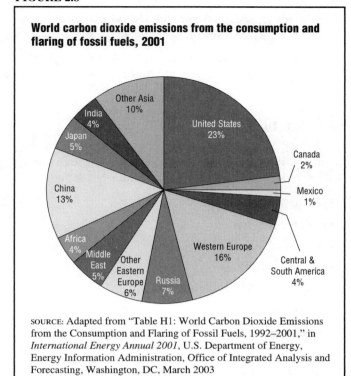

**World carbon dioxide emissions from the consumption and flaring of fossil fuels, 2001**

SOURCE: Adapted from "Table H1: World Carbon Dioxide Emissions from the Consumption and Flaring of Fossil Fuels, 1992–2001," in *International Energy Annual 2001*, U.S. Department of Energy, Energy Information Administration, Office of Integrated Analysis and Forecasting, Washington, DC, March 2003

FIGURE 2.9

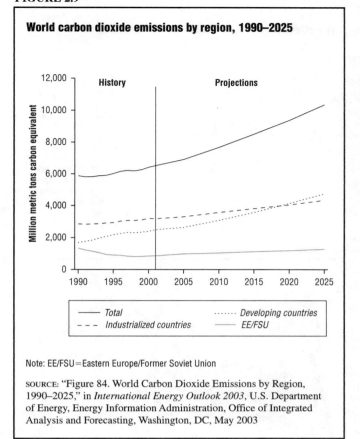

**World carbon dioxide emissions by region, 1990–2025**

Note: EE/FSU=Eastern Europe/Former Soviet Union

SOURCE: "Figure 84. World Carbon Dioxide Emissions by Region, 1990–2025," in *International Energy Outlook 2003*, U.S. Department of Energy, Energy Information Administration, Office of Integrated Analysis and Forecasting, Washington, DC, May 2003

PFCs are a class of chemicals containing fluorine and carbon. They also are increasingly used by industry as substitutes for ozone-depleting CFCs. SF6 is a colorless, odorless gas commonly used as an insulating medium in electrical equipment and as an etchant (an etching agent) in the semiconductor industry.

Although emissions of these chemicals are very small in comparison to other greenhouse gases, they are of particular concern because of their long life in the atmosphere. PFCs and SF6 have atmospheric lifetimes of thousands of years and are actually far more potent greenhouse gases than $CO_2$ per unit of molecular weight.

## THE INTERNATIONAL PANEL ON CLIMATE CHANGE—A SCIENTIFIC CONSENSUS?

Global warming was only acknowledged as an international problem at the end of the twentieth century. At the world's first ecological summit, the 1972 Stockholm Conference, climate change was not even listed among the threats to society.

Climate-change science has been developing rapidly, leading observers to recognize the complexity of the issues. In order to understand the issue of climate change and any possible global warming, scientists worldwide gathered together in 1990 to exchange information. They formed the International Panel on Climate Change (IPCC), which is sponsored jointly by the United Nations Environment Program (UNEP) and the World Meteorological Organization.

The panel's mission was to advise the parties to the 1992 Global Warming Treaty. It received the input of nearly 400 of the world's climate experts from 25 nations.

The IPCC report was the most comprehensive summary of climate-change science to date. With regard to various statements about climate change, the scientists responded based on how sure they were of global warming: "virtually certain" (nearly unanimous agreement and no credible alternative), "very probable" (roughly a 90 percent likelihood of occurring), "probable" (a two out of three chance of happening), and "uncertain" (evidence does not support). The outcome indicated an international consensus on three basic statements related to global warming: that greenhouse gases do produce climate change, that there is a natural greenhouse effect, and that emissions resulting from human activities are contributing to the proliferation and effects of greenhouse gases.

### A Landmark Judgment—The 1995 IPCC Report

In 1995 the IPCC reassessed the state of knowledge about climate change. The panel reaffirmed its earlier conclusions and updated its forecasts, predicting that, if no further action is taken to curb emissions of greenhouse gases, temperatures will increase 1.44 degrees to 6.30 degrees Fahrenheit by 2100. The panel concluded that the evidence suggested a human influence on global climate.

**FIGURE 2.10**

### The effect of forests on carbon dioxide concentrations

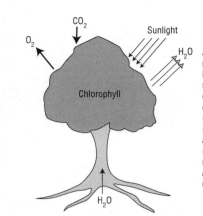

As plants and trees grow, photosynthesis — involving the interaction of sunlight, chlorophyll in green leaves, carbon dioxide ($CO_2$) and water ($H_2O$) — results in a net removal of $CO_2$ from the air and the release of oxygen ($O_2$) as a by-product. Also, moisture is released to the air through evapotranspiration.

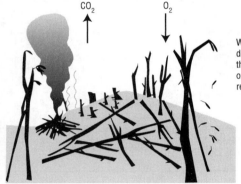

When forests die and decay, or are burned, the biomass is oxidized and $CO_2$ is returned to the air.

SOURCE: "Figure 2a" and "Figure 2b," in *Biosphere, NASA Facts*, National Aeronautics and Space Administration, Goddard Space Flight Center, Greenbelt, MD, April 1998

The cautiously worded statement was a compromise following intense discussions. Nonetheless, it was a landmark conclusion because the panel, until then, had always said that global warming and climate changes could have been the result of natural variability. Despite acrimonious debate and challenges to the report, the conclusions have largely held up to scientific scrutiny.

### The 2001 Update

The IPCC issued its third assessment report in 2001. It actually comprises four reports: *Climate Change 2001: The Scientific Basis, Climate Change 2001: Impacts, Adaptation and Vulnerability, Climate Change 2001: Mitigation,* and *Climate Change 2001: The Synthesis Report.* The IPCC's assessment covered the adaptability and vulnerability of North America to climate change impacts likely to occur from global warming. Among the suggested possible effects of global warming are:

- Expansion of some diseases in North America

- In coastal areas, increased erosion, flooding, and loss of wetlands

- Risk to "unique natural ecosystems"

- Changes in seasonal snowmelts, which would have effects on water users and aquatic ecosystems

- Some initial benefits for agriculture, but those benefits would decline over time and possibly "become a net loss"

### Next IPCC Update Due in 2007

The IPCC plans to publish its fourth major assessment report in 2007. At meetings held in Paris during 2003 seven themes were selected for focus in the report:

- risk and uncertainty
- regional integration
- water
- key vulnerabilities
- adaptation and mitigation
- sustainable development
- technology

## SOME RESEARCHERS STILL QUESTION THE THEORY OF CLIMATE CHANGE

In the *Swiss Review of World Affairs* (July 1994), Dr. Richard Lindzen, professor of meteorology at the Massachusetts Institute of Technology, observed that a solid scientific foundation is lacking from claims that global warming is imminent. He believes that Earth's climate depends on more than the $CO_2$ content. Instead, Earth cools by atmospheric movements upward and poleward rather than only by radiation; and climate changes since 1850 are indistinguishable from natural variability. He further claims that no large-scale model includes all the known major factors in the water vapor cycle. In 1997 Dr. Lindzen added that the natural system has built-in resilience and that the observed changes are insignificant and not urgent.

In October 2003 Anders Sivertsson, Kjell Aleklett, and Colin Campbell, researchers from the University of Uppsala in Sweden, disputed the IPCC's predictions of extreme temperature change due to $CO_2$ emissions. According to Andy Coghlan in his article "'Too Little' Oil for Global Warming," in the journal *New Scientist,* the researchers noted that their data show that the world's oil and gas supplies will be depleted long before atmospheric $CO_2$ concentrations build to sufficient levels to make a major difference in Earth's climate.

Among the claims of critics of global climate warming are:

- The increase in the Earth's surface temperature during the past 150 years is less than the best existing climate models can explain. As the models improve, however, they predict less and less warming.

FIGURE 2.11

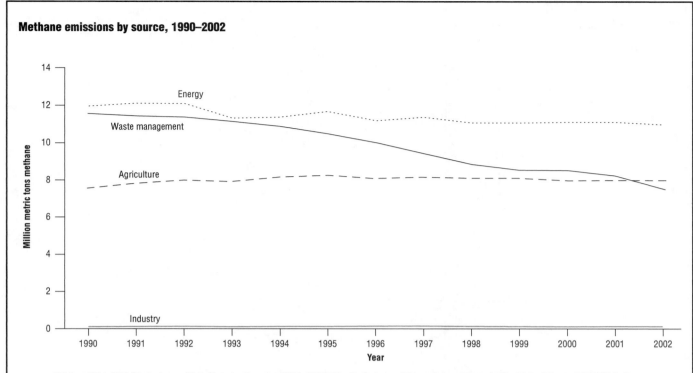

**Methane emissions by source, 1990–2002**

SOURCE: "Figure ES4. U.S. Emissions of Methane by Source, 1990–2002," in *Emissions of Greenhouse Gases in the United States 2002*, U.S. Department of Energy, Energy Information Administration, Office of Integrated Analysis and Forecasting, Washington, DC, October 2003

- Climate has been known to change dramatically within a relatively short period without any human influence.

- Temperature readings already showed increased temperatures before $CO_2$ levels rose significantly (before 1940).

- Natural variations in climate may exceed any human-caused climate change.

- Some of the increase in temperatures can be attributed to sunspot activity.

- If warming should occur, it will not stress Earth; it may even have benefits, such as for agriculture, and may delay the next ice age.

- Reducing emissions will raise energy prices, reduce gross domestic product, and produce job losses.

- While clouds are crucial to climate predictions, so little is known about them that computer models cannot produce accurate predictions.

In 2000 a handful of scientists at the Oregon Institute of Science and Medicine began a drive to collect the signatures of other scientists on a petition reading in part: "There is no convincing scientific evidence that human release of carbon dioxide, methane, or other greenhouse gasses is causing or will, in the foreseeable future, cause catastrophic heating of the Earth's atmosphere and disruption of the Earth's climate. Moreover, there is substantial scientific evidence that increases in atmospheric carbon dioxide produce many beneficial effects on the natural plant and animal environments of the Earth." As of April 2004 the so-called Petition Project claims to have collected more than 19,500 signatures, including 17,100 from scientists, mostly with advanced degrees.

## THE RESPONSE OF THE NATIONS

### United States

Many industrialized countries have committed themselves to stabilizing or reducing $CO_2$ emissions. President George H. W. Bush's administration (1989–1993) opposed precise deadlines for $CO_2$ limits, arguing that the extent of the problem was too uncertain to justify painful economic measures. However, in 1989 the U.S. Global Change Research Program (USGCRP) was established and later authorized by Congress in the Global Change Research Act of 1990.

When President Bill Clinton took office in 1993, he joined the European Community in calling for overall emissions to be stabilized at 1990 levels by 2000, but this goal was not met. In October 1993 the United States, under the United Nations (UN) Framework Convention on Climate Change, released *The Climate Change Action Plan* detailing the nation's response to climate change.

FIGURE 2.12

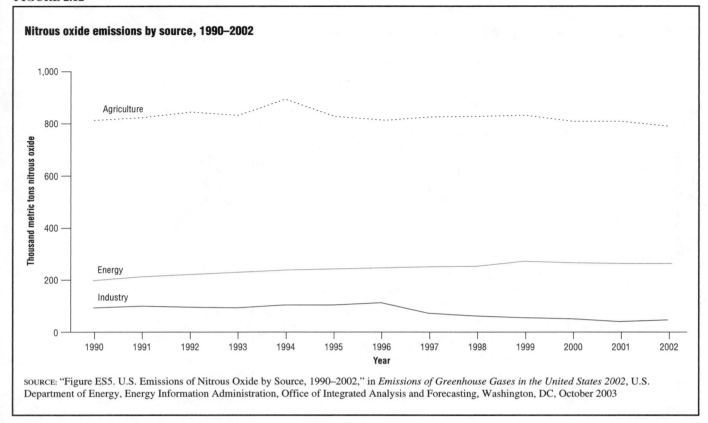

**Nitrous oxide emissions by source, 1990–2002**

SOURCE: "Figure ES5. U.S. Emissions of Nitrous Oxide by Source, 1990–2002," in *Emissions of Greenhouse Gases in the United States 2002*, U.S. Department of Energy, Energy Information Administration, Office of Integrated Analysis and Forecasting, Washington, DC, October 2003

The plan included a set of measures by both government and the private sector to lay a foundation for the nation's participation in world response to the climate challenge.

The measures called for under the *Action Plan* would reduce emissions for all greenhouse gases to 1990 levels by 2000. However, since the time the projections were prepared and the *Action Plan* was published, the economy grew at a more robust rate than anticipated, which led to increased emissions. Furthermore, the U.S. Congress did not provide full funding for the actions contained in the plan.

At the time, the U.S. global warming program was coordinated through the Committee on Earth and Environmental Science within the Office of Science and Technology Policy of the White House. Some eighteen federal agencies were represented in the multiyear, multibillion-dollar research program, nine of which received the bulk of funding. NASA received 66 percent, and the National Science Foundation (NSF) was provided 10 percent. Rounding out the top nine were the Smithsonian Institution and a half dozen federal agencies, including the EPA and the Departments of Agriculture, Commerce, Defense, Energy, and the Interior.

Even though the United States had a comprehensive global warming program in place, Congress was reluctant to take steps to reduce emissions. However, the Clinton administration implemented some policies that did not require congressional approval. In 1999 President Clinton made a number of executive orders and began several new initiatives. These included the Climate Change Technology initiative, a five-year, $6.3 billion program of tax incentives and investments focusing on improving energy efficiency and renewable energy technologies; the Wind Powering America initiative, seeking to supply 5 percent of the nation's electricity through wind technology by 2020; and a Brightfields initiative, aimed at using former industrial sites contaminated with toxic waste for producing pollution-free solar energy. Executive Order 13134 aimed at coordinating federal efforts to develop technology to grow the economy while simultaneously solving some environmental problems, such as converting crops, trees, and other "biomass" into fuels, power, and products. Executive Order 13123 required all federal government agencies to reduce greenhouse gas emissions below 1990 levels by 2010. The federal government was the largest energy consumer in the nation, and the Clinton administration wanted it to serve as an example to the nation's businesses and consumers, who ultimately could reap benefits from energy improvements. President Clinton also established the U.S. Climate Change Research Initiative to study areas of uncertainty about global climate change science and identify priorities for public investments.

After President George W. Bush took office in 2001 he established a new cabinet-level management structure

FIGURE 2.13

**Sources of nitrous oxide (N₂O) emissions from agricultural soils**

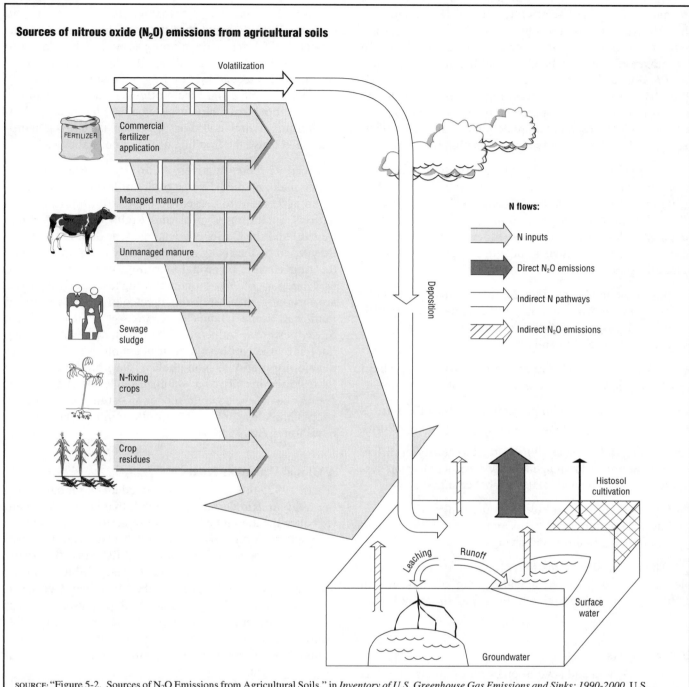

SOURCE: "Figure 5-2. Sources of N₂O Emissions from Agricultural Soils," in *Inventory of U.S. Greenhouse Gas Emissions and Sinks: 1990-2000,* U.S. Environmental Protection Agency, Washington, DC, April 15, 2002

to oversee government investments in climate change science and technology. Both the U.S. Climate Change Research Initiative and the USGCRP were placed under the oversight of the interagency Climate Change Science Program (CCSP), which reports integrated research sponsored by thirteen federal agencies. The CCSP is overseen by the Office of Science and Technology Policy, the Council on Environmental Quality, and the Office of Management and Budget.

In June 2002 the George W. Bush administration released the third National Communication of the United States of America under the UN Framework Convention on Climate Change. The *U.S. Climate Action Report— 2002* acknowledges that greenhouse gases resulting from human activities are accumulating in the atmosphere and that they are causing air and ocean temperatures to increase. However, it does not rule out the still unknown role of natural variability in global warming. In addition, the report reiterates that the administration plans to reduce the nation's greenhouse gas intensity by 18 percent over the following decade through a combination of existing regulations and voluntary, incentive-based measures. In November of that year the CCSP publication *Our Changing Planet* was issued as a supplement to the President's 2003 fiscal year budget. According to this document the administration's overall approach to climate change is "to achieve both environmental protection and a healthy economy, based on the best possible information for action."

In July 2003 the CCSP published two major reports: *Strategic Plan for the U.S. Climate Change Science Program* and *The U.S. Climate Change Science Program: Vision for the Program and Highlights of the Scientific Strategic Plan*. Together these documents outline the approach the CCSP plans to take to achieve its five main scientific goals:

- Improve knowledge of the Earth's past and present climate and environment, including its natural variability, and improve understanding of the causes of observed variability and change

- Improve quantification of the forces bringing about changes in the Earth's climate and related systems

- Reduce uncertainty in projections of how the Earth's climate and related systems may change in the future

- Understand the sensitivity and adaptability of different natural and managed ecosystems and human systems to climate and related global changes

- Explore the uses and identify the limits of evolving knowledge to manage risks and opportunities related to climate variability and change

The CCSP illustrates that climate change is a complex issue and involves a wide range of natural and anthropogenic (human-related) factors as shown in Figure 2.14.

## Other Nations

Saudi Arabia and several other oil producing nations have resisted the setting of new targets for emissions because they fear this will reduce the demand for their oil. On the other hand, many environmentalists believe that reducing pollution depends on increasing energy efficiency and gradually switching from fossil fuels to renewable energy, an idea most industrialized countries have been slow to embrace.

China's ongoing economic revolution may lead that nation to become the world's largest contributor to global warming. China is burning increasing amounts of coal and is expected to become the single greatest creator of greenhouse gases in the coming decades. China's leaders indicate that they are well aware that coal burning causes pollution, but the Chinese government has made it clear that it will not sacrifice development for the environment. Chinese leaders argue that it is unfair to impose constraints on China or developing nations when Western countries have been willfully polluting the environment for more than a century, pointing out that China has only recently become a significant offender. Critics of China's argument note that China itself would be harmed by global warming, which could dry up crops, shrink water supplies, and cause the flooding of major coastal cities.

In 1992 representatives of thirty-seven island nations, which make up the Alliance of Small Island States, drafted an agreement to present to the Earth Summit. These nations, including Cyprus and Malta (in the Mediterranean) and the Caribbean islands, fear their existence is threatened by the rising sea level. They claim they will be the first victims of global warming, becoming a whole new category of environmental refugees. Tourist beaches are shrinking, dikes are being erected to protect reclaimed land, and some islands have already been evacuated. Since they are so vulnerable, these nations feel an urgency not felt by the Northern industrial countries and the larger developing nations. Nonetheless, larger countries may have similar incentives to stabilize sea levels. A three-foot rise in ocean levels would render an estimated 72 million people homeless in China, 11 million in Bangladesh, and 8 million in Egypt.

## A Global Warming Treaty

In 1992, 143 countries approved a UN global warming treaty in Rio de Janeiro, Brazil, that committed them to reduce the amount of greenhouse gases emitted into the atmosphere. Many environmentalists criticized the treaty as too weak because it did not establish specific targets that governments must meet. The treaty did not include specific targets mainly because then-President George H. W. Bush, representing the United States, refused to accept them. President Clinton signed the treaty in 1994, although adherence to the environmental measures has largely been disappointing even among those nations that originally signed.

Supporters of a global warming strategy advocate limiting the emissions of the four main greenhouse gases and recommend a gradual transition away from fossil fuels, which currently provide about three-quarters of the world's energy.

In 1995, 120 parties to the global warming treaty met in Berlin in what is known as the Berlin Mandate to determine the success of existing treaties and to embark on discussions of emissions after 2000. Differences persisted along North-South lines, with developing countries making essentially a moral argument for requiring more of the richer nations. They pointed out that the richer nations are responsible for most of the pollution. The Berlin talks

FIGURE 2.14

**Major factors affecting climate and climate change**

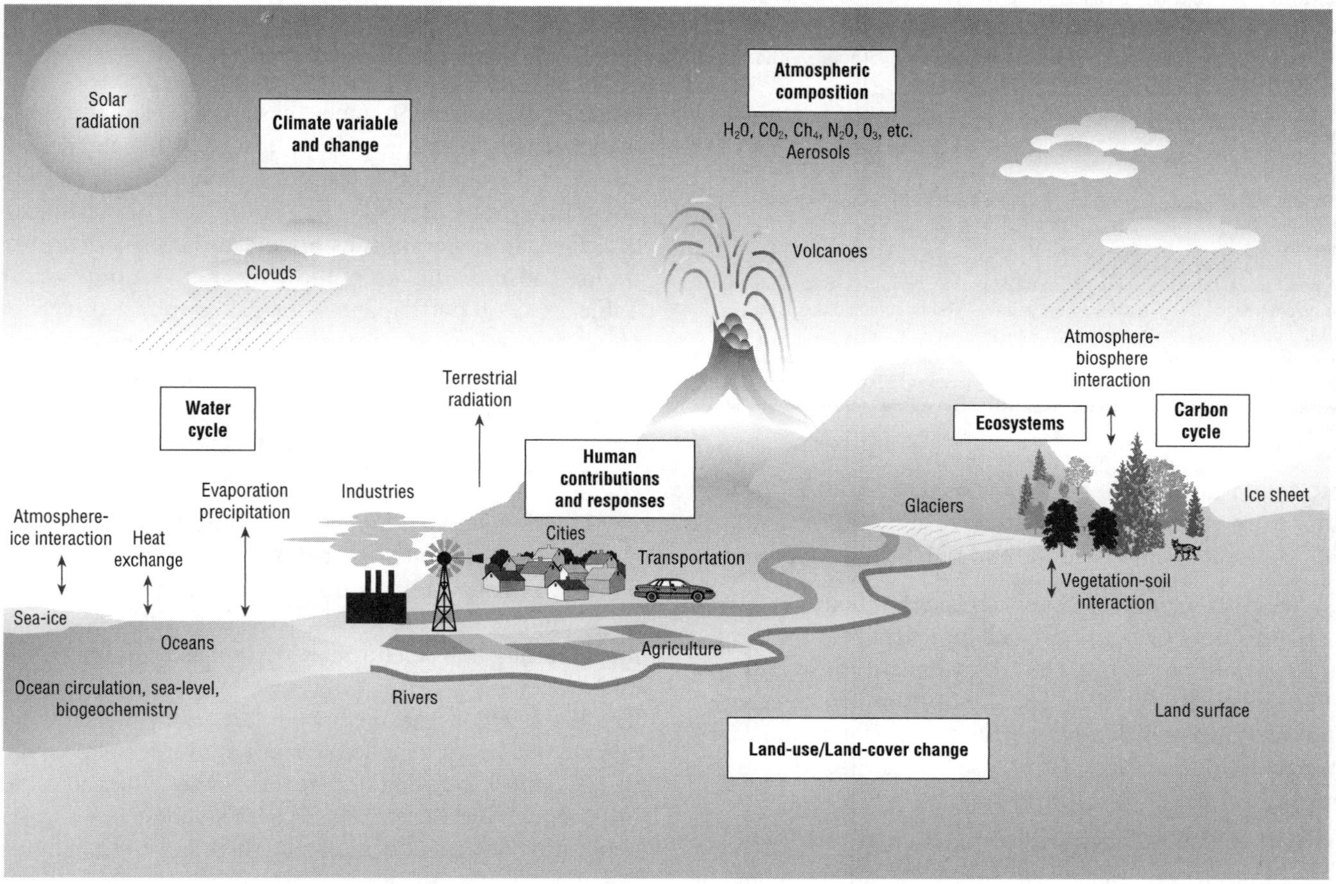

SOURCE: "Figure 1: Major Components Needed to Understand the Climate System and Climate Change," in *The U.S. Climate Change Science Program Vision for the Program and Highlights of the Scientific Strategic Plan,* Climate Change Science Program and Subcommittee on Global Change Research, Washington, DC, July 2003

essentially failed to endorse binding timetables for reductions in greenhouse gases.

### The Kyoto Protocol—Rich versus Poor Countries

In December 1997 delegates from 166 countries met in Kyoto, Japan, at the UN Climate Change Conference to negotiate actions to reduce global warming. The task was more complicated and difficult than was envisioned in 1995, when parties to the 1992 Rio de Janeiro treaty on climate change decided that stronger action was needed. What was originally envisioned as a matter of deciding on a reduction target and a timetable for industrialized countries once again broadened into a contentious debate between developed and developing countries as to the proper role of each.

Some developed nations, including the United States, wanted to require all countries to reduce their emissions. However, developing countries felt the industrialized nations had caused, and were still causing, most global warming and therefore should bear the brunt of economic sacrifices to clean up the environment. Even within the industrialized community, the European Union criticized the United States for lagging behind in reducing emissions, as it had previously pledged.

Different targets for different countries, tailored to their economic and social circumstances, emerged as a possible way to get around the impasse between nations. The final compromise, signed by the parties, was for the industrialized nations to cut emissions by an average of 5.2 percent between 2008 and 2012. The treaty also set up an emission trading system that would allow countries that exceed their pollution limits to purchase on an open market "credits" from countries that pollute less. This provision was viewed as necessary to U.S. congressional approval. The developing nations feared that such a trading system would allow rich countries to buy their way

**TABLE 2.3**

**Public understanding of the global warming issue, 2004**

NEXT, THINKING ABOUT THE ISSUE OF GLOBAL WARMING, SOMETIMES CALLED THE "GREENHOUSE EFFECT," HOW WELL DO YOU FEEL YOU UNDERSTAND THIS ISSUE— WOULD YOU SAY VERY WELL, FAIRLY WELL, NOT VERY WELL, OR NOT AT ALL?

| | Very well % | Fairly well % | Not very well % | Not at all % | No opinion % |
|---|---|---|---|---|---|
| 2004 Mar 8–11 | 18 | 50 | 26 | 6 | * |
| 2003 Mar 3–5 | 15 | 53 | 27 | 5 | – |
| 2002 Mar 4–7 | 17 | 52 | 25 | 6 | * |
| 2001 Mar 5–7 | 15 | 54 | 24 | 6 | 1 |
| 1997 Nov 6–9 | 16 | 45 | 28 | 10 | 1 |
| 1992 Jan | 11 | 42 | 22 | 22 | 3 |

SOURCE: "Next, thinking about the issue of global warming, sometimes called the 'greenhouse effect,' how well do you feel you understand this issue— would you say very well, fairly well, not very well, or not at all?," in *Poll Topics and Trends: Environment*, The Gallup Organization, Princeton, NJ, March 17, 2004 [Online] www.gallup.com [accessed March 30, 2004]

into compliance rather than make unpopular emissions cuts. Enforcement mechanisms were not agreed to, nor did developing nations commit to binding participation.

The U.S. Congress never ratified the treaty. In March 2001 President George W. Bush indicated that the United States would pull out of the treaty because it would cost an estimated $400 billion and 4.9 million jobs to comply. The treaty was ratified by the European Union and Japan in June 2002.

## PUBLIC OPINION ABOUT GLOBAL WARMING

In March 2004 the Gallup Organization conducted its annual poll on topics related to the environment. Participants were asked several questions about global warming and the Kyoto Protocol.

As shown in Table 2.3 more than a quarter of those asked admitted they did not have a good understanding of the global warming issue. Half of those asked said they understood the issue fairly well. Another 18 percent felt they had a very good understanding about the issue. In 1992, 22 percent of people responded that they did not understand the issue at all, but this had dropped to 6 percent in 2004.

Table 2.4 shows how poll participants view the timeliness of global warming. Just over half (51 percent) of those asked in 2004 believe that global warming has already begun, while 17 percent feel that global warming will occur within the next few years or within their lifetimes. Another 18 percent consider global warming a problem for the distant future, and 11 percent believe that

it will never happen at all. This breakdown of opinions on the subject has remained fairly constant through the years.

Concern about global warming has wavered over the years but is generally lower in 2004 than in previous years. (See Table 2.5). In a poll from March 2004, 26 percent of respondents expressed a great deal of concern about global warming and 25 percent expressed a fair amount of concern. Together these opinions represent just over half of all the people polled. In 1989 nearly two-thirds of those asked felt a great deal or fair amount of concern about this issue. In 2000 nearly three-quarters did. Overall concern does appear to be on a downward trend. These results coincide with the data presented in Table 2.6. Poll participants were asked to rate their views about the seriousness of global warming based on the information they have seen in the media. In 2004 more people (38 percent) considered the seriousness to be exaggerated by the media rather than reported correctly (25 percent) or underestimated (33 percent).

In 2004 Gallup asked people about the response of the Bush administration to global warming. Pollsters pointed out that a group of "prominent" scientists have criticized the administration's approach to dealing with global warming, claiming that scientific evidence has been ignored and distorted. Only 8 percent of the poll respondents indicated they had heard a great deal about this criticism. Another 26 percent had heard a moderate amount of information about it. The largest fraction (40 percent) claimed they knew very little about the criticism and 26 percent knew nothing at all about it. When asked which side they tended to believe in this dispute (the Bush administration or the scientists) a majority (59 percent) sided with the scientists, while 32 percent sided with the Bush administration.

Despite the administration's well-publicized opposition to the Kyoto treaty, Gallup's 2004 poll shows that more people support the treaty than oppose it. Figure 2.15 illustrates that 42 percent of poll respondents indicated that the United States should agree to abide by the provisions of the treaty, compared to 22 percent who disagreed. A sizable percentage of those asked (36 percent) had no opinion on the matter.

People participating in Gallup's 2003 environment poll were asked whether increases in the Earth's temperature over the last century were due more to natural causes or to pollution associated with human activities. As shown in Figure 2.16, human activities received nearly twice the blame than did natural causes. This breakdown is identical to that reported by Gallup when the same question was asked in 2001.

TABLE 2.4

**Public opinion about when the effects of global warming will begin, 2004**

WHICH OF THE FOLLOWING STATEMENTS REFLECTS YOUR VIEW OF WHEN THE EFFECTS OF GLOBAL WARMING WILL BEGIN TO HAPPEN—[ROTATED: THEY HAVE ALREADY BEGUN TO HAPPEN, THEY WILL START HAPPENING WITHIN A FEW YEARS, THEY WILL START HAPPENING WITHIN YOUR LIFETIME, THEY WILL NOT HAPPEN WITHIN YOUR LIFETIME, BUT THEY WILL AFFECT FUTURE GENERATIONS, (OR) THEY WILL NEVER HAPPEN]?

| | Already begun % | Within a few years % | Within your lifetime % | Not within lifetime, but affect future % | Will never happen % | No opinion % |
|---|---|---|---|---|---|---|
| 2004 Mar 8–11 | 51 | 5 | 12 | 18 | 11 | 3 |
| 2003 Mar 3–5 | 51 | 6 | 12 | 17 | 10 | 4 |
| 2002 Mar 4–7 | 53 | 5 | 13 | 17 | 9 | 3 |
| 2001 Mar 5–7 | 54 | 4 | 13 | 18 | 7 | 4 |
| 1997 Nov 6–9 | 48 | 3 | 14 | 19 | 9 | 7 |

SOURCE: "Which of the following statements reflects your view of when the effects of global warming will begin to happen—[ROTATED: they have already begun to happen, they will start happening within a few years, they will start happening within your lifetime, they will not happen within your lifetime, but they will affect future generations, (or) they will never happen]?," in *Poll Topics and Trends: Environment*, The Gallup Organization, Princeton, NJ, March 17, 2004 [Online] www.gallup.com [accessed March 30, 2004]

TABLE 2.5

**Public concern about global warming, 2004**

PLEASE TELL ME IF YOU PERSONALLY WORRY ABOUT THIS PROBLEM A GREAT DEAL, A FAIR AMOUNT, ONLY A LITTLE, OR NOT AT ALL. THE "GREENHOUSE EFFECT" OR GLOBAL WARMING?

| | Great deal % | Fair amount % | Only a little % | Not at all % | No opinion % |
|---|---|---|---|---|---|
| 2004 Mar 8–11 | 26 | 25 | 28 | 19 | 2 |
| 2003 Mar 3–5 | 28 | 30 | 23 | 17 | 2 |
| 2002 Mar 4–7 | 29 | 29 | 23 | 17 | 2 |
| 2001 Mar 5–7 | 33 | 30 | 22 | 13 | 2 |
| 2000 Apr 3–9 | 40 | 32 | 15 | 12 | 1 |
| 1999 Apr 13–14 | 34 | 34 | 18 | 12 | 2 |
| 1999 Mar 12–14 | 28 | 31 | 23 | 16 | 2 |
| 1997 Oct 27–28 | 24 | 26 | 29 | 17 | 4 |
| 1991 Apr 11–14 | 35 | 27 | 22 | 12 | 5 |
| 1990 Apr 5–8 | 30 | 27 | 20 | 16 | 6 |
| 1989 May 4–7 | 35 | 28 | 18 | 12 | 7 |

SOURCE: "Please tell me if you personally worry about this problem a great deal, a fair amount, only a little, or not at all. The 'greenhouse effect' or global warming?," in *Poll Topics and Trends: Environment*, The Gallup Organization, Princeton, NJ, March 17, 2004 [Online] www.gallup.com [accessed March 30, 2004]

TABLE 2.6

**Public opinion about the seriousness of global warming, 2004**

THINKING ABOUT WHAT IS SAID IN THE NEWS, IN YOUR VIEW IS THE SERIOUSNESS OF GLOBAL WARMING—[ROTATED: GENERALLY EXAGGERATED, GENERALLY CORRECT, OR IS IT GENERALLY UNDERESTIMATED]?

| | Generally exaggerated % | Generally correct % | Generally underestimated % | No opinion % |
|---|---|---|---|---|
| 2004 Mar 8–11 | 38 | 25 | 33 | 4 |
| 2003 Mar 3–5 | 33 | 29 | 33 | 5 |
| 2002 Mar 4–7 | 31 | 32 | 32 | 5 |
| 2001 Mar 5–7 | 30 | 34 | 32 | 4 |
| 1997 Nov 6–9* | 31 | 34 | 27 | 8 |

*Based on half sample.

SOURCE: "Thinking about what is said in the news, in your view is the seriousness of global warming—[ROTATED: generally exaggerated, generally correct, or is it generally underestimated]?," in *Poll Topics and Trends: Environment*, The Gallup Organization, Princeton, NJ, March 17, 2004 [Online] www.gallup.com [accessed March 30, 2004]

**FIGURE 2.15**

**Public opinion about the Kyoto agreement, March 2004**

BASED ON WHAT YOU HAVE HEARD OR READ, DO YOU THINK THE UNITED STATES SHOULD— OR SHOULD NOT—AGREE TO ABIDE BY THE PROVISIONS OF THE KYOTO AGREEMENT ON GLOBAL WARMING?

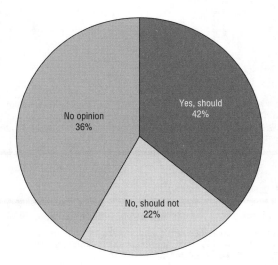

SOURCE: "Next, turning to the agreement on global warming that was drawn up at a world conference in Kyoto, Japan several years ago, based on what you have heard or read, do you think the United States should or should not agree to abide by the provisions of the Kyoto agreement on global warming?," in *Poll Topics and Trends: Environment*, The Gallup Organization, Princeton, NJ, March 17, 2004 [Online] www.gallup.com [accessed March 30, 2004]

**FIGURE 2.16**

**Public opinion regarding Earth's increasing temperature, March 2003**

DO YOU BELIEVE INCREASES IN THE EARTH'S TEMPERATURE OVER THE LAST CENTURY ARE DUE MORE TO

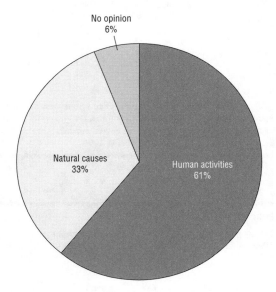

SOURCE: "Do you believe increases in the Earth's temperature over the last century are due more to—[ROTATED: the effects of pollution from human activities (or) natural changes in the environment that are not due to human activities]?," in *Poll Topics and Trends: Environment*, The Gallup Organization, Princeton, NJ, March 17, 2004 [Online] www .gallup.com [accessed March 30, 2004]

# CHAPTER 3
# A HOLE IN THE SKY: OZONE DEPLETION

## EARTH'S PROTECTIVE OZONE LAYER

Ozone is a gas naturally present in Earth's atmosphere. Unlike regular oxygen, which contains two oxygen atoms ($O_2$), ozone contains three oxygen atoms ($O_3$). A molecule of regular oxygen can be converted to ozone by ultraviolet (UV) radiation, electrical discharge (such as from lightning), or complex chemical reactions. These processes split apart the two oxygen atoms, which are then free to bind with other loose oxygen atoms to form ozone.

Ozone exists in Earth's atmosphere at two levels—the troposphere and the stratosphere. (See Figure 3.1.) Tropospheric (or ground-level) ozone accounts for only a small portion of Earth's total ozone, but it is a potent air pollutant with serious health consequences. Ground-level ozone is the primary component in smog and is formed via complex chemical reactions involving emissions of industrial chemicals and through fossil fuel combustion. Ozone formation is intensified during hot weather, when more radiation reaches the ground. Smog retards crop and tree growth, impairs health, and limits visibility.

Approximately 90 percent of the Earth's ozone lies in the stratosphere at altitudes greater than about twenty miles. (See Figure 3.2.) Ozone molecules at this level protect life on Earth by absorbing UV radiation from the sun and preventing it from reaching the ground. The so-called ozone layer is actually a scattering of molecules constantly undergoing change from oxygen to ozone and back. Although most of the ozone changes back to oxygen, a small amount of ozone persists. As long as this natural process stays in balance, the overall ozone layer remains thick enough to do its job. The amount of ozone in the stratosphere varies greatly depending upon location, altitude, and temperature.

## EVIDENCE OF OZONE DEPLETION

Many scientists believe that the introduction of certain chemicals into the stratosphere alters the natural ozone balance by depleting ozone molecules. Chlorine and bromine atoms are particularly destructive. They can bind to loose oxygen atoms and prevent them from reforming either oxygen or ozone. Chlorine and bromine are found in the sea salt from ocean spray. Chlorine is also present in the form of hydrochloric acid emitted with volcanic gases. These are natural sources of ozone-depleting chemicals.

In the mid-1970s scientists first began to speculate that the ozone layer was rapidly being destroyed by reactions involving industrial chemicals that contained chlorine and bromine. Two chemists, F. Sherwood Rowland and Mario Molina, discovered that chlorofluorocarbons (CFCs) could break down in the stratosphere, releasing chlorine atoms that could destroy thousands of ozone molecules. This discovery led to a ban on CFCs as a propellant in aerosols in the United States and other countries.

In 1984 British scientists at Halley Bay in Antarctica measured the ozone in the air column above them and discovered alarmingly low concentrations. Measurements indicated ozone levels about 50 percent lower than they had been in the 1960s.

Scientists report ozone concentrations in units called Dobson units. The unit is named after G.M.B. Dobson (1889–1976), a British scientist who invented an instrument for measuring ozone concentrations from the ground. One Dobson unit (DU) corresponds to a layer of atmospheric ozone that would be 0.001 millimeters thick if it was compressed into a layer at standard temperature and pressure at the Earth's surface. Atmospheric ozone is considered "thin" if its concentration falls below 220 DU. A thin spot in the ozone layer is commonly called an "ozone hole."

Since 1982 an ozone hole has appeared over Antarctica in the springtime (our autumn) for several months. (See Figure 3.3.) The hole varies in size throughout each of its appearances, as indicated by the vertical lines on the graph. The average area calculated for each year is indicated by the

FIGURE 3.1

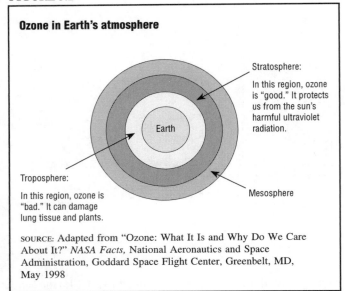

**Ozone in Earth's atmosphere**

Stratosphere:

In this region, ozone is "good." It protects us from the sun's harmful ultraviolet radiation.

Earth

Troposphere:

In this region, ozone is "bad." It can damage lung tissue and plants.

Mesosphere

SOURCE: Adapted from "Ozone: What It Is and Why Do We Care About It?" *NASA Facts,* National Aeronautics and Space Administration, Goddard Space Flight Center, Greenbelt, MD, May 1998

FIGURE 3.2

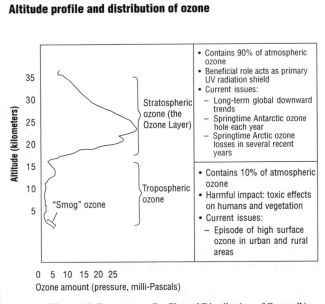

**Altitude profile and distribution of ozone**

Altitude (kilometers)

35

30

25

20

15

10

5

"Smog" ozone

Stratospheric ozone (the Ozone Layer)

Tropospheric ozone

- Contains 90% of atmospheric ozone
- Beneficial role acts as primary UV radiation shield
- Current issues:
  – Long-term global downward trends
  – Springtime Antarctic ozone hole each year
  – Springtime Arctic ozone losses in several recent years

- Contains 10% of atmospheric ozone
- Harmful impact: toxic effects on humans and vegetation
- Current issues:
  – Episode of high surface ozone in urban and rural areas

0  5  10  15  20  25
Ozone amount (pressure, milli-Pascals)

SOURCE: "Figure 14. Temperature Profile and Distribution of Ozone," in *Reporting on Climate Change: Understanding the Science*, National Safety Council, Environmental Health Center, Washington, DC, June 2000

heavy dots on the graph. As shown in Figure 3.3, the hole's average area has increased dramatically since the early 1980s. By the early 1990s the hole was consistently larger than the area of Antarctica. Throughout most of the early 2000s the hole has been larger in size than the continent of North America. In 2002 the hole size dropped dramatically due to unusually warm weather at the South Pole. The hole rebounded to its largest area yet in the autumn of 2003.

Ozone depletion has also been detected over other parts of the world. In 1993 satellite measurements indicated a 10 to 20 percent reduction in ozone levels over parts of Canada, Scandinavia, Russia, and Europe compared to 1992 levels. In 1995 scientists first detected thinning of the ozone over the North Pole. The extreme cold and unique climate conditions over the poles are thought to make the ozone layers there particularly susceptible to thinning. In colder areas of the planet, such as Antarctica, where cloud and ice particles are present, reactions that hasten ozone destruction also occur on the surface of ice particles.

In January 2002 the European Space Agency (ESA) announced that ozone thinning had been detected over Europe by the Global Ozone Monitoring Experiment instrument aboard the ESA's *ERS-2* satellite. The thinning in the ozone layer lasted for only three days. Similar thinning was observed previously in November 1999 and November 2001. Scientists believe that unusual air currents in the stratosphere, rather than chemical depletion, may be responsible for the thinning.

## CONSEQUENCES OF OZONE DEPLETION

Ironically, destruction of the ozone layer at the upper levels could increase the amount of ozone at the Earth's surface. In addition, the decline in stratospheric ozone is thought to increase hydrogen peroxide in the stratosphere, contributing to acid rain. The ozone layer acts as a protective shield against UV radiation. As ozone diminishes in the upper atmosphere, the Earth could receive more UV radiation. Increased radiation, especially of the frequency known as ultraviolet-B, the most damaging wavelength, promotes skin cancers and cataracts, suppresses the human immune system, and produces wrinkled, leather-like skin. It also reduces crop yields and fish populations, damages some materials such as plastics, and intensifies smog creation.

### Threats to Human Health

The skin is the largest organ in the human body. It covers and protects the organs inside the body. Globally, the incidence of skin cancer is rising. There are two types of skin cancer: melanoma and nonmelanoma.

The American Cancer Society reported in 2004 that more than 1 million cases of nonmelanoma skin cancer are diagnosed in the United States each year and estimated that between 1,000 and 2,000 people would die from the disease. The incidence of cancer is closely tied to cumulative exposure to UV radiation. Each 1 percent drop in ozone is projected to result in a 4 to 6 percent increase in these types of skin cancer.

Melanoma skin cancer is less common, but far more deadly, accounting for just 4 percent of skin cancer cases but a staggering 79 percent of skin cancer deaths, according to the American Cancer Society in 2004. Melanoma is more likely to metastasize (spread to other parts of the body, particularly major organs such as the lungs, liver, and brain).

FIGURE 3.3

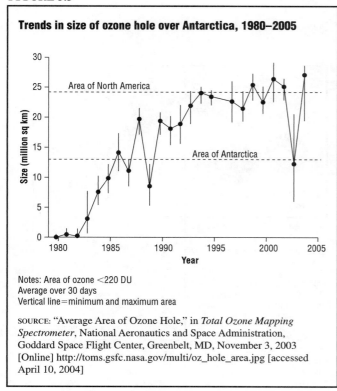

**Trends in size of ozone hole over Antarctica, 1980–2005**

Notes: Area of ozone <220 DU
Average over 30 days
Vertical line=minimum and maximum area

SOURCE: "Average Area of Ozone Hole," in *Total Ozone Mapping Spectrometer*, National Aeronautics and Space Administration, Goddard Space Flight Center, Greenbelt, MD, November 3, 2003 [Online] http://toms.gsfc.nasa.gov/multi/oz_hole_area.jpg [accessed April 10, 2004]

In June 2003 the American Cancer Society predicted that in 2004 there would be 55,100 new cases of melanoma in the United States and that 7,910 Americans would die from the disease. The Environmental Protection Agency (EPA) estimates that ozone depletion will result in an additional 31,000 to 126,000 cases and 7,000 to 30,000 fatalities among white Americans born before 2075. Melanoma appears to be associated with acute radiation exposure, such as severe sunburns, which are more likely to occur when the ozone hole is larger.

The EPA estimates that, among Americans born before 2075, depletion of the ozone layer could be responsible for 555,000 to 2.8 million additional cases of cataracts. Cataracts cause clouding of the eye's lens, which results in blurred vision and—if left untreated—blindness. It is also expected that victims will be stricken at younger and younger ages.

Some medical researchers theorize that UV radiation also depresses the human immune system, lowering the body's resistance to tumors and infectious diseases.

### Damage to Plant and Animal Ecosystems

Terrestrial and aquatic ecosystems are also affected by the depletion of the ozone layer. UV radiation alters photosynthesis, plant yield, and growth in plant species. Phytoplankton (one-celled organisms found in the ocean) are the backbone of the marine food web. Studies have found that a 25 percent reduction in ozone would decrease their productivity by about 35 percent. Fish species that live solely on phytoplankton would likely disappear. As ecosystems are altered, important fish species will become vulnerable. These fish are a vital component in feeding the increasing human population. In addition, scientists have identified the rise in UV radiation caused by the thinning of the ozone layer as the culprit behind the decline in the number of frogs and other amphibians.

### Deterioration of Materials

Increased UV radiation also affects synthetic materials. Plastics are especially vulnerable, tending to weaken, become brittle and discolored, and break.

### OZONE-DEPLETING CHEMICALS

Most ozone destruction in the atmosphere is believed to be anthropogenic (caused by humans). In 1999 the World Meteorological Organization in Geneva, Switzerland, estimated that only 18 percent of the sources contributing to ozone depletion were natural. The remaining 82 percent of sources contributing to ozone depletion were industrial chemicals. The blame is largely placed on chemicals developed by modern society for use as refrigerants, air conditioning fluids, solvents, cleaning agents, and foam-blowing agents. These chemicals can persist for years in the atmosphere. Thus, there is a significant lag between the time that emissions decline at the Earth's surface and the time at which ozone levels in the stratosphere recover.

Table 3.1 lists the chemicals of particular concern to scientists worried about thinning ozone levels. Each chemical is assigned a value called an ozone depletion potential (ODP) based on its harmfulness to the ozone layer. The most common depleters are the chlorofluorocarbons known as CFC-11, CFC-12, and CFC-13. Each of these chemicals is arbitrarily assigned an ODP of 1. The ODPs for other chemicals are determined by comparing their relative harmfulness to that of CFC-11. Class I chemicals are those with an ODP value greater than or equal to 0.2. Class II chemicals have ODP values less than 0.2.

### Class I Chemicals

Although a number of chemicals can destroy stratospheric ozone, CFCs are the main offenders because they are so prevalent. When CFCs were invented in 1930, they were welcomed as chemical wonders. Discovered by Thomas Midgley, Jr., they were everything the refrigeration industry needed at the time—nontoxic, nonflammable, noncorrosive, stable, and inexpensive. Their artificial cooling provided refrigeration for food and brought comfort to warm climates. The compound was originally marketed under the trademark Freon.

Over time, new formulations were discovered and the possibilities for use seemed endless. CFCs could be used as coolants in air conditioners and refrigerators, as propellants in aerosol sprays, in certain plastics such as poly-

TABLE 3.1

**Lifetime and ozone depletion potential of various chemicals**

| | Lifetime in years | Ozone depletion potential |
|---|---|---|
| **CLASS I** | | |
| CFC-11 | 45 | 1 |
| CFC-12 | 100 | 1 |
| CFC-13 | 640 | 1 |
| CFC-113 | 85 | 0.8–1 |
| CFC-114 | 300 | 0.94–1 |
| CFC-115 | 1700 | 0.44–0.6 |
| Halon 1211 | 16 | 3–6 |
| Halon 1301 | 65 | 10–12 |
| Halon 2402 | 20 | 6–8.6 |
| Carbon tetrachloride | 26 | 0.73–1.1 |
| Methyl bromide | 0.7 | 0.38–0.6 |
| Methyl chloroform | 5 | 0.1–0.12 |
| **CLASS II** | | |
| HCFC-21 | 1.7 | 0.04 |
| HCFC-22 | 1.2 | 0.05–0.055 |
| HCFC-123 | 1.3 | 0.02–0.06 |
| HCFC-124 | 5.8 | 0.02–0.04 |
| HCFC-141b | 9.3 | 0.1–0.12 |
| HCFC-142b | 17.9 | 0.06–0.07 |
| HCFC-225ca | 1.9 | 0.02–0.025 |
| HCFC-225cb | 5.8 | 0.03–0.033 |

The ODP is the ratio of the impact on ozone of a chemical compared to the impact of a similar mass of CFC–11. Thus, the ODP of CFC–11 is defined to be 1.0. Other CFCs and HCFCs have ODPs that range from 0.01 to 1.0.

SOURCE: Adapted from "Class I Ozone-Depleting Substances," and "Class II Ozone-Depleting Substances," in *Ozone Depletion Chemicals*, U.S. Environmental Protection Agency, Washington, DC, February 12, 2004 [Online] http://www.epa.gov/ozone/ods.html [accessed April 10, 2004]

styrene, in insulation, in fire extinguishers, and as cleaning agents. World production doubled every five years through 1970, and another growth spurt occurred in the 1980s as new uses were discovered—primarily as a solvent to clean circuit boards and computer chips.

CFCs are extremely stable; it is this stability that allows them to float intact through the troposphere (the layer of air nearest the Earth's surface) and into the ozone layer. They reach the stratosphere after six to eight years. Once there, some can survive for hundreds of years. CFCs do not degrade in the lower atmosphere but, on entering the stratosphere, they encounter the sun's UV radiation and eventually break down into chlorine, fluorine, and carbon. Many scientists believe it is the chlorine that damages the ozone layer. (See Figure 3.4.)

While CFCs are primarily blamed for ozone loss, other gases are also at fault. One of those gases is halon, which contains bromine. As shown in Table 3.1, halons have much higher ODP values than do CFCs. The bromine atoms in halons destroy ozone in a manner similar to that shown in Figure 3.4 for chlorine, but they are chemically more powerful. This means that the impact to ozone of a particular mass of halon is more destructive than a similar mass of a CFC. Halons are relatively long-

FIGURE 3.4

**Destruction of ozone**

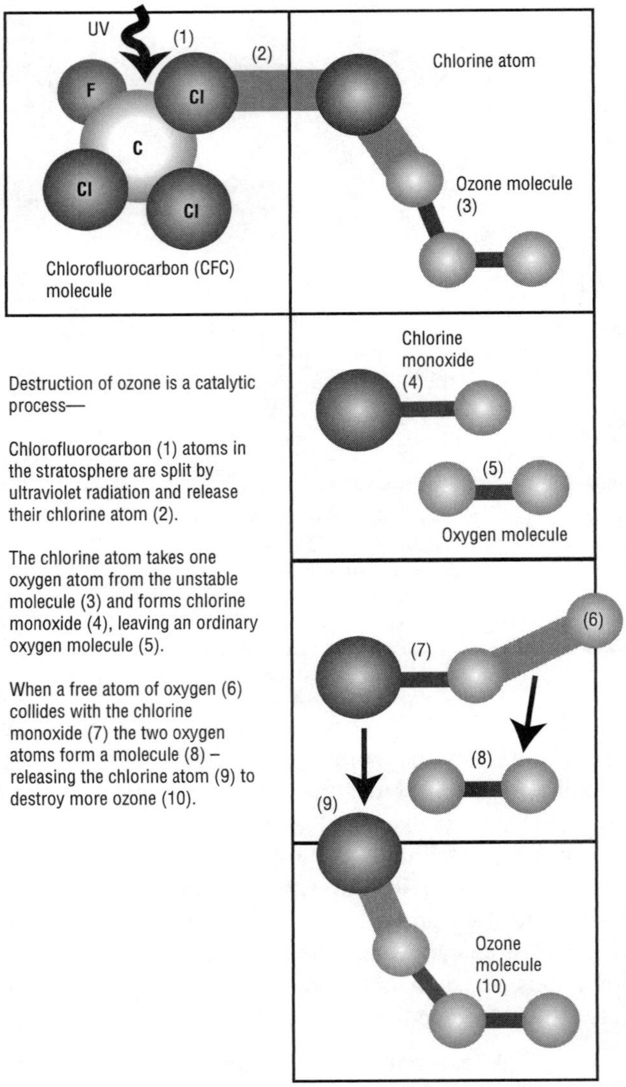
UV
(1)
(2)
F
Cl
C
Cl
Cl
Chlorine atom
Ozone molecule
(3)
Chlorofluorocarbon (CFC)
molecule
Chlorine
monoxide
(4)
Destruction of ozone is a catalytic process—
Chlorofluorocarbon (1) atoms in the stratosphere are split by ultraviolet radiation and release their chlorine atom (2).
The chlorine atom takes one oxygen atom from the unstable molecule (3) and forms chlorine monoxide (4), leaving an ordinary oxygen molecule (5).
When a free atom of oxygen (6) collides with the chlorine monoxide (7) the two oxygen atoms form a molecule (8) – releasing the chlorine atom (9) to destroy more ozone (10).
(5)
Oxygen molecule
(6)
(7)
(8)
(9)
Ozone
molecule
(10)

SOURCE: Adapted from "Ozone: What It Is and Why Do We Care About It?" *NASA Facts,* National Aeronautics and Space Administration, Goddard Space Flight Center, Greenbelt, MD, May 1998

lived in the atmosphere, lingering for up to sixty-five years before being broken down. Halon is used primarily for fighting fires. Civilian and military fire-fighting training accounts for much of the halon emission.

Other Class I ozone destroyers include carbon tetrachloride, methyl bromide, and methyl chloroform. These chemicals are commonly used by industry as solvents and cleaning agents.

### Class II Chemicals

The most common Class II ozone-depleting chemicals are hydrochlorofluorocarbons (HCFCs). HCFCs contain hydrogen. This makes them more susceptible to atmospheric breakdown than CFCs. As shown in Table 3.1, most HCFCs have a lifetime of less than six years. The most long-lived, HCFC-142b, lasts for only 17.9 years. HCFCs have much lower ODP values than CFCs, halons, and industrial ozone depleters. HCFCs are considered good short-term replacements for CFCs. Although HCFCs are less destructive to ozone than the chemicals they are replacing, scientists believe that HCFC use must also be phased out to allow the ozone layer to fully recover.

## A LANDMARK IN INTERNATIONAL DIPLOMACY: THE MONTREAL PROTOCOL

CFCs and halons were widely used in thousands of products and represented a significant share of the international chemical industry, with billions of dollars in investment and hundreds of thousands of jobs. Ozone depletion was a global problem that necessitated international cooperation, but nations mistrusted one another's motives. As with the issues of global warming and pollution, developing countries resented being asked to sacrifice their economic development for a problem they felt the industrialized nations had created. To complicate matters, gaps in scientific proof led to disagreements over whether a problem actually existed.

In 1985, as the first international response to the ozone threat, twenty nations signed an agreement in Vienna, Austria, calling for data gathering, cooperation, and a political commitment to take action at a later date. In 1987 negotiators meeting in Montreal, Canada, finalized a landmark in international environmental diplomacy: the Montreal Protocol on Substances That Deplete the Ozone Layer. It is generally referred to as the Montreal Protocol. The agreement was signed by twenty-nine countries including the United States, Canada, Mexico, Japan, Australia, all western European nations, the Russian Federation, and a handful of other countries around the world.

The agreement called for industrial countries to cut CFC emissions in half by 1998 and to reduce halon emissions to 1986 levels by 1992. Developing countries were granted deferrals to compensate for their low levels of production. Industrial countries agreed to reimburse developing countries that complied with the protocol for "all agreed incremental costs," meaning all additional costs above any they would have expected to incur had they developed their infrastructure in the absence of the accord. And, very importantly, the protocol also called for further amending as new data became available.

Although the ozone agreement was a major achievement in international negotiations, the world community could not enjoy the success for long. In 1991 new scientific information revealed that ozone depletion was occurring twice as fast as expected. This news spurred a call to again revise the treaty.

In 1992 representatives from eighty-seven nations met in Copenhagen, Denmark, to advance to January 1996 the deadline for halting production of CFCs. They agreed to halt the manufacture of new halon as of January 1994. (Developing nations were given a ten-year grace period to phase out manufacture of CFCs and halon.) A 1997 meeting in Montreal established deadlines for the phase-out of methyl bromide—2005 for developed nations and 2015 for developing countries. A deadline of 2015 was also set for developing nations to end production of methyl chloroform, which had been phased out by developed countries in 1996. The delegates also set a deadline for eliminating HCFCs. Their phaseout was to begin in 2004 and end no later than 2030. (See Table 3.2.)

The Montreal Protocol was hailed as an historic event—the most ambitious attempt ever to combat environmental degradation on a global scale. It ushered in a new era of environmental diplomacy. Some historians view the signing of the accord as a defining moment, the point at which the definition of international security was expanded to include environmental issues as well as military matters. In addition, an important precedent was established—that science and policymakers had a new relationship. Many observers thought that the decision to take precautionary action in the absence of complete proof of a link between CFCs and ozone depletion was an act of foresight that would now be possible with other issues.

## MONTREAL PROTOCOL PROBLEMS

Implementation and enforcement of the Montreal Protocol have been aggravated by two factors—the reluctance of many countries to ratify amendments to the Protocol and the thriving black market in banned chemicals.

### Amendment Ratification Moves Slowly

The United Nations reported that 186 countries had ratified the original Montreal Protocol as of January 2004. In total, four amendments to the Protocol have been adopted. These are known as the London Amendment (effective 1992), the Copenhagen Amendment (effective 1994), the Montreal Amendment (effective 1999), and the Beijing Amendment (effective 2002). Nations have been slow to ratify the amendments. For example, as of early 2004 only 68 nations have ratified the Beijing Amendment. The United States has ratified all amendments to the Montreal Protocol.

### Black-Market CFCs

Under the Montreal Protocol, developing countries are permitted to produce CFCs until January 1, 2010. It is illegal, however, to import CFCs into the United States. The ban on CFCs—including Freon, which was widely used for air conditioning automobiles—created a black market for the product. Demand by the approximately 80

**TABLE 3.2**

**Comparison of the Montreal Protocol and United States phaseout schedules**

| Montreal Protocol | | United States | |
| --- | --- | --- | --- |
| Year to be implemented | % Reduction in consumption using the cap as a baseline | Year to be implemented | Implementation of HCFC phaseout through Clean Air Act regulations |
| 2004 | 35.0% | 2003 | No production and no importing of HCFC-141b |
| 2010 | 65.0% | 2010 | No production and no importing of HCFC-142b and HCFC-22, except for use in equipment manufactured before 1/1/2010 (so no production or importing for NEW equipment that uses these refrigerants) |
| 2015 | 90.0% | 2015 | No production and no importing of any HCFCs, except for use as refrigerants in equipment manufactured before 1/1/2020 |
| 2020 | 99.5% | 2020 | No production and no importing of HCFC-142b and HCFC-22 |
| 2030 | 100.0% | 2030 | No production and no importing of any HCFCs |

HCFC = hydrochlorofluorocarbon

SOURCE: "Comparison of the Montreal Protocol and United States Phaseout Schedules," in *HCFC Phaseout Schedule*, U.S. Environmental Protection Agency, Washington, DC, May 7, 2002 [Online] http://www.epa.gov/ozone/title6/phaseout/hcfc.html [accessed May 15, 2002]

million American owners of older cars, which still use the refrigerant in their air-conditioning systems, caused prices to spiral upward.

Several U.S. government agencies—the EPA, the Customs Service, the Departments of Commerce and Justice, and the Internal Revenue Service—began intensive anti-smuggling efforts. In 2000 the EPA reported that approximately two million pounds of CFCs and other ozone-depleting substances had been seized and impounded since the 1996 phaseout of CFCs began and more than ninety individuals and businesses had been charged with smuggling. Early arrests for CFC smuggling were made primarily in Miami; authorities then began to notice these illegal imports creeping into other port cities, such as Houston and Baltimore. They also made their way into Europe, which banned CFCs one year before the United States.

In late 2003 a London-based nonprofit organization called the Environmental Investigation Agency (EIA) reported that international trade in illegal ozone-depleting substances is very detrimental to achieving progress under the Montreal Protocol (Tom Maliti, "Illegal Trade in Ozone-Depleting Substances Is Thriving over Three Con-

tinents, Says Report," Associated Press, November 11, 2003). According to the EIA, the port cities of Dubai in the United Arab Emirates and Singapore are "major transit points" in the smuggling efforts. The report notes that demand for illegal CFCs remains high in the United States, Russia, China, Vietnam, Cambodia, and Nepal.

Successful smuggling of CFCs typically involves false labeling, counterfeit paperwork, and fake export corporations. The EIA claims to have evidence that several companies based in Singapore illegally export CFCs to the United States. The organization says the companies make a 75–225 percent profit on each kilogram sold. The EIA estimates worldwide trade in illegal CFCs to be 18–27 million kilograms per year.

In November 2003 the U.S. Department of Justice (DOJ) announced its latest conviction in a massive Freon smuggling ring based in Panama that operated during the 1990s to supply Freon to customers in southern Florida. This is the third major conviction achieved in the case. In 2001 a U.S. district judge sentenced two men to up to two years in prison for their role in the smuggling scheme and for tax evasion. In 2003 their business partner was found guilty of tax fraud and money laundering. Authorities claim that the man laundered more than $8 million in profits from the smuggling operation by moving the money via wire transfers through a series of bank accounts and corporations in Panama.

## MONTREAL PROTOCOL SCIENCE

Article 6 of the Montreal Protocol requires that the ratifying nations base their decision-making on scientific information assessed and presented by an international panel of ozone experts. This panel includes the World Meteorological Organization (WMO), the United Nations Environment Programme (UNEP), the European Commission, and two U.S. organizations—the National Oceanic and Atmospheric Administration (NOAA) and the National Aeronautics and Space Administration (NASA).

In March 2003 the WMO published the panel's latest findings in *Scientific Assessment of Ozone Depletion: 2002*. This is the fifth scientific assessment of the world's ozone condition. Previous assessments were published in 1989, 1991, 1994, and 1998. The reports are based on analysis of data collected from satellites, aircraft, balloons, and ground-based instruments and the results of laboratory investigations and computer modeling.

The scientific assessment published in 2003 presents the following major findings:

- For the period 1997–2001 Earth's average stratospheric ozone concentration was approximately 3 percent less than the concentration measured prior to 1980 (the baseline condition).

- Ozone loss has occurred primarily at Earth's poles and over mid-latitude regions of the Northern and Southern hemispheres. No statistically significant ozone loss has been recorded over tropical regions near Earth's equator.

- Ozone depletion over the Southern hemisphere occurs year-round. Depletion over the Northern hemisphere is more seasonal, worsening during winter and spring months. Ozone depletion over Antarctica (the "ozone hole") has historically been a springtime phenomenon. However, data collected in recent years show that the ozone hole is persisting for a longer time each year, not disappearing until early summer.

- Ozone concentrations measured over Antarctica during occurrences of the ozone hole consistently average 40–50 percent less than concentrations measured prior to 1980.

- The average size of the Antarctic ozone hole increased during the 1980s and 1990s. Although the growth rate was slower during the 1990s than it was during the 1980s, scientists are not sure yet that the size of the hole has reached a maximum.

- Ozone depletion has been recorded at certain times over the North Pole during the last decade. However, scientists do not believe that an Arctic ozone hole that returns on a regular basis is likely to form due to the highly variable meteorological conditions at the North Pole.

- Measurements taken at various sites around the world indicate an increase of 6–14 percent in the amount of ultraviolet radiation reaching the ground since the early 1980s. However, lack of long-term data and an abundance of other factors besides ozone that affect UV levels (such as cloudiness) make it difficult to determine the statistical significance of the recorded UV increases.

- The total concentration of ozone-depleting chemicals in the troposphere (lower atmosphere) has declined since peaking in 1992–1994. Chlorine concentrations measured in 2000 were 5 percent lower than those recorded in 1992–1994. Bromine concentrations increased by 3 percent per year over the same time period. The increase in bromine is blamed on halon emissions. Atmospheric concentrations of methyl chloroform have declined significantly. CFC-11 and CFC-113 concentrations are also decreasing. CFC-12 concentrations continue to increase, although at a slower rate each year than in previous decades.

- Continued reductions in ozone-depleting chemicals in the stratosphere should lead to complete recovery of Earth's ozone layer during the twenty-first century. Computer models suggest that falling atmospheric ozone concentrations should level off by the year 2012 and then begin to increase to their normal levels. The annual occurrence of the Antarctic ozone hole is expected to cease by the year 2050.

## PRODUCTION AND DEMAND OF OZONE-DEPLETING CHEMICALS

In 1950 the worldwide production of CFCs was approximately 42,000 tons annually; in 1988, it peaked at 1.3 million tons. By the late 1990s, production had dropped to an estimated 300,000 tons.

In the industrial world, many countries did more than was required by the protocol. As a result, when the official CFC phaseout date arrived, most industrial nations were ready, and some had phased out ozone-depleting substances before they were required to. By the end of 1994, the European Union no longer permitted CFCs. In addition to banning Freon in 1997, the Clean Air Act required the United States to end the use of methyl bromide by 2001—nine years ahead of protocol requirements.

Although CFCs are no longer used in new applications, existing users can continue using them provided they are maintained under strict regulation, such as being replenished and "reclaimed" by authorized technicians. There are other exceptions to the ban, including medical inhalers, which commonly use CFCs as propellants.

### U.S. Emissions and Markets

Title VI of the 1990 Clean Air Act Amendments (PL 101-549) is the United States' primary response to ozone depletion. As part of the act, the U.S. Congress approved a provision requiring the president to speed up the schedule for chemical phaseout if new evidence warranted it. In fact, new data did become available, including worrisome evidence that showed ozone depletion was occurring over the Northern Hemisphere.

U.S. emissions of ozone-depleting substances for 1990 and 1995–2001 are shown in Table 3.3. Emissions of CFCs, carbon tetrachloride, and methyl chloroform have all decreased, some significantly. Halon-1301 emissions have also decreased since 1990. However, emissions of halon-1211 actually increased by a small amount.

Figure 3.5 compares annual emissions of major ozone-depleting substances from 1990 to 2001. Although emissions of CFCs have declined dramatically, emissions of HCFC-22 (the most widely used CFC substitute) have risen due to increased production. Figure 3.6 compares emissions of major ozone-depleting substances for 2001.

In 1999 the EPA published an analysis of the market for CFC-12 over the next decade (*Report on the Supply and Demand of CFC-12 in the United States*). The agency estimated that 67 million units using CFC-12 would still

**TABLE 3.3**

## Emissions of ozone-depleting substances, 1990 and 1995–2001

[in gigagrams (Gg)]

| Compound | 1990 | 1995 | 1996 | 1997 | 1998 | 1999 | 2000 | 2001 |
|---|---|---|---|---|---|---|---|---|
| **Class I** | | | | | | | | |
| CFC-11 | 53.5 | 36.2 | 26.6 | 25.1 | 24.9 | 24.0 | 22.8 | 22.8 |
| CFC-12 | 112.6 | 51.8 | 35.5 | 23.1 | 21.0 | 14.0 | 17.2 | 21.3 |
| CFC-113 | 52.7 | 17.1 | + | + | + | + | + | + |
| CFC-114 | 4.7 | 1.6 | + | + | + | + | + | + |
| CFC-115 | 4.2 | 3.0 | 3.2 | 2.9 | 2.7 | 2.6 | 2.3 | 1.5 |
| Carbon tetrachloride | 32.3 | 4.7 | + | + | + | + | + | + |
| Methyl chloroform | 316.6 | 92.8 | + | + | + | + | + | + |
| Halon-1211 | 1.0 | 1.1 | 1.1 | 1.1 | 1.1 | 1.1 | 1.1 | 1.1 |
| Halon-1301 | 1.8 | 1.4 | 1.4 | 1.3 | 1.3 | 1.3 | 1.3 | 1.2 |
| **Class II** | | | | | | | | |
| HCFC-22 | 34.0 | 39.3 | 41.0 | 42.4 | 43.8 | 74.1 | 79.1 | 80.5 |
| HCFC-123 | + | 0.6 | 0.7 | 0.8 | 0.9 | 1.0 | 1.1 | 1.2 |
| HCFC-124 | + | 5.6 | 5.9 | 6.2 | 6.4 | 6.5 | 6.5 | 6.5 |
| HCFC-141b | 1.3 | 9.9 | 9.9 | 8.8 | 9.7 | 10.9 | 10.9 | 10.7 |
| HCFC-142b | 0.8 | 3.6 | 4.0 | 4.3 | 4.7 | 5.0 | 5.4 | 5.8 |
| HCFC-225ca/cb | + | + | + | + | + | + | + | + |

Note: + Does not exceed 0.05 Gg

SOURCE: "Table ES-13: Emissions of Ozone Depleting Substances (Gg)," in *Inventory of U.S. Greenhouse Gas Emissions and Sinks: 1990–2001*, U.S. Environmental Protection Agency, Office of Atmospheric Programs, Washington, DC, April 15, 2003

be in use in the United States in the year 2005. Nearly 90 percent of these units are refrigerated appliances. Mobile air conditioners comprise most of the remainder. The number of units using CFC-12 is expected to fall steadily through 2010.

Demand for CFC-12 is expected to decline as substitutes increase and aging equipment is replaced. The EPA estimated demand for CFC-12 in 1999 at 23 million pounds. By 2005 U.S. consumers will demand approximately 3 million pounds of CFC-12. Mobile air conditioning accounts for 80 percent of this demand. The remainder includes commercial refrigeration at supermarkets and cold storage warehouses, industrial processes, refrigerated appliances, and refrigerated transport.

Recycled halon and inventories produced before January 1, 1994 are the only supplies now available. It is legal under the Montreal Protocol and the U.S. Clean Air Act to import recycled halon, but each shipment requires approval from the EPA. Certain uses, such as fire protection, are classified as "critical use" and are permitted as long as supplies remain. The EPA also maintains a list of acceptable substitutes for halon.

## SUBSTITUTES AND NEW TECHNOLOGIES

As pressure increased to discontinue use of CFCs and halons, substitute chemicals and technologies began to be developed. One of the most popular substitutes is a class of compounds called hydrofluorocarbons (HFCs). HFCs do not contain chlorine, a potent ozone destroyer. They are also relatively short-lived in the atmosphere. Most survive intact for less than twelve years. This means that HFCs do not directly impact Earth's protective ozone layer. HFCs have ODP values of zero.

During the 1990s use of HFCs increased dramatically. NASA reports that atmospheric levels of HFCs also surged over this time period. This is a concern to scientists studying global warming because HFCs are believed to enhance atmospheric heating. Also, HFC breakdown in the atmosphere produces a chemical called trifluoroacetic acid, large concentrations of which are known to be harmful to certain plants (particularly in wetlands). Continued heavy use of HFCs during the twenty-first century could introduce or aggravate other environmental problems.

Development of effective chemical substitutes with acceptable health and environmental effects is an enormous challenge. Some experts propose returning to the refrigerant gases used before the invention of CFCs. These include sulfur dioxide, ammonia, and various hydrocarbon compounds. However, these chemicals have their own issues; for example, most are highly toxic.

The EPA's Significant New Alternative Policy program evaluates alternatives to ozone-depleting substances and determines their acceptability for use. Submissions for evaluation include those that could be used in a variety of industrial applications, including refrigeration and air conditioning, foam blowing, and fire suppression and protection.

Many industrial engineers are pursuing new technologies for cooling, including semiconductors that cool down

FIGURE 3.5

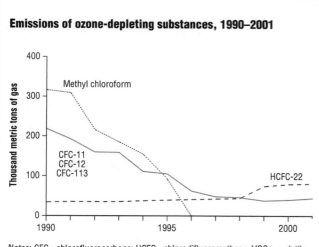

**Emissions of ozone-depleting substances, 1990–2001**

Notes: CFC=chlorofluorocarbons; HCFC=chlorodifluoromethane; VOCs=volatile organic compounds. Because vertical scales differ, graphs should not be compared.

SOURCE: "Figure 12.6. Ozone Depleting Substances and Criteria Pollutants: Ozone Depleting Substances, 1990–2001," in *Annual Energy Review 2002*, U.S. Department of Energy, Energy Information Administration, Washington, DC, October 2003

FIGURE 3.6

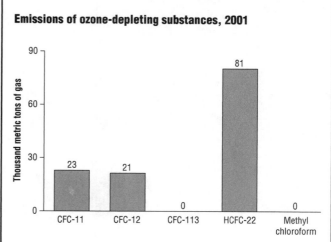

**Emissions of ozone-depleting substances, 2001**

Notes: CFC=chlorofluorocarbons; HCFC=chlorodifluoromethane

SOURCE: "Figure 12.6. Ozone Depleting Substances and Criteria Pollutants: Ozone Depleting Substances, 2001," in *Annual Energy Review 2002*, U.S. Department of Energy, Energy Information Administration, Washington, DC, October 2003

when charged with electricity, refrigeration that uses plain water as a refrigerant, and the use of thermoacoustics (sound energy). Extensive investment in research and development of new technologies will be required to produce cooling methods acceptable to industry and environmentalists.

## PUBLIC OPINION ABOUT THE OZONE

In March 2004 the Gallup Organization conducted its annual poll of Americans' beliefs and attitudes about environmental issues. The results show that damage to the Earth's ozone layer ranks very low on the list of environmental problems about which Americans are worried. (See Figure 1.8 in Chapter 1.) Only 33 percent of those asked in 2004 expressed a great deal of worry about damage to the ozone layer. As shown in Table 3.4, this percentage is down from a peak of 51 percent in 1989. In 2004 another 27 percent expressed a fair amount of concern about the problem, while 26 percent felt only a little concern and 14 percent felt no concern at all.

**TABLE 3.4**

**Public concern about damage to the Earth's ozone layer, 2004**

PLEASE TELL ME IF YOU PERSONALLY WORRY ABOUT THIS PROBLEM A GREAT DEAL, A FAIR AMOUNT, ONLY A LITTLE, OR NOT AT ALL. DAMAGE TO THE EARTH'S OZONE LAYER?

|  | Great deal % | Fair amount % | Only a little % | Not at all % | No opinion % |
|---|---|---|---|---|---|
| 2004 Mar 8–11 | 33 | 27 | 26 | 14 | * |
| 2003 Mar 3–5 | 35 | 31 | 21 | 12 | 1 |
| 2002 Mar 4–7 | 38 | 29 | 21 | 11 | 1 |
| 2001 Mar 5–7 | 47 | 28 | 16 | 8 | 1 |
| 2000 Apr 3–9 | 49 | 29 | 14 | 7 | 1 |
| 1999 Apr 13–14 | 44 | 32 | 15 | 8 | 1 |
| 1997 Oct 27–28 | 33 | 27 | 25 | 13 | 2 |
| 1991 Apr 11–14 | 49 | 24 | 16 | 8 | 4 |
| 1990 Apr 5–8 | 43 | 28 | 15 | 10 | 4 |
| 1989 May 4–7 | 51 | 26 | 13 | 8 | 2 |

SOURCE: "Please tell me if you personally worry about this problem a great deal, a fair amount, only a little, or not at all. Damage to the Earth's ozone layer?," in *Poll Topics and Trends: Environment*, The Gallup Organization, Princeton, NJ, March 17, 2004 [Online] www.gallup.com [accessed March 30, 2004]

# CHAPTER 4
# WASTE DISPOSAL

One of the consequences of a modern society is the generation of enormous amounts of waste. The scale of materials use by industrialized countries dwarfs that of a century ago. By 2000 the stock of materials drew from all ninety-two naturally occurring elements in the periodic table compared with just twenty in 1900. The U.S. Geological Survey (USGS) estimates that, in the United States alone, consumption of metal, glass, wood, cement, and chemicals has grown eighteen-fold since 1900 and that the nation accounts for one-third of all materials used throughout the world.

The production and processing of almost any material generates by-products (which may or may not be useful) and releases them to into the air and water. Manufacturing, mining, oil and gas drilling, chemical processing, and coal-burning power plants produce many billions of tons of waste each year. Generation of radioactive and hazardous wastes has grown as society has advanced technologically. Even agriculture generates about a billion tons of waste annually, primarily crop residuals. Finally, residential and commercial generation of municipal solid waste (garbage) is at 230 million tons per year.

So where does it all go? In the past, worries about waste disposal were eased by the apparent ability of the environment (land, air, and water) to absorb that waste. The old saying "out of sight, out of mind" ruled the day. Today, we realize that any type of waste disposal has significant environmental consequences.

## LAWS GOVERNING WASTE DISPOSAL

The Resource Conservation and Recovery Act (RCRA; PL 94-580), the major federal law on waste disposal, was passed in 1976. Its primary goal was to "protect human health and the environment from the potential hazards of waste disposal." RCRA is also concerned with reducing the amount of waste generated, ensuring that wastes are managed properly, and conserving natural resources and energy. The RCRA regulates solid waste, hazardous waste, and underground storage tanks containing petroleum products or certain chemicals.

The RCRA definition of solid waste includes garbage and other materials we would ordinarily consider "solid," as well as sludges, semisolids, liquids, and even containers of gases. These wastes can come from industrial, agricultural, commercial, and residential sources. The RCRA primarily covers hazardous waste, which is only a small part of all waste generated. State and local governments are mainly responsible for passing laws concerning nonhazardous waste, although the federal government will supply money and guidance to local governments so they can better manage their garbage systems.

Other federal laws cover other areas of waste disposal. For example, the Clean Water Act (PL 95-217) regulates wastewater disposal; the Safe Drinking Water Act (PL 93-523) controls underground injections (when wastewater is injected into deep wells); and the Clean Air Act (PL 95-95) governs air pollution.

## INDUSTRIAL WASTES

Prior to the 1970s most industrial waste was dumped in landfills, stored on-site, burned, or discharged to surface waters with little or no treatment. Since the Pollution Prevention Act of 1990, industrial waste management follows a hierarchy introduced by the Environmental Protection Agency (EPA). (See Figure 4.1.) Source reduction is the preferred method for waste management. This is an activity that prevents the generation of waste initially, for example, a change in operating practices or raw materials. The second choice is recycling, followed by energy recovery. If none of these methods is feasible, then treatment prior to disposal is recommended.

For example, a paper mill that changes its pulping chemicals might reduce the amount of toxic liquid left

FIGURE 4.1

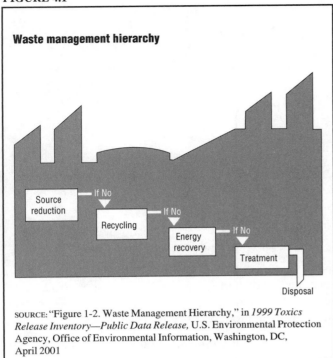

**Waste management hierarchy**

SOURCE: "Figure 1-2. Waste Management Hierarchy," in *1999 Toxics Release Inventory—Public Data Release,* U.S. Environmental Protection Agency, Office of Environmental Information, Washington, DC, April 2001

over after the paper is produced. If that is not possible, perhaps the pulping liquid could be recycled and reused in the process. If not, perhaps the liquid can be burned for fuel to recover energy. If not, and the liquid requires disposal, it should be treated as necessary to reduce its toxicity before being released into the environment.

Industrial waste is categorized based on its relative harm to the environment, chiefly to human health. Most wastes produced by industry are nonhazardous. However, the potential danger from hazardous wastes is so severe that the disposal of such wastes is heavily regulated.

### Hazardous Industrial Wastes

Hazardous waste is the inevitable by-product of industrialization. Manufacturers use many chemicals to create their products. Hazardous waste is generated by big industries like automobile and computer manufacturers and by small businesses like neighborhood photo shops and cleaners. Although people can reduce quantities of hazardous waste through careful management, it is not possible to eliminate hazardous residues entirely because of the continual demand for goods.

Industrial wastes are usually a combination of compounds, one or more of which may be hazardous; for example, used pickling solution from a metal processor can also contain residual acids and metal salts. A mixture of waste produced regularly as a result of the industrial process is called a waste stream, and it generally consists of diluted rather than full-strength compounds. Often, the hazardous components are diluted in a mixture of dirt, oil, or water.

Officially, hazardous waste is defined as a waste that is either listed as such in EPA regulations or exhibits one or more of the following characteristics: ignitability, corrosivity, reactivity, or toxicity (containing constituents in excess of federal standards). In 2004 the EPA had a list of more than 500 hazardous wastes. Hazardous wastes are regulated under subtitle C of the RCRA. The EPA has the primary responsibility for permitting facilities that treat, store, and dispose of hazardous waste. States can adopt more stringent regulations if they wish. Because of their potential dangers, hazardous wastes require special care when being stored, transported, or discarded.

Contamination of the air, water, and soil with hazardous wastes can frequently lead to serious health problems. The EPA estimates that roughly 1,000 cases of cancer annually, as well as degenerative diseases, mental retardation, birth defects, and chromosomal changes, can be linked to public exposure to hazardous waste. While most scientists agree that exposure to high levels of hazardous waste is dangerous, there is less agreement on the danger of exposure to low levels.

Every two years the EPA, in partnership with the states, publishes *The National Biennial RCRA Hazardous Waste Report.* The latest report available was published in 2003 and includes data from 2001. The report only includes wastewaters that are disposed via deep well/underground injection. All other wastewaters that are managed in wastewater treatment systems are not included.

The EPA distinguishes between large-quantity and small-quantity generators of hazardous waste. A large-quantity generator is one that

- generates at least 1,000 kilograms (2,200 pounds) of RCRA hazardous waste in any single month,

- generates in any single month or accumulates at any time at least 1 kilogram (2.2 pounds) of RCRA acute hazardous waste, or

- generates or accumulates at any time at least 100 kilograms (220 pounds) of spill cleanup material contaminated with RCRA acute hazardous waste.

Ninety percent of all hazardous waste in the United States is produced by large-quantity generators. The chemical industry is by far the largest producer, followed by petroleum refiners and the metal-processing industry.

In 2001 there were 19,024 large-quantity generators that reported the generation of 40.8 million tons of RCRA hazardous waste. The five states with the largest generation of hazardous waste were Texas (7.5 million tons), Louisiana (3.9 million tons), New York (3.5 million tons), Kentucky (2.7 million tons), and Mississippi (2.2 million tons). Together, these states accounted for 49 percent of the total quantity generated.

The remaining 10 percent of hazardous waste comes from more than 100,000 small-quantity generators—businesses that produce less than 1,000 kilograms of hazardous waste per month. Table 4.1 shows a list of typical small-quantity generators and the types of hazardous waste they produce. Hazardous wastes from small-quantity generators and households are regulated under subtitle D of the RCRA.

Household hazardous wastes include solvents, paints, cleaners, stains, varnishes, pesticides, motor oil, and car batteries. The EPA reports that Americans generate 1.6 million tons of household hazardous waste every year. The average home can have as much as 100 pounds of these wastes in basements, garages, and storage buildings. Many communities hold special collection days for household hazardous waste to ensure that it is disposed of properly.

## Methods of Dealing with Hazardous Waste

A variety of techniques exist for safely managing hazardous wastes:

- Reduction—This approach reduces the waste stream at the outset. Waste generators change their manufacturing and materials in order to produce less waste. For example, a food packaging plant might replace solvent-based adhesives used to seal packages with water-based adhesives.

- Recycling—Some waste materials become raw material for another process or can be recovered, reused, or sold.

- Treatment—A variety of chemical, biological, and thermal processes can be applied to neutralize or destroy toxic compounds. For example, microorganisms or chemicals can remove hydrocarbons from contaminated water.

- Incineration—Hazardous waste can also be burned. Unfortunately incineration has a flaw—as waste is burned, hot gases spew into the atmosphere, carrying toxic materials not consumed by the flames. In 1999 the Clinton administration imposed a ban on new hazardous waste incinerators.

- Land disposal—Some hazardous wastes are buried in landfills. State and federal regulations require the pretreatment of most hazardous wastes before they can be disposed of in landfills. These treated materials can only be placed in specially designed land disposal facilities.

In 2001 there were 2,479 treatment, storage, and disposal facilities that treated and/or disposed of 46 million tons of RCRA hazardous waste. Table 4.2 shows the man-

**TABLE 4.1**

### Typical hazardous waste generated by small businesses

| Type of business | How generated | Typical wastes |
|---|---|---|
| Drycleaning and laundry plants | Commercial drycleaning processes | Still residues from solvent distillation, spent filter cartridges, cooked powder residue, spent solvents, unused perchloroethylene |
| Furniture/wood manufacturing and refinishing | Wood cleaning and wax removal, refinishing/stripping, staining, painting, finishing, brush cleaning and spray brush cleaning | Ignitable wastes, toxic wastes, solvent wastes, paint wastes |
| Construction | Paint preparation and painting, carpentry and floor work, other specialty contracting activities, heavy construction, wrecking and demolition, vehicle and equipment maintenance for construction activities | Ignitable wastes, toxic wastes, solvent wastes, paint wastes, usedoil, acids/bases |
| Laboratories | Diagnostic and other laboratory testing | Spent solvents, unused reagents, reaction products, testing samples, contaminated materials |
| Vehicle maintenance | Degreasing, rust removal, paint preparation, spray booth, spray guns, brush cleaning, paint removal, tank cleanout, installing lead-acid batteries, oil and fluid replacement | Acids/bases, solvents, ignitable wastes, toxic wastes, paint wastes, batteries, used oil, unused cleaning chemicals |
| Printing and allied industries | Plate preparation, stencil preparation for screen printing, photoprocessing, printing, cleanup | Acids/bases, heavy metal wastes, solvents, toxic wastes, ink, unused chemicals |
| Equipment repair | Degreasing, equipment, cleaning rust removal, paint preparation, painting, paint removal, spray solvents booth, spray guns, and brush cleaning | Acids/bases, toxic wastes, ignitable wastes, paint wastes |
| Pesticide end-users/application services | Pesticide application and cleanup | Used/unused pesticides, solvent wastes, ignitable wastes, contaminated soil (from spills), contaminated rinsewater, empty containers |
| Educational and vocational shops | Automobile engine and body repair, metalworking, graphic arts-plate preparation, woodworking | Ignitable wastes, solvent wastes, acids/bases, paint wastes |
| Photo processing | Processing and developing negatives/prints, stabilization system cleaning | Acid regenerants, cleaners, ignitable wastes, silver |
| Leather manufacturing | Hair removal, bating, soaking, tanning, buffing, and dyeing | Acids/bases, ignitables wastes, toxic wastes, solvent wastes, unused chemicals |

SOURCE: Adapted from "Typical Hazardous Waste Generated by Small Businesses," in *Managing Your Hazardous Waste: A Guide for Small Businesses,* U.S. Environmental Protection Agency, Office of Solid Waste, Washington, DC, December 2001

agement methods used by these facilities. More than 17 million tons of hazardous wastewater were injected into deep underground wells. Figure 4.2 shows a typical Class I (deep) well. There are 163 of these wells located around the country; most are in Texas (78) and Louisiana (18). Eleven of the wells are for commercial hazardous waste injection. They are the only facilities allowed to accept hazardous waste generated off-site. Ten of these wells are

**TABLE 4.2**

**Management method, by quantity of RCRA (Resource Conservation and Recovery Act) hazardous waste managed, 2001**

| Management method | Tons managed | Percentage of quantity | Number of facilities* | Percentage of facilities* |
|---|---|---|---|---|
| Deepwell or underground injection | 17,681,650 | 38.3 | 48 | 1.9 |
| Other disposal | 6,429,341 | 13.9 | 206 | 8.3 |
| Aqueous organic treatment | 4,501,963 | 9.8 | 97 | 3.9 |
| Aqueous inorganic treatment | 3,672,052 | 8.0 | 322 | 13.0 |
| Other treatment | 2,355,272 | 5.1 | 562 | 22.7 |
| Landfill/surface impoundment | 2,089,701 | 4.5 | 69 | 2.8 |
| Energy recovery | 1,700,078 | 3.7 | 105 | 4.2 |
| Incineration | 1,646,217 | 3.6 | 174 | 7.0 |
| Metals recovery | 1,461,606 | 3.2 | 191 | 7.7 |
| Stabilization | 1,269,609 | 2.8 | 243 | 9.8 |
| Other recovery | 1,026,255 | 2.2 | 97 | 3.9 |
| Fuel blending | 923,332 | 2.0 | 117 | 4.7 |
| Storage and/or transfer | 717,785 | 1.6 | 639 | 25.8 |
| Solvents recovery | 425,459 | 0.9 | 564 | 22.8 |
| Sludge treatment | 178,975 | 0.4 | 99 | 4.0 |
| Land treatment/application/farming | 65,508 | 0.1 | 14 | 0.6 |
| **Total** | **46,144,802** | **100.0** | **2479** | |

*Columns may not sum because facilities may have multiple handling methods.

SOURCE: "Exhibit 2.6. Management Method, by Quantity of RCRA Hazardous Waste Managed, 2001," in *The National Biennial RCRA Hazardous Waste Report (Based on 2001 Data)*, U.S. Environmental Protection Agency, Office of Solid Waste and Emergency Response, Washington, DC, 2003

located in Gulf Coast states. The other well is in the Great Lakes region. Various types of land disposal and combustion were used to dispose of most of the remaining hazardous wastes from large-quantity generators.

**LAND DISPOSAL AND CONTAMINATION.** Groundwater is a major source of drinking water for many parts of the United States. If not properly constructed, land disposal facilities for hazardous waste may leak contaminants into the underlying groundwater. The RCRA imposed control over such disposal facilities to minimize their adverse environmental impacts. The EPA, in order to implement the act, requires that owners/operators of hazardous waste sites install wells to monitor the groundwater under their facilities.

**SHIPPING ELSEWHERE.** Many states have refused to accept toxic trash from states that have not developed their own disposal programs. Their position has been undermined, however, by the U.S. Supreme Court's determination that waste is a commodity in interstate commerce and is subject to federal, not state, regulation.

Because of the problems in finding disposal sites, the United States is sending larger and larger amounts of toxic waste out of the country. Mexico and Central and South America have become preferred spots for disposing of sludge and incinerator ash. However, toxic waste is sometimes mislabeled nontoxic by the time it arrives in South American countries. Until 1988 Africa had been a favorite location for dumping toxic waste. At that time, however, most African countries signed agreements that restricted importation of dangerous materials.

**Toxic Chemicals in Industrial Waste**

The Toxics Release Inventory (TRI) was established under the Emergency Planning and Community Right-to-Know Act of 1986 (PL 99-499). Under the program, certain industrial facilities using specific toxic chemicals must report annually on their waste management activities and toxic chemical releases. More than 650 toxic chemicals are on the TRI list.

Manufacturing facilities (called "original" industries) have had to report under the TRI program since 1987. In 1998 the TRI requirements were extended to a second group of industries called the "new" industries. These include metal and coal mining, electric utilities burning coal or oil, chemical wholesale distributors, petroleum terminals, bulk storage facilities, RCRA subtitle C hazardous water treatment and disposal facilities, solvent recovery services, and federal facilities. However, only facilities with 10 or more full-time employees that use certain thresholds of toxic chemicals are included.

The *2001 Toxics Release Inventory (TRI) Public Data Release Report* was published in July 2003. The report states that 26.7 billion pounds (13.4 million tons) of TRI chemicals in production-related waste were managed during 2001. The chemical manufacturing industry accounted for 40 percent of the waste, followed by the primary metals industry with 12 percent and metal mining with 11 percent. The largest amounts of production-related waste were managed by facilities in Texas, Louisiana, and Illinois.

Approximately 36 percent of the waste was recycled, while 13 percent was burned for energy recovery, and 28 percent was treated. The remainder (23 percent) was released to the environment in some way. Figure 5.18 in chapter 5 shows the distribution of TRI releases in 2001. More than half of the releases were to land disposal.

FIGURE 4.2

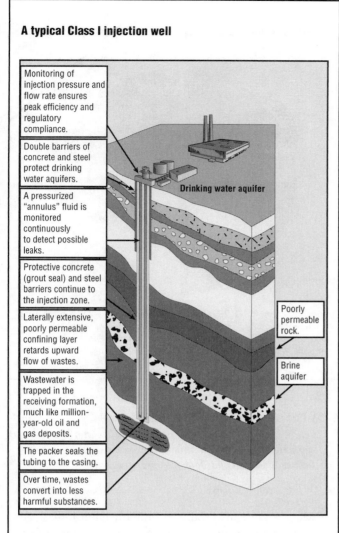

**A typical Class I injection well**

Monitoring of injection pressure and flow rate ensures peak efficiency and regulatory compliance.

Double barriers of concrete and steel protect drinking water aquifers.

A pressurized "annulus" fluid is monitored continuously to detect possible leaks.

Protective concrete (grout seal) and steel barriers continue to the injection zone.

Laterally extensive, poorly permeable confining layer retards upward flow of wastes.

Wastewater is trapped in the receiving formation, much like million-year-old oil and gas deposits.

The packer seals the tubing to the casing.

Over time, wastes convert into less harmful substances.

Drinking water aquifer

Poorly permeable rock.

Brine aquifer

SOURCE: "Exhibit 3. A Typical Class I Injection Well," in *Class I Underground Injection Well Control Program: Study of the Risks Associated with Class I Underground Injection Wells,* U.S. Environmental Protection Agency, Office of Water, Washington, DC, March 2001

## Nonhazardous Industrial Waste

Nonhazardous industrial wastes are neither hazardous wastes nor municipal wastes. Figure 4.3 shows a general breakdown of nonhazardous industrial waste by material, as reported by the Office of Industrial Technology.

Manufacturing produces huge amounts of nonhazardous waste. The paper industry, which uses many chemicals to produce paper, accounts for a very large proportion of manufacturing waste. The metal and chemical industries are also large waste producers. Many big manufacturing plants have sites on their own property where they dispose of waste or treat it so it will not become dangerous. Still others ship it to private disposal sites for dumping or for treatment. Smaller manufacturers might use private waste disposal companies or even the city garbage company.

FIGURE 4.3

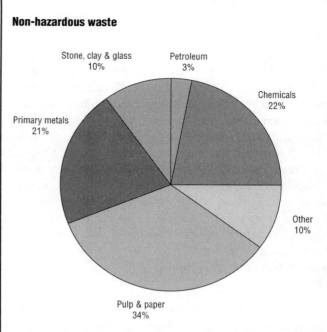

**Non-hazardous waste**

Stone, clay & glass 10%

Petroleum 3%

Chemicals 22%

Primary metals 21%

Other 10%

Pulp & paper 34%

SOURCE: Adapted from "Figure W-1. Non-Hazardous Waste," in "Waste Generation," *Energy, Environmental and Economics (E³) Handbook,* U.S. Department of Energy, Office of Energy Efficiency and Renewable Energy, Office of Industrial Technology, Washington, DC, September 1997

Mining also produces much waste, most of it rock and tailings. Normally, miners have to move rock to retrieve the ore or minerals. Tailings are left over after miners have sifted through the rocks and dirt for the ore or minerals. Chemicals used to remove minerals from ore become waste after they have done their job. Sometimes these chemical wastes are liquid, and sometimes solid. Either way, they must be disposed of appropriately so as not to pollute the environment.

Almost all (96 to 98 percent) of the waste from gas and oil drilling is water. Water is either pumped out of the ground before the oil is found, or it is found mixed with oil. This water is often salt water and it must be separated from the oil and gas before these natural products can be turned into refined products for use in automobiles or home heating. Other waste comes from mud and rock extracted by the drilling process. Most oil and gas companies dispose of their own waste.

One method is called "surface impoundment," which consists of a large pond in which liquid wastes can be stored and then treated so they can be disposed of safely. Almost all wastes from oil and gas production, mining, and agriculture end up in surface impoundments.

When an electric company burns coal to heat water to make electricity, about 90 percent of the coal is burned up, but it leaves about 10 percent in the form of ash. This waste must then be discarded somewhere.

State and local governments have regulatory responsibility for the management of most nonhazardous wastes. Different states have different regulatory schemes. For example, Texas categorizes nonhazardous industrial wastes into three classes based on their potential harm to the environment and human health. Class 1 wastes include asbestos, ash, and various solids, sludges, and liquids contaminated with nonhazardous chemicals. Every four years the state evaluates its disposal capacity for Class 1 wastes. The last report available was published in 2000 and includes data for 1997. According to *Needs Assessment for Industrial Class 1 Nonhazardous Waste Commercial Disposal Capacity in Texas (2000 Update)* nearly 83 million tons of Class 1 waste were generated in Texas that year. The vast majority of the waste (96 percent) was liquid.

Class 2 wastes include containers that held Class 1 wastes, depleted aerosol cans, some medical wastes, paper, food wastes, glass, aluminum foil, plastics, Styrofoam, and food packaging resulting from industrial processes. Class 3 wastes include all other chemically inert and insoluble substances such as rocks, brick, glass, dirt, and some rubbers and plastics.

Industries do not have to report how much Class 2 or Class 3 wastes they generate or how they dispose of it. However, municipal solid waste landfills are required to report the receipt of all industrial waste.

## MUNICIPAL SOLID WASTE

Most of the waste that people see is produced by ordinary households throwing out their uneaten food, yesterday's newspapers, packaging materials, lawn clippings, and branches from bushes and trees. This is the type of garbage that the EPA calls municipal solid waste (MSW).

Since 1960 the EPA has collected data on MSW generation and disposal in the United States. According to its report *Municipal Solid Waste in the United States: 2001 Facts and Figures* (October 2003), Americans produced 229 million tons of MSW in 2001, down slightly from 230 million tons in 1999, but up from 205 million tons in 1990 and 88.1 million tons in 1960. The largest category of MSW was paper, accounting for 35.7 percent of the total. (See Figure 4.4.)

Per capita generation of MSW was 4.4 pounds per person per day, up from 2.7 pounds per day in 1960. (See Figure 4.5.) However, per capita generation was down slightly from 1990, indicating that the rate has leveled off. This is due in large part to growing rates of recovery, which is using trash for some other purpose rather than discarding it. (See Figure 4.6.)

**FIGURE 4.4**

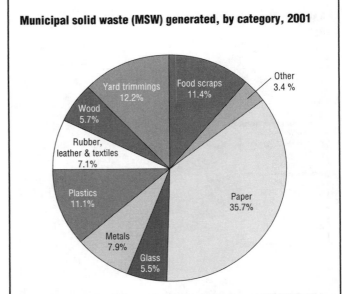

**Municipal solid waste (MSW) generated, by category, 2001**

SOURCE: "Figure ES-3: 2001 Total MSW Generation—229 Million Tons (Before Recycling)," in *Municipal Solid Waste in the United States: 2001 Facts and Figures*, U.S. Environmental Protection Agency, Office of Solid Waste, Washington, DC, October 2003

## Why Is There So Much Garbage?

*The widespread human appetite for all materials has defined this century in much the same way that stone, bronze, and iron characterized previous eras.*

— U.S. Geological Survey, *Mineral Commodity Summaries*, 1998

Most Americans produced little trash until the twentieth century. Food scraps were boiled into soups or fed to animals, which were themselves part of the food chain, providing milk, eggs, or meat for household use or for sale. Durable items were passed on to the next generation or to people more in need. Objects that were of no further use to adults became toys for children. Broken items were repaired or dismantled for reuse. Many Americans possessed the skills required for repairing items. Things that could no longer be used were burned for fuel, especially in the homes of the poor. Even middle-class Americans traded rags to peddlers in exchange for buttons or tea kettles. These "ragmen" worked the streets, begging for or buying for pennies items such as bones, paper, old iron, rags, and bottles. They then sold the "junk" to dealers who marketed it to manufacturers.

Spending time to prolong the useful lives of items and to use scraps saved money. Besides giving away clothes, mending and remaking them, and using them as rags for work, women reworked textiles into useful household furnishings such as quilts, rugs, and upholstery. Rags were also important materials collected for recycling in factories: Paper mills used rags to make paper, and a growing

FIGURE 4.5

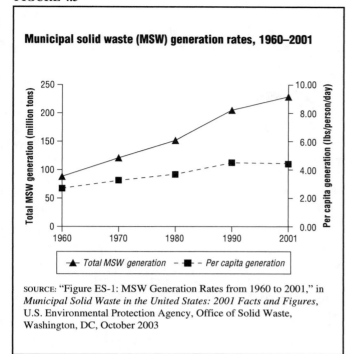

**Municipal solid waste (MSW) generation rates, 1960–2001**

SOURCE: "Figure ES-1: MSW Generation Rates from 1960 to 2001," in *Municipal Solid Waste in the United States: 2001 Facts and Figures,* U.S. Environmental Protection Agency, Office of Solid Waste, Washington, DC, October 2003

FIGURE 4.6

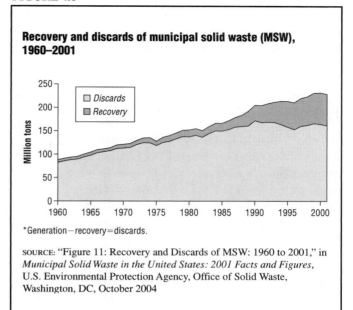

**Recovery and discards of municipal solid waste (MSW), 1960–2001**

*Generation − recovery = discards.

SOURCE: "Figure 11: Recovery and Discards of MSW: 1960 to 2001," in *Municipal Solid Waste in the United States: 2001 Facts and Figures,* U.S. Environmental Protection Agency, Office of Solid Waste, Washington, DC, October 2004

paper industry made it profitable for thrifty homemakers to save rags.

This trade in used goods provided crucial resources for early industrialization, but these early systems of recycling began to pass into history around the turn of the twentieth century. Sanitary reformers and municipal trash collection did away with scavenging. Technology made available cheap and new alternatives. People made fewer things themselves, and they bought more than previous generations had. They saved and repaired less and threw out more.

People of the growing middle class learned to throw things away, attracted by convenience and wanting to avoid any association with scavenging and poverty. Success often meant that one did not have to use secondhand things. As municipalities became responsible for collecting and disposing of refuse, Americans found it easier to throw things out.

**THE SORTING PROCESS.** Nothing is inherently trash. Trash is produced by a human behavior called sorting. Items in people's lives eventually require a decision—to keep or to discard. Some things go here, and some things go there.

The sorting process varies from person to person, from place to place, and changes over time. What is considered rubbish changes from decade to decade. Some societies value saving things more than others: Nomadic people, who must travel light, save less. During times of war, people often have to conserve and reuse materials, a situation that is not as common during peacetime. Age

also plays a role. The youth of the late twentieth and early twenty-first century have more readily adopted the notions of convenience and disposability than their parents and grandparents, and are less likely to conserve.

Sorting is also an issue of class. Trash making creates social differences based on economic status. The wealthy can more readily afford to replace older items with new ones, and discarding things can be a way of demonstrating affluence and power.

At the turn of the millennium, people in developed nations discarded things for reasons unheard of in developing nations or in earlier times—because they no longer wanted them. Disposing of out-of-style clothes and outmoded equipment reflected a worship of newness that was not widespread before the twentieth century. Dumpsters filled with "perfectly good stuff" that was simply not new anymore—stuff of which the owner had tired.

Economic growth during the twentieth century was fueled—in part—by a growing consumer culture that demanded a continual supply of new products, disposables, and individually packaged consumer items. This, combined with "planned obsolescence" on the part of manufacturers, produced increasing volumes of garbage. Colored plastic trash bags represent the contemporary attitude about trash, far from the homemade soup, darned underwear, and flour-sack dresses of an earlier time.

As the United States became richer, the nation produced more garbage and pollution. Between 1960 and 2001 America's population grew by 55 percent, but according to the EPA in *Municipal Solid Waste in the United States: 2001 Facts and Figures,* the amount of garbage produced increased 160 percent.

Growing populations, rising incomes, and changing consumption patterns have combined to complicate the waste management problem. Garbage generation expands as a city grows in size and, as consumers earn more money, their demand for consumer goods increases. This includes "convenience foods" with packaging that immediately becomes waste. Other convenience items, such as disposable diapers, also add to the mountains of waste.

ONE-TIME USE. Today, most consumer goods are designed for short-term use. This contrasts sharply with the practices of earlier eras when materials were reused or transformed for other uses. In *A Social History of Trash* (New York: Henry Holt and Co., 1999), Susan Strasser notes that more and more things today are being made and sold with the understanding that they will soon become worthless or obsolete.

The volume of waste increases with income—poor neighborhoods generate lower amounts of solid wastes per capita than richer neighborhoods. An inventory of what Americans throw away would reveal valuable metals, paper representing millions of acres of trees, and plastics incorporating highly refined petrochemicals.

## Garbage Disposal

The United States is facing a problem with its ever-growing mountains of garbage. America generates more garbage than any other nation on Earth, twice as much per person as in Europe. As with most environmental issues, waste disposal has grown to crisis proportions. The cost of handling garbage is the fourth biggest item—after education and police and fire protection—in many city budgets. Most of the nation's solid waste is dumped in landfills, but sites are rapidly filling up and many are leaking toxic substances into the nation's water supply.

The more that is learned about garbage, the more apparent it is that trucking garbage to landfills does not necessarily eliminate it. As a result, municipal governments worldwide are struggling to find the best methods for managing waste.

## The History of Garbage Disposal

Around 500 B.C.E. ("before the common era"), Athens, Greece, issued the first known edict against throwing garbage into the streets and organized the first municipal dumps by mandating that scavengers transport wastes to no less than one mile from the walls of the city. This method was not practiced in medieval Europe (circa 500–1485). Parisians in France and Londoners in England continued to toss trash and sewage out their windows until the 1800s. The west end of London and the west side of Paris became fashionable in the late seventeenth and eighteenth centuries because the prevailing winds blew west to east, carrying the smell of rotting garbage with them.

Industrialization brought with it a greater need for collection and disposal of refuse. Garbage was transported beyond the city limits and dumped in piles in the countryside. As cities grew, the noxious odors and rat infestations at the dumps became intolerable. Freestanding piles gave way to pits, but that solution soon became unsatisfactory.

In 1874 the first systematic incineration (burning) of municipal waste was tested in England. Burning reduced waste volume by 70 to 90 percent, but the expense of building incinerators and the reduced air quality caused many cities to abandon the method. Waste burial remained the most widely practiced form of disposal.

Throughout the nineteenth century many cities passed antidumping ordinances, but they were largely ignored. Some landowners and merchants resented ordinances, which they considered infringements of their rights. As cities grew so did garbage, becoming not only a public eyesore but a threat to public health.

City leaders began to recognize that they had to do something about the garbage. By the turn of the twentieth century, most major cities had set up garbage collection systems. Many cities introduced incinerators to burn some of the garbage. By the time World War I began in Europe in 1914, about 300 incinerators were operating in the United States and Canada.

Cities located downstream from other cities that were pouring their garbage into rivers sued the upstream cities because the water was polluted. As a result, more and more cities stopped dumping their garbage into rivers and began to build landfills or garbage dumps to get rid of their waste. Many coastal cities began to take their refuse out into the ocean and dump it, although much of the garbage they poured into the sea washed back to pollute the beaches.

Some health officials and reformers knew that pollution was unhealthy and could lead to sickness, but most people were not concerned about the long-term effects of garbage and pollution on the environment. As the twentieth century progressed, however, more and more Americans became concerned that garbage and pollution were harming the environment.

## How Is Garbage Disposed of Today?

The EPA's *Municipal Solid Waste in the United States: 2001 Facts and Figures* reported that 55.7 percent of MSW goes to land disposal, while 29.7 percent is recovered and 14.7 percent is incinerated. (See Figure 4.7.)

LAND DISPOSAL. Land disposal includes landfills, land application, and underground injection into deep wells. Landfilling is the most widely used method. Landfills are areas set aside specifically for garbage dumping.

FIGURE 4.7

FIGURE 4.8

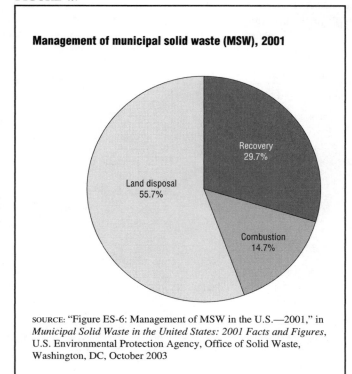

**Management of municipal solid waste (MSW), 2001**

SOURCE: "Figure ES-6: Management of MSW in the U.S.—2001," in *Municipal Solid Waste in the United States: 2001 Facts and Figures*, U.S. Environmental Protection Agency, Office of Solid Waste, Washington, DC, October 2003

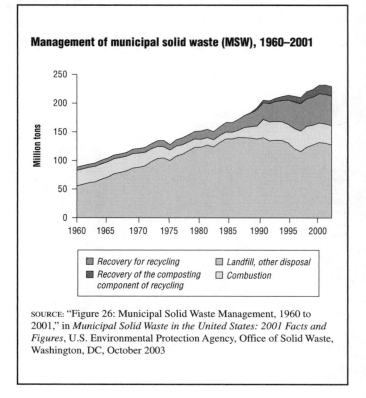

**Management of municipal solid waste (MSW), 1960–2001**

SOURCE: "Figure 26: Municipal Solid Waste Management, 1960 to 2001," in *Municipal Solid Waste in the United States: 2001 Facts and Figures*, U.S. Environmental Protection Agency, Office of Solid Waste, Washington, DC, October 2003

There are several types of landfills. In the most common type, garbage is dumped into a large pit and ultimately buried with earth. In land application, waste is taken to a designated area and spread over the surface of the land. Garbage may also be dumped onto a waste pile on the ground where it is stored and may eventually be treated.

Landfills are popular because, when compared with the cost of alternative disposal methods, dumping waste in the ground is a relatively cheap solution to an immediate problem. MSW discards to land disposal grew steadily from the 1960s to the 1980s and then declined through the early 1990s as use of other disposal methods increased. (See Figure 4.8.) However, land disposal rebounded somewhat in the late 1990s before leveling off again in the early 2000s.

In January 2004 *Biocycle* magazine published the results of its 14th annual State of Garbage in America survey. The magazine reported that in 2002 there were 1,767 landfills operating in the United States, down from 2,142 in 2000 and approximately 8,000 in 1988. (See Table 4.3.) Estimates of remaining landfill capacity vary greatly from state to state.

The growing amount of waste has led to a depletion of landfill capacity. Many of the nation's active landfills will reach capacity and have to close over the coming decades. Cities that exhaust their landfills are forced to find sites elsewhere, usually in more remote areas, which increases cost for transportation and landfill fees.

New landfills are becoming harder to find. Many local officials blame the lack of new facilities on public opposition, the "not in my backyard" syndrome, rather than a lack of available space. Landfills, like prisons and interstate highways, have never been welcome in neighborhoods.

During the late twentieth century, interest in recycling grew, and most states began making recycling an important part of waste collection and disposal. Nonetheless, the nation's landfills began filling up and new ones had to be constructed. Most states began sending some of their garbage to other states that would accept it, or even to other countries. By the beginning of the twenty-first century, some states and countries no longer accepted other people's garbage.

Some states have tried to enact bans on importing garbage into their states, but the U.S. Supreme Court, in *Chemical Waste Management v. Hunt* (112 S. Ct. 2009, 1992), ruled that such shipments are protected by the constitutional right to conduct commerce across state borders, because states that accept garbage charge fees for garbage dumping.

**ENVIRONMENTAL CONSEQUENCES OF LAND DISPOSAL.** Regulations passed under the RCRA, which took effect in 1993, required landfill operators to do several things to lessen the chance of pollution. The most important standard requires all landfills to monitor groundwater for contaminants. According to the EPA, less than one-third of the nation's toxic waste dumps in 2000 met

TABLE 4.3

**Number of municipal solid waste (MSW) landfills and waste to energy (WTE) plants, average tip fees, and capacity by state, 2000**

| State | Number of MSW landfills | Average landfill tip fee ($/ton) | Total landfill capacity remaining (tons) | Number of WTE plants | Average WTE tip fee ($/ton) |
|---|---|---|---|---|---|
| Arizona | 41 | n/a | n/a | 0 | – |
| Arkansas | 24 | 28.45 | n/a | 2 | n/a |
| California | 161 | 13.63 | 410,501,190 | 3 | n/a |
| Colorado | 65 | n/a | n/a | 0 | – |
| Connecticut | 2 | n/a | n/a | 6 | 65 |
| Delaware | 3 | 58.50 | 20,000,000 | 0 | – |
| Florida | 100 | 42.47 | n/a | 13 | 59 |
| Georgia | 60 | 33.50 | 135,349,274[1] | 1 | 45 |
| Hawaii | 9 | n/a | n/a | 1 | n/a |
| Idaho | 29 | n/a | n/a | 0 | – |
| Illinois | 51 | n/a | 212,393,636[1] | 0 | – |
| Indiana | 35 | n/a | 52,231,795[1] | 1 | n/a |
| Iowa | 59 | 33.25 | 40,182,628 | 1 | 53 |
| Kansas | 51 | 28 | n/a | 0 | – |
| Kentucky | 25 | 27.57 | 36,363,636[1] | 1 | n/a |
| Louisiana | 24 | 25 | n/a | 0 | – |
| Maine | 8 | 55 | 3,030,303[1] | 4 | 65 |
| Maryland | 20 | 50 | n/a[2] | 3 | 49 |
| Massachusetts | 19 | 72.60 | n/a | 7 | 71 |
| Michigan | 52 | n/a | 143,939,394[1] | 4 | 76 |
| Minnesota | 21 | 50 | 18,700,000 | 15 | 50 |
| Mississippi | 17 | 26 | n/a | 0 | – |
| Missouri | 24 | 33.54 | 41,432,836[1] | 0[3] | – |
| Montana[4] | 30 | 32 | 32,727,273 | 0 | – |
| Nebraska | 24 | 25 | n/a | 0 | – |
| Nevada | 23 | 30 | 60,742,056[1] | 0 | – |
| New Hampshire | 10 | 68 | 15,000,000 | 2 | 81 |
| New Jersey | 12 | 60 | 40,000,000 | 5 | 60 |
| New Mexico | 35 | n/a | 190,966,142[1] | 0 | – |
| New York | 26 | 50 | 90,000,000 | 10 | 65 |
| North Carolina | 41 | 30 | 100,000,000 | 1 | 50 |
| North Dakota | 14 | 26.56 | n/a | 0 | – |
| Ohio | 44 | 32.20 | 124,079,624[1] | 0 | – |
| Oklahoma | 40 | 20 | n/a | 1 | n/a |
| Oregon | 30 | 34.50 | n/a | 1 | 68 |
| Pennsylvania | 49 | 48 | 298,585,524 | 6 | 74 |
| Rhode Island | 2 | 41.50 | n/a | 0 | – |
| South Carolina | 19 | 27 | 109,534,023 | 4 | n/a |
| South Dakota | 15 | 30 | 16,757,576[1] | 0 | – |
| Tennessee | 34 | 28.38 | n/a | 1 | n/a |
| Texas | 175 | 27 | 970,000,000 | 2 | n/a |
| Utah | 38 | n/a | n/a | 1 | n/a |
| Vermont | 5 | 80 | 1,453,778 | 0 | – |
| Virginia | 67 | n/a | 251,810,045 | 5 | n/a |
| Washington | 21 | 46.48 | 180,002,767 | 4 | n/a |
| West Virginia | 18 | 43 | >5,674,330 | 0 | – |
| Wisconsin | 42 | 36.43 | 30,440,024[1] | 2 | n/a |
| Wyoming | 53 | n/a | n/a | 0 | – |
| **Totals** | **1,767** | | | **107** | |

[1]Tonnage based on conversion from cubic yards reported (conversion of 3.3 cubic yards/ton)
[2]Landfill capacity remaining exceeds ten years
[3]Waste-to-energy plant burns tires for fuel
[4]2001 data from MSW Management

SOURCE: Scott M. Kaufman, Nora Goldstein, Karsten Millrath, and Nickolas J. Themelis, "Table 7. Number of Municipal Solid Waste Landfills and Waste to Energy Plants, Average Tip Fees, and Capacity by State for 2002," in "The State of Garbage in America," in *BioCycle*, vol. 45, no. 1, January 2004

requirements under disposal laws for monitoring underground water supplies near their sites.

The rules also require plastic liners for dump sites, and all debris must be covered with soil to prevent odors and trash from being blown away. Methane gas must be monitored, and the owner is responsible for cleanup of any contamination. To prevent pollution of the environment, these rules must be observed for a 30-year period after the landfill is closed.

Although the government plans to close all dumps that fail to meet requirements, and many landfills have been shut down at least in part because of noncompliance, the process has been slow due to a lack of resources to prosecute violators. The virtual disappearance of affordable environment-impairment liability insurance has also forced many dumps to shut down.

Newer, state-of-the-art landfills are now being built with multiple liners to prevent leaks and with equipment to treat emissions. This is very expensive. Experts point out that many of these landfills will have to accept waste from a wide region to be financially viable.

Experts agree that even the most advanced landfills may eventually leak, releasing hazardous materials into surface or underground water. Methane, a flammable gas, is produced when organic matter decomposes in the absence of oxygen. If not properly vented or controlled, it can cause explosions and underground fires that smolder for years. Increasingly, this gas is being recovered through pipes inserted into landfills and distributed or used to generate energy.

## Landfills of the Twenty-First Century

Landfills are, and will continue to be, the cornerstone of the nation's waste services system. However, a number of changes will occur in site design and function. Sites will become more standardized, especially in the areas of liners and in the collection of landfill gas, which is expected to stimulate the development of new landfill gas-to-energy plants.

The number of landfills is expected to decrease due to the stricter standards and the need for operators to provide assurance that they can fund closure, cleanup, and security in the event of contamination. In many cases, it will be difficult to justify small-scale sites economically. The result will be fewer, but larger and more regional, operations. Most waste will move away from its point of generation, resulting in increased interdependence among communities and states in waste disposal. More waste will cross state lines.

The volume of waste traditionally handled at landfills will also decrease. Landfills will provide diverse services—burial of waste, bioremediation, recycling facilities, leachate collection (contaminants picked up through the leaching of soil), and gas recovery. To make landfills more pleasing to neighborhoods, operators will establish larger buffer zones and more green space and will show more sensitivity to land-use compatibility and landscaping.

## Source Reduction

Many experts believe that the primary solution to the world's mounting garbage problem is "source reduction," or waste prevention. The less waste people create, the less there is to throw away. Source reduction involves minimizing the amount or toxicity of materials in products and/or reducing the amount of wastes produced during manufacturing. Manufacturing, packaging, or processing goods in certain ways "up front" will generate less refuse to be discarded.

Nearly $1 out of every $12 Americans spend for food and beverages pays for packaging. According to the U.S. Department of Agriculture (USDA), packaging is the sec-

ond largest portion of the cost of marketing food (advertising is the largest portion). The increasing numbers of women in the workforce and changes in family structure have resulted in greater demand for convenience products—carry-out meals and frozen and vacuum-packed foods. One way to reduce waste is to trim the amount of packaging.

Soft drink consumption has also risen, increasing waste in the form of cans and plastic containers. Aluminum, the most abundant metal manufactured on Earth, was first refined into a valuable product in the 1820s. Its use has continually escalated and beverage cans are the largest single use of aluminum. They are also a source of much waste.

The advent of low-priced petrochemicals in the early twentieth century ushered in the age of plastics. Several times more plastics—a family of more than forty-six types—are now produced in the United States than aluminum and all other nonferrous metals combined. Most of these plastics are nonbiodegradable and, once discarded, remain relatively intact for many years.

Many Americans claim they would pay more for a product with environmental benefits. Marketing that considers these consumer preferences is known as "green marketing." "Green design" can make products more environmentally safe.

COMPOSTING. Composting is a form of source reduction that involves the mixing of vegetable and organic refuse in order to speed the natural decomposition into fiber and micronutrients. For example, a compost pile in one's backyard recycles food scraps and lawn clippings that are deposited and mixed periodically. The decomposed product can then be used as fertilizer for the garden. Organic waste gives off energy in the form of heat when microorganisms metabolize the waste, causing it to lose between 40 and 75 percent of its original volume. After decontamination and refinement processes have been completed, the finished product is often used in landscaping, land reclamation, landfill cover, farming, and for nurseries. About one-quarter of household trash is organic material—food and yard waste.

Composting is a particularly promising method of disposal of household wastes in developing countries and is also very advanced in Europe. Yard waste is a prime candidate for composting due to its high moisture content. Because compost contains moisture and micronutrients, slows soil erosion, and improves water retention, it is an alternative to the use of environmentally dangerous chemical fertilizers. According to the EPA report *Municipal Solid Waste in the United States: 2001 Facts and Figures,* approximately 50 percent of the U.S. population is subject to state legislation discouraging the disposal of yard trimmings. The tonnage of yard trimmings dumped in landfills has declined, most likely because such legislation prompts more people to use backyard composting and mulching lawn mowers.

## Recovery

The terms "recovery" and "recycling" are often used interchangeably. Both mean that an item is not discarded as trash, but reused in some way. Recovery reduces the amount of garbage requiring disposal, thus saving landfill space and conserving the energy that would be used for incineration. Recovery also reduces environmental degradation and chemicals that pollute water resources, generates jobs and small-scale enterprise, reduces dependence on foreign imports of metals, and conserves water. Some analysts claim that more than half of consumer waste could be economically recycled.

However, recycling sometimes requires more energy and water consumption than waste disposal. It depends on how far the materials must be transported and what is necessary to "clean" them before they can be reused. Demand for some recyclable materials is weak, making them economically unfeasible to recycle in a market-driven society.

According to the EPA's report 68 million tons of MSW was recycled in 2001, up from 33.2 million tons in 1990. (See Figure 4.9.) This was an average recycling rate of 1.3 pounds per person per day. Recycling rates for various materials are shown in Figure 4.10 for 2001 and previous years dating back to 1970. Automobile batteries are the most frequently recycled item with a recycling rate of 94 percent in 2001. Recycling for all other materials shown has increased dramatically since 1970.

Every state has some type of recycling program. The oldest recycling law is the Oregon Recycling Opportunity Act, passed in 1983 and put into effect in 1986. A growing number of states require that many items sold must be made from recycled products. According to the American Forest and Paper Association Web site (http://www.afandpa.org/) at least 12 states require that recycled paper be used to make newspapers; many require that recycled materials be used in making telephone directories, trash bags, glass, and plastic containers. Most states have goals to recycle from 25 to 70 percent of MSW. Rhode Island (70 percent) and New Jersey (60 percent) have the highest recycling goals of all the states; Maryland had the lowest (20 percent). More than 30 states bar some recyclable materials from being thrown into landfills. These include car and boat batteries, grass cuttings, tires, used oil, glass, plastic containers, and newspapers. Almost all books and pamphlets printed by the U.S. Government Printing Office are printed on recycled paper.

## Incineration

Some observers think incinerators are the best alternative to landfills. The EPA reported that, in 2001, nearly

FIGURE 4.9

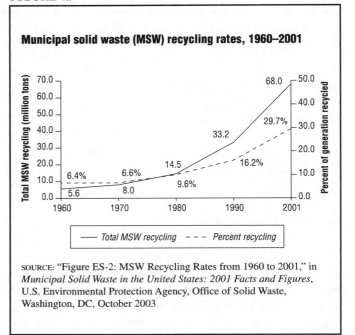

SOURCE: "Figure ES-2: MSW Recycling Rates from 1960 to 2001," in *Municipal Solid Waste in the United States: 2001 Facts and Figures*, U.S. Environmental Protection Agency, Office of Solid Waste, Washington, DC, October 2003

15 percent of MSW in the United States was burned. (See Figure 4.7.) When an incinerator burns waste, it reduces the amount of garbage. About 75 percent of the weight of the garbage burns off. Some incinerators do more than burn garbage; by using the heat from the burning garbage to make energy, they can also be waste-to-energy (WTE) facilities. WTE incinerators are preferred over older incinerator models because they use an improved combustion process, have better pollution-control technology, and produce energy from trash. Table 4.3 shows the number of WTE facilities in each state in 2002. There were 107 WTE plants in the United States that year, most in Minnesota, Florida, and New York.

Incinerators are very expensive to build. The country's largest incinerator, in Detroit, Michigan, cost $438 million in 1988. This huge incinerator produces enough steam to heat half of Detroit's central business district and enough electricity to supply 40,000 homes. However, most experts agree that energy recovery from MSW has the potential for making only a limited contribution to the nation's overall energy production. The U.S. Department of Energy (DOE) has set a goal for waste-derived energy at 2 percent of the total supply by 2010.

During the operation of a typical incinerator, trucks dump waste into a pit; the waste is moved to the furnace by a crane; and the furnace burns the waste at a very high temperature, heating a boiler that produces steam for generating electricity and heat. Ash collects at the bottom of the furnace where it is later removed and dumped in a landfill. (See Figure 4.11.)

Most experts believe that incineration can never serve as a primary method of garbage disposal because it

FIGURE 4.10

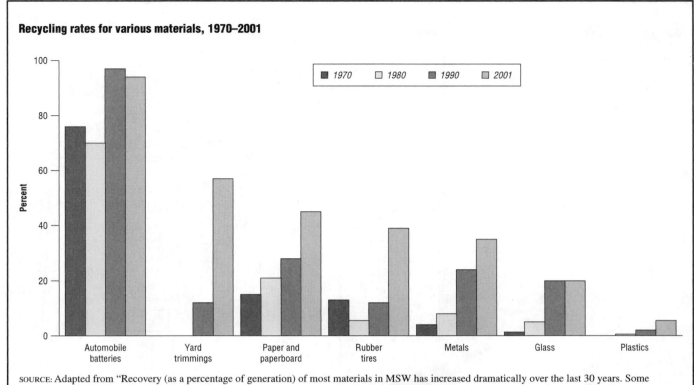

**Recycling rates for various materials, 1970–2001**

SOURCE: Adapted from "Recovery (as a percentage of generation) of most materials in MSW has increased dramatically over the last 30 years. Some examples," in *Municipal Solid Waste in the United States: 2001 Facts and Figures*, U.S. Environmental Protection Agency, Office of Solid Waste, Washington, DC, October 2003

produces (1) residue that must then be transferred to a landfill, and (2) poisonous gases, primarily dioxin and mercury, which are increasingly being found to be dangerous. Incineration may, however, be useful to augment landfill and recycling. WTE plants also have these problems. Mercury is largely impossible to screen with pollution-control devices such as scrubbers (an air pollution device that uses a spray of water or reactant to trap pollutants). In the process of burning paints, fluorescent lights, batteries, or electronics, mercury is released as a gaseous vapor that is poisonous to humans and to the environment. Most of the first incinerators built have been retired because they failed to meet subsequent air quality standards. Some analysts are not satisfied that the emissions problems have been solved, especially the problems of burning materials containing chlorine. Chlorine molecules, when burned, create dioxin, a known carcinogen (cancer-causing agent).

Regulators are also concerned about the acid gases and heavy metals released from WTE plants. Scrubbers reduce but do not eliminate these emissions. Even when the toxic elements are largely removed from emissions, the resulting ash is still toxic and, when put in landfills, can leach into the groundwater. Thus, toxic compounds in incinerator ash are simply removed from one environmental medium to enter another. Toxic compounds still end up in the soil. By law, toxic residue created by burning waste in incinerators must be treated as hazardous waste and must not be dumped in ordinary landfills.

## THE NATIONAL PRIORITIES LIST— THE SUPERFUND

The Comprehensive Environmental Response, Compensation, and Liability Act of 1980 (CERCLA; PL 96-510) established the Superfund to pay for cleaning up abandoned disposal sites. The Superfund—initially a $1.6 billion, five-year program—was intended to clean up leaking dumps that jeopardized groundwater and posed public health risks. During the act's original mandate, only six sites were cleaned up, and when it expired in 1985, many observers viewed the program as a billion-dollar fiasco rampant with scandal and mismanagement. The negative publicity surrounding the program increased public awareness of the magnitude of the cleanup job required in America to reduce the risk to public health. Consequently, the Superfund has been reauthorized several times since its establishment.

### A Huge Project

CERCLA requires the government to maintain a National Priorities List (NPL) of sites that pose the highest potential threat to human health and the environment. The NPL is constantly changing as new sites are officially

**FIGURE 4.11**

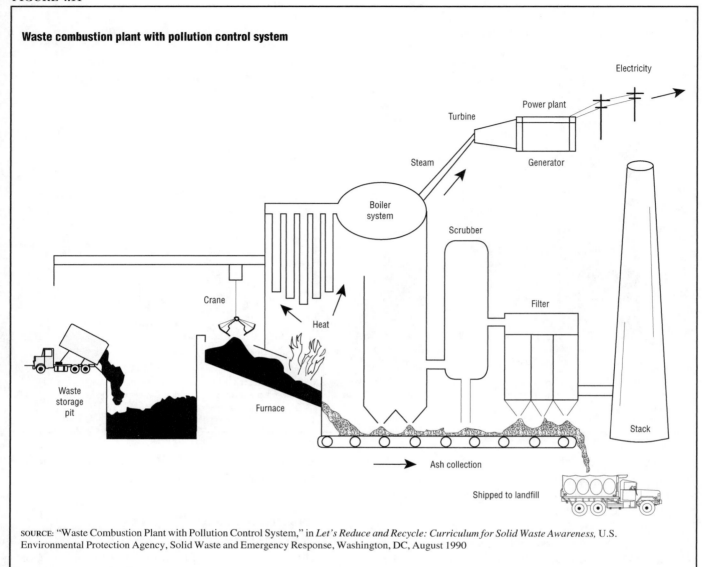

Waste combustion plant with pollution control system

SOURCE: "Waste Combustion Plant with Pollution Control System," in *Let's Reduce and Recycle: Curriculum for Solid Waste Awareness,* U.S. Environmental Protection Agency, Solid Waste and Emergency Response, Washington, DC, August 1990

added (finalized) and other sites are deleted. During the early 1980s hundreds of sites were added to the NPL each year. By 1992 a total of more than 1,200 sites had been added. Over the years sites were deleted as clean-ups were accomplished. Table 4.4 shows NPL site listings and clean-up milestones achieved by fiscal year (October through September) for 1992 through 2004. These data were reported in April 2004, meaning that only data for October 2003 through April 2004 are included under fiscal year 2004.

As of April 2004 the NPL, which is maintained by the EPA, includes 1,238 sites (158 federal sites and 1080 non-federal sites). The states containing the most NPL sites are New Jersey (113), California (96), Pennsylvania (92), New York (90), Michigan (67), and Florida (51). These six states account for just over 40 percent of all NPL final sites.

In 2003 the EPA proposed 14 new sites be added to the NPL based on preliminary investigations. Officials estimate that about one-fourth of these sites will actually be added to the NPL. As shown in Table 4.4, in general, the number of sites proposed for the NPL each year is outpacing the number deleted each year.

According to the EPA, more than three times as many Superfund sites were cleaned up between 1993 and 2000 than in all of the prior years of the program combined. However, many NPL sites are still years away from being cleaned up. The EPA estimates that 85 percent of the NPL sites will be cleaned up by 2008. Completion for the remaining 15 percent of the sites may take well beyond 2008.

When the Superfund was created, the program was expected to deal with a limited number of sites over a relatively short time. It eventually became clear that the number of sites needing attention was much larger than

TABLE 4.4

**Number of sites associated with the National Priorities List, 1992–2004**

by Fiscal year

| Action | 1992 | 1993 | 1994 | 1995 | 1996 | 1997 | 1998 | 1999 | 2000 | 2001 | 2002 | 2003 | 2004 |
|---|---|---|---|---|---|---|---|---|---|---|---|---|---|
| Sites proposed to the NPL | 30 | 52 | 36 | 9 | 27 | 20 | 34 | 37 | 40 | 45 | 9 | 14 | 11 |
| Sites finalized on the NPL | 0 | 33 | 43 | 31 | 13 | 18 | 17 | 43 | 39 | 29 | 19 | 20 | 0 |
| Sites deleted from the NPL | 2 | 12 | 13 | 25 | 34 | 32 | 20 | 23 | 19 | 30 | 17 | 9 | 6 |
| **Milestone** | | | | | | | | | | | | | |
| Partial deletions* | – | – | – | – | 0 | 6 | 7 | 3 | 5 | 4 | 7 | 7 | 6 |
| Construction completions | 88 | 68 | 61 | 68 | 64 | 88 | 87 | 85 | 87 | 47 | 42 | 40 | 10 |

A fiscal year is October 1 through September 30.
Fiscal year 2004 includes actions and milestones achieved from October 1, 2003 to the present.
Partial deletion totals are not applicable until fiscal year 1996, when the policy was first implemented.
*These totals represent the total number of partial deletions by fiscal year and may include multiple partial deletions at a site. Currently, there are 45 partial deletions at 37 sites.

SOURCE: "Number of NPL Site Actions and Milestones," in *National Priorities List*, U.S. Environmental Protection Agency, Washington, DC, April 27, 2004 [Online] http://www.epa.gov/superfund/sites/query/queryhtm/nplfy.htm [accessed May 5, 2004]

originally believed and that the program could run several more decades. The number of sites reported has declined steadily since 1985. The percentage of sites that the EPA believes warrant further consideration after initial investigation has remained relatively constant since 1984, with about 40 percent of the recommended sites needing further action. The average site cleanup takes approximately 12 years to complete.

### What Is the Cost and Who Pays the Bill for Superfund?

FUNDING FOR SUPERFUND. Funding for the Superfund program is derived through two major sources: the Superfund Trust Fund and monies appropriated from the federal government's general fund.

The Superfund Trust Fund was set up as part of the original Superfund legislation of 1980. It was designed to help the EPA pay for cleanups and related program activities. Table 4.5 shows the revenue going into the Superfund Trust Fund each year from 1993 through 2002. Until 1995 the Superfund Trust Fund was financed primarily by dedicated taxes collected from companies in the chemical and crude oil industries. The system was extremely unpopular with many corporations arguing that environmentally responsible companies should not have to pay for the mistakes of others. In 1995 the tax was eliminated.

The Superfund Trust Fund is also financed through cost recoveries—money the EPA recovers through legal settlements with responsible parties. The EPA is authorized to compel parties responsible for creating hazardous pollution, such as waste generators, waste haulers, site owners, or site operators, to clean up the sites. If these parties cannot be found, or if a settlement cannot be reached, the Superfund program finances the cleanup. After completing a cleanup, the EPA can take action against the responsible parties to recover costs and replenish the fund. The average cost of cleanup is about $30 million, large enough to make it worthwhile for parties to pursue legal means to spread the costs among large numbers of responsible parties. Many cleanups involve dozens of parties.

Disputes have arisen between industries and cities over who is responsible for a cleanup, and numerous lawsuits have been filed by industries against cities over responsibility for what is usually a huge expense. Many businesses and municipalities may be unable to assume such expense. The EPA reports that the government currently collects only one-fifth of the cleanup costs that could be recovered from polluters under the Superfund law. According to the EPA, in many cases, the polluters have disappeared or are unable to pay. In other cases, the agency lacks the staff or evidence to proceed with lawsuits.

All of these factors have resulted in only modest amounts of money being collected for the Superfund Trust Fund through cost recoveries. Total revenue into the Fund dropped from 2.4 billion dollars in 1993 to $368 million in 2002 as shown in Table 4.5. However, the EPA has continued to add sites to the NPL that require cleanup. According to a U.S. General Accounting Office (GAO) analysis conducted in 2003, the EPA consistently spent between $1.3 and $1.7 billion each year from 1993 to 2002 to operate the Superfund Program. The GAO reports that the unexpended balance of the Superfund Trust Fund stood at only $3.4 billion at the end of fiscal year 2002. At current rates of spending the Fund is expected to be depleted in a short amount of time.

In recent years the EPA has increasingly relied on money appropriated from the federal government's general fund to pay for NPL cleanups. During the early 2000s the general fund accounted for roughly half of all appropriations to the Superfund Program as shown in Figure 4.12. This means that all American taxpayers are increasingly paying to clean up hazardous waste sites under the Superfund Program. The GAO estimates that the general fund will supply about 80 percent of the

**TABLE 4.5**

**Revenue into the Superfund trust fund, fiscal years 1993–2002**

Constant 2002 dollars in millions

| Revenue source | Fiscal year | | | | | | | | | |
|---|---|---|---|---|---|---|---|---|---|---|
| | **1993** | **1994** | **1995** | **1996** | **1997** | **1998** | **1999** | **2000** | **2001** | **2002** |
| Taxes | $2,019 | $1,685 | $1,672 | $705 | $82 | $85 | $22 | $5 | $6 | $7 |
| Cost recoveries | 214 | 231 | 285 | 276 | 341 | 343 | 338 | 239 | 205 | 248 |
| Interest on unexpended balance | 165 | 202 | 359 | 388 | 359 | 313 | 233 | 245 | 223 | 111 |
| Fines and penalties | 4 | 3 | 3 | 4 | 3 | 5 | 4 | 1 | 2 | 1 |
| Total | $2,403 | $2,121 | $2,318 | $1,372 | $785 | $745 | $597 | $490 | $437 | $368 |

SOURCE: "Table 1: Revenue into the Superfund Trust Fund, Fiscal Years 1993 Through 2002," in *Superfund Program: Current Status and Future Fiscal Challenges*, GAO-03-850, U.S. General Accounting Office, Washington, DC, July 31, 2003

monies needed for the Superfund Program in EPA's fiscal year 2004 budget.

Some critics have called for the federal government to reinstate dedicated taxes against petroleum and chemical corporations to fund the Superfund Program, instead of burdening taxpayers. The GAO notes that Congress is reluctant to appropriate more general fund monies to the Superfund Program and fears that costs will continue to escalate as the EPA adds more sites to the NPL. A major complaint is that the Superfund Program lacks an effective system for indicating the progress that it is making toward cleaning up the nation's hazardous waste sites. Congress has asked the EPA to develop performance indicators that could help them make better funding decisions for the Superfund Program. An EPA advisory council is expected to make its recommendations during 2004.

## BROWNFIELDS

Many former industrial sites have become eyesores of urban scenery. These trash-strewn plots, concentrated mostly in the northeast and Midwest, are called brownfields and, as of EPA estimates in early 2004, numbered more than 400,000 sites. The EPA defines brownfields as abandoned, idled, or underused industrial or commercial sites where expansion or redevelopment is complicated by real or perceived environmental contamination.

Many of the properties are polluted. They are shunned by developers, often stalling efforts to revive poor, inner-city neighborhoods. Until 1995, developers and buyers had avoided some 38,000 sites listed as possible targets under the Superfund Law, which says that anyone involved in the management of a property can be held liable for the entire cost of cleanup. Many of those sites had, in fact, been passed over by the EPA as not contaminated enough for Superfund action. Nonetheless, many of those properties were deemed untouchable by the real estate industry. A 1995 survey of the American Bankers Association showed that 83 percent of smaller banks had refused to make loans to projects because of concerns about environmental liability.

To help the reclamation effort, in 1995 the EPA removed 25,000 of the least-polluted sites from the list. The sites required some type of cleanup but would not be subjected to the tougher Superfund standards. In addition to restoring the environment, the purpose of reclamation programs is to encourage the reuse of abandoned sites, revitalize cities, create jobs, and generate municipal tax revenues. Redevelopment of polluted sites is becoming a thriving business. Experts estimate that about one-third of real estate sales involve sifting databases of environmental agencies for records of toxic spills before a real estate transaction can take place. Sensing a new business possibility, several insurance companies have created divisions offering policies that protect developers of polluted real estate against unforeseen cleanup costs or lawsuits.

In 1997 U.S. President Bill Clinton signed the Taxpayer Relief Acts (PL 105-34 and PL 105-32), both of which included new tax incentives to spur the cleanup and redevelopment of brownfields. The acts enable taxpayers to consider any qualified environmental remediation expenditure as tax deductions in the year paid rather than having to be capitalized over time.

## NUCLEAR WASTE

Nuclear waste includes a wide range of materials with varying levels of radioactivity. There is currently no agreed-upon safe way to dispose of nuclear waste. None of the current options guarantees protection of the biosphere from radiation, which can linger for many thousands of years. Because of the scientific and political difficulties with geologic burial and other methods, aboveground "temporary" storage, despite its dangers, may remain the only option well into the twenty-first century.

### Low-Level Radioactive Waste

Low-level waste includes items that have become contaminated with radioactive material or have become radioactive through exposure to neutron radiation. This waste typically consists of contaminated protective shoe covers and clothing, wiping rags, mops, filters, reactor

FIGURE 4.12

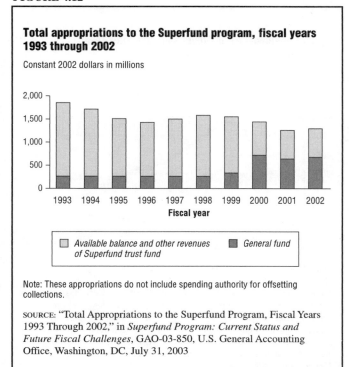

**Total appropriations to the Superfund program, fiscal years 1993 through 2002**

Constant 2002 dollars in millions

Legend:
- □ Available balance and other revenues of Superfund trust fund
- ■ General fund

Note: These appropriations do not include spending authority for offsetting collections.

SOURCE: "Total Appropriations to the Superfund Program, Fiscal Years 1993 Through 2002," in *Superfund Program: Current Status and Future Fiscal Challenges*, GAO-03-850, U.S. General Accounting Office, Washington, DC, July 31, 2003

FIGURE 4.13

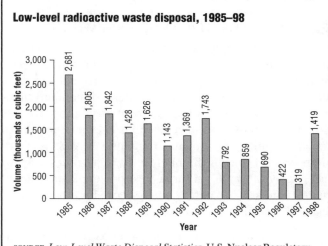

**Low-level radioactive waste disposal, 1985–98**

SOURCE: *Low-Level Waste Disposal Statistics,* U.S. Nuclear Regulatory Commission, Rockville, MD, December 21, 2001 [Online] http://www .nrc.gov/waste/llw-disposal/statistics.html [accessed May 5, 2004]

water treatment residues, equipment and tools, luminous dials, medical tubes, swabs, injection needles, syringes, and laboratory animal carcasses and tissues. The radioactivity can range from just above background levels found in nature, to very highly radioactive as in the case of, for example, parts from inside the reactor vessel in a nuclear power plant.

Low-level waste is typically stored on-site by licensees, either until it has decayed away and can be disposed of as ordinary trash, or until amounts are large enough for shipment to a low-level waste disposal site in containers approved by the U.S. Department of Transportation. As shown in Figure 4.13 approximately 1.4 million cubic feet of commercial low-level waste was shipped to disposal facilities in 1998, the latest year for which data are available.

Most low-level radioactive waste decays in less than 50 years. Until the 1960s, the United States dumped low-level wastes into the ocean. The first commercial site to house such waste was opened in 1962, and by 1971 six sites were licensed for disposal. By 1979 only three commercial low-level waste sites were still operating—Hanford, Washington; Beatty, Nevada; and Barnwell, South Carolina. The volume of low-level waste increased during the initial years (1963–80) of commercially generated waste disposal; this, coupled with the threatened closing of the South Carolina site, prompted Congress to pass the Low Level Radioactive Waste Policy Act of 1980 (PL 96-573), calling for the establishment of a national system of

such facilities. Since then, volume has decreased. The act made every state responsible for finding a low-level disposal site by 1986 for wastes generated within its borders. It also gave states the right to bar imports of low-level wastes if they were engaged in regional compacts for waste disposal. The disposal of high-level wastes, however, remains a federal responsibility.

COMPACTS. The 1980 law encouraged states to organize themselves into compacts to develop new low-level waste facilities. As of 2004 ten compacts serving forty-three states had been approved by Congress. Compacts and unaffiliated states have confronted significant barriers to developing disposal sites, however, including: public health and environmental concerns, antinuclear sentiment, substantial financial requirements, political issues, and "not in my backyard" campaigns by some citizen activists.

No compact or state had successfully developed a new disposal facility for low-level wastes by early 2004. Certain conditions have led some states to remain uncommitted to disposal development and to consider other options. The reopening of the Barnwell, South Carolina, facility in 1995 eased some of the pressure on the states. The emergence in 1995 of new private-sector nuclear waste handlers—such as Envirocare of Utah, Inc.—has increased interest in the possibility of privately operated waste disposal facilities. Collectively, the Barnwell; Richland, Washington (operated by American Ecology Corporation); and Envirocare facilities provide disposal capacity for almost all types of low-level wastes.

The number of DOE shipments of low-level waste is projected to decrease throughout the remainder of the decade. (See Figure 4.14.) According to the DOE, shipments of low-level radioactive wastes accounted for less than 0.5 percent of all hazardous material shipments in 2001.

## Spent Fuel and High-Level Radioactive Waste

The most dangerous radioactive waste is irradiated uranium from commercial nuclear power plants. Spent fuel, the used uranium fuel removed from a nuclear reactor, is far from being completely "spent." It contains highly penetrating and toxic radioactivity and requires isolation from living things for thousands of years. It still contains significant amounts of uranium, as well as plutonium created during the nuclear fission process. Spent fuel is a serious problem for nuclear power plants that will be decommissioned before a long-term, high-level waste disposal repository is available.

Unless a temporary site becomes available, decommissioned plants have the following options:

- The fuel can be left in place.

- On-site storage casks can be used. This is not an option for hot fuel (fuel that is less than five years out of the core).

- The spent fuel can be shipped abroad for reprocessing. France, which is heavily dependent on nuclear power, developed the technology to reprocess spent fuel, something not available in the United States. In 1993 the British government also opened a nuclear fuel reprocessing plant that reprocesses spent fuel from nuclear power generators around the world. Nuclear watch groups and some Americans fear that shipping spent fuel abroad will undermine efforts to halt the spread of nuclear arms because the process of transporting such materials increases the possibility for theft or accident.

- The unit can continue to operate.

- The fuel can be shipped to a monitored retrievable storage facility, if there is one available.

Figure 4.15 shows the major sites storing spent nuclear fuel as of December 2003. The Idaho National Engineering and Environmental Laboratory is the main storage site, accounting for 54 percent of the total stored.

## The Vestiges of Nuclear Disarmament

Nuclear disarmament resulted in the dismantling of much of the U.S. nuclear arsenal and the resulting need to store tons of plutonium. The federal government has proceeded to take apart as many as 15,000 warheads with intentions of eventually storing them at one of two former nuclear weapons–making plants—Pantex, near Amarillo, Texas, and Savannah River, South Carolina. DOE officials predict that dismantling will continue through 2003. The government must then decontaminate buildings used at those facilities, dispose of millions of gallons of boiling

**FIGURE 4.14**

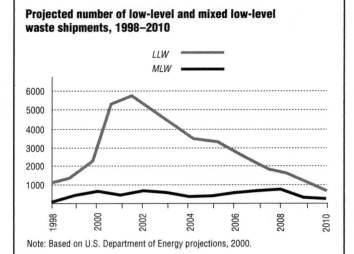

**Projected number of low-level and mixed low-level waste shipments, 1998–2010**

Note: Based on U.S. Department of Energy projections, 2000.

SOURCE: "Figure 7. DOE Low-Level and Mixed Low-Level Waste Projected Number of Shipments through 2010," in *A Guide to the U.S. Department of Energy's Low-Level Radioactive Waste,* National Safety Council, Environmental Health Center, Washington, DC, January 2002. Reproduced from *A Guide to the U.S. Department of Energy's Low-Level Radioactive Waste* with permission from the National Safety Council's Environmental Health Center, January 2002.

radioactive water, and decontaminate hundreds of square miles of desert at the Nevada nuclear test site.

In 2000 the United States and Russia agreed that each country would dispose of 34 metric tons of surplus weapons-grade plutonium by 2019. The Clinton administration had intended to immobilize permanently most of the plutonium in glass or ceramic to prevent its potential use in nuclear weapons. The remainder of the plutonium was to be converted to a mixed-oxide fuel (MOX) for use in commercial nuclear power reactors. MOX fuel has been used in existing reactors in Europe, but never tried in the United States.

In April 2002 the Bush administration announced plans to convert all of the plutonium to MOX fuel. The DOE proposed construction of a MOX Fuel Fabrication Facility at the Savannah River site in South Carolina. This facility was to be operational by 2007. The governor of the state fought the plan in court and threatened to use the state police to stop waste shipments from entering South Carolina. He feared that the MOX Fuel Fabrication Facility would never be funded and that the waste would be permanently stored in South Carolina.

As of 2004 the Nuclear Regulatory Commission has not granted final approval for construction of the facility. In addition the project is being fought in court by environmental groups including Georgians Against Nuclear Energy. Even if the project is approved, federal budget cuts are expected to delay the operational date to 2008. The utility

## FIGURE 4.15

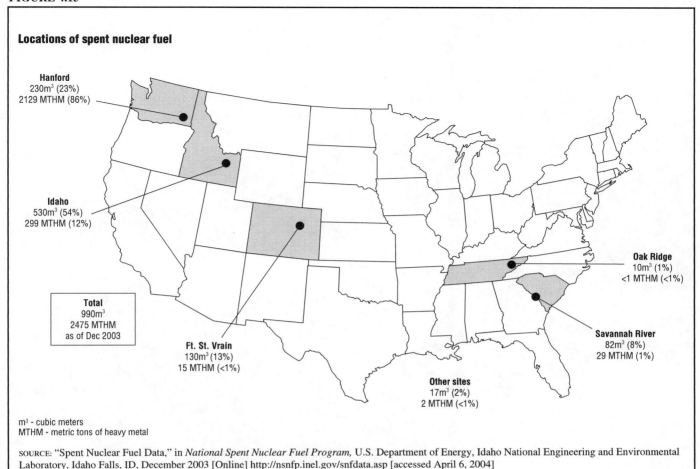

**Locations of spent nuclear fuel**

**Hanford**
230m³ (23%)
2129 MTHM (86%)

**Idaho**
530m³ (54%)
299 MTHM (12%)

**Total**
990m³
2475 MTHM
as of Dec 2003

**Ft. St. Vrain**
130m³ (13%)
15 MTHM (<1%)

**Oak Ridge**
10m³ (1%)
<1 MTHM (<1%)

**Savannah River**
82m³ (8%)
29 MTHM (1%)

**Other sites**
17m³ (2%)
2 MTHM (<1%)

m³ - cubic meters
MTHM - metric tons of heavy metal

SOURCE: "Spent Nuclear Fuel Data," in *National Spent Nuclear Fuel Program,* U.S. Department of Energy, Idaho National Engineering and Environmental Laboratory, Idaho Falls, ID, December 2003 [Online] http://nsnfp.inel.gov/snfdata.asp [accessed April 6, 2004]

company Duke Power has announced that it plans to test MOX combustion at one of its power plants in 2005 and begin full-scale use of MOX fuel by 2010.

## FEDERAL NUCLEAR WASTE REPOSITORIES

In the United States the federal government is focusing on two locations as eventual long-term nuclear waste repositories. The Waste Isolation Pilot Plant in southeastern New Mexico would store defense-generated transuranic waste (waste left over from research and production of nuclear weapons), while Nevada's Yucca Mountain would house civilian nuclear waste.

### The Waste Isolation Pilot Plant

The Waste Isolation Pilot Plant (WIPP) became the world's first deep depository for nuclear waste when it received its first shipment of waste on March 26, 1999. This large facility is located near Carlsbad, New Mexico. WIPP is 655 meters below the Earth's surface in the salt beds of the Salado Formation and is intended to house up to 6.25 million cubic feet of transuranic waste for more than 10,000 years.

Under congressional mandate, the WIPP facility does not accept commercial or high-level waste; it only accepts transuranic waste. More than 99 percent of transuranic waste is temporarily stored in drums at nuclear defense sites around the country. Waste shipped to WIPP is tracked by satellite and moved only at night, when traffic is lighter. It can be transported only in good weather and must be routed around major cities. Figure 4.16 shows a tractor trailer container transporting radioactive waste to WIPP.

As of May 1, 2004, the WIPP facility has received 2,543 shipments of radioactive waste and disposed of 19,793 cubic meters. By 2035, barring court challenges, almost 40,000 truckloads of nuclear waste will be trucked across the country to WIPP.

### Yucca Mountain

The centerpiece of the federal government's plan to dispose of highly radioactive waste is a proposed facility at Yucca Mountain in Nevada. (See Figure 4.17.) The Nuclear Waste Policy Act of 1982 was amended in 1987 to require the secretary of energy to investigate the site and, if suitable, recommend to the president that the site

**FIGURE 4.16**

Using specially designed containment canisters, a truck hauls radioactive waste to the Waste Isolation Pilot Plant in New Mexico, where it can be safely stored. *(U.S. Department of Energy.)*

be established. The investigation of Yucca Mountain has taken a long time and cost more than $3 billion. In February 2002 DOE secretary Spencer Abraham recommended the site to President Bush as a storage facility for nuclear waste. Despite intense protests from environmentalists and Nevada politicians, the president accepted the recommendation. The DOE hopes to bury 77,000 tons of radioactive waste at the Yucca site beginning in 2009.

Yucca Mountain is a 1,200-foot-high volcanic ridge located on federally owned land in the desert approximately 100 miles northwest of Las Vegas. The repository would be constructed 1,200 feet below the land surface and 800 feet above the groundwater table.

**STANDARDS FOR CONTAINMENT.** For the Yucca Mountain Repository to be built, the DOE must satisfactorily demonstrate to the Nuclear Regulatory Commission that the combination of the site and the repository design complies with the standards set forth by the EPA. The EPA's standard is based on a new approach of using numerical probabilities to establish requirements for containing radioactivity within the repository. Their quantitative terms are as follows:

- Cumulative releases of radioactivity from a repository must have a likelihood of less than 1 chance in 10 of exceeding limits established in the standard, and a likelihood of less than 1 chance in 1,000 of exceeding 10 times the limits, for a period of 10,000 years.

- Exposures of radiation to individual members of the public for 1,000 years must not exceed specified limits.

- Limits are placed on the concentration of radioactivity for 1,000 years after disposal from the repository to a nearby source of groundwater that currently supplies drinking water for thousands of persons, and is irreplaceable.

- Prescribed technical or institutional procedures or steps must provide confidence that the containment requirements are likely to be met.

**FIGURE 4.17**

An aerial view of Yucca Mountain, Nevada. Yucca Mountain is the proposed site for a major, long-term, nuclear waste storage facility. *(U.S. Department of Energy, Office of Civilian Radioactive Waste Management.)*

## Crisis in the Industry

The long delay in providing disposal sites for nuclear wastes, coupled with the accelerated pace at which nuclear plants are being retired, has created a crisis in the industry. Several aging plants are being maintained—at a cost of $20 million a year for each reactor—simply because there is no place to send the waste once the plants are decommissioned. Under the Nuclear Waste Policy Act of 1982, the DOE was scheduled to begin picking up waste on January 31, 1998.

The utilities have been paying $.01 per 10 kilowatt-hours of nuclear power generated by the reactors to finance a repository. Although the 1987 waste amendment designated Yucca Mountain as the site, little progress was made in approving the project. In 1996 the U.S. Court of Appeals, in *Indiana Michigan Power Co. v. Department of Energy* (88F3d1272 [D.C. Ctr., 1996]), ruled that the nuclear waste act created an obligation for the DOE to start disposing of utilities' waste no later than January 31, 1998. Because the DOE missed the deadline, more than twenty utilities have sued.

In February 1999 the DOE announced that because it was unable to receive nuclear waste for permanent storage, it would take ownership of the waste and pay temporary storage costs with money the utilities have paid to develop the permanent repository. The waste will stay where it currently is being stored, and the DOE will pay the storage costs. Even without the expense of temporary storage, the nuclear waste fund (the money collected from the utilities) is many billions short of what Yucca Mountain is expected to cost.

## A Serious Leak of Radioactive Waste—Hanford

In 1997 scientists discovered that about 900,000 gallons of radioactive waste had leaked into the soil from 68 of the 149 tanks at the nuclear weapons plant in Hanford, Washington. Eventually, all the tanks are expected to leak. The leak contaminated underground water moving toward the nearby Columbia River. Managers at the plant maintained that the leaks were insignificant because radioactive materials would be trapped by the area above the water table (the "vadose zone"). Furthermore, officials had been saying for decades that no waste from the tanks would reach the groundwater in the next 10,000 years.

However, as of early 2004 government officials estimate that groundwater contamination affects approximately eighty square miles of the site, prompting comparisons of the situation to the Chernobyl disaster. (In April 1986 an explosion at the nuclear power plant located in the Soviet [now the Ukraine] town of Chernobyl killed thirty people.) A threatened lawsuit by the State of Washington against the DOE over the leaks at the Hanford site resulted in an agreement to clean up the two

TABLE 4.6

**Public concern about contamination of soil and water by toxic waste, 2004**

PLEASE TELL ME IF YOU PERSONALLY WORRY ABOUT THIS PROBLEM A GREAT DEAL, A FAIR AMOUNT, ONLY A LITTLE, OR NOT AT ALL. CONTAMINATION OF SOIL AND WATER BY TOXIC WASTE?

| | Great deal % | Fair amount % | Only a little % | Not at all % | No opinion % |
|---|---|---|---|---|---|
| 2004 Mar 8–11 | 48 | 26 | 21 | 5 | * |
| 2003 Mar 3–5 | 51 | 28 | 16 | 5 | * |
| 2002 Mar 4–7 | 53 | 29 | 15 | 3 | * |
| 2001 Mar 5–7 | 58 | 27 | 12 | 3 | * |
| 2000 Apr 3–9 | 64 | 25 | 7 | 4 | * |
| 1999 Apr 13–14 | 63 | 27 | 7 | 3 | * |
| 1999 Mar 12–14 | 55 | 29 | 11 | 5 | * |
| 1991 Apr 11–14 | 62 | 21 | 11 | 5 | 1 |
| 1990 Apr 5–8 | 63 | 22 | 10 | 5 | * |
| 1989 May 4–7 | 69 | 21 | 6 | 3 | * |

SOURCE: "Please tell me if you personally worry about this problem a great deal, a fair amount, only a little, or not at all. Contamination of soil and water by toxic waste?," in *Poll Topics and Trends: Environment*, The Gallup Organization, Princeton, NJ, March 17, 2004 [Online] www.gallup.com [accessed March 30, 2004]

indoor pools near the Columbia River by 2007. Progress reports on site clean-up are available at http://www.hanford.gov/cp/gpp.

## PUBLIC OPINION ABOUT WASTE DISPOSAL

In March 2004 the Gallup Organization conducted its annual poll regarding environmental issues. Participants were asked to express the level of concern they feel about various environmental problems. As shown in Table 4.6 less than half of those asked felt a great deal of concern about the possibility of contamination of soil and water by toxic waste. This percentage has, in general, fallen steadily since its high point of 69 percent in 1989. Nearly as many people (47 percent) felt either a fair amount or a little amount of concern about this problem. A small percentage (5 percent) expressed no concern at all.

## CHAPTER 5
# AIR QUALITY

## THE AIR PEOPLE BREATHE

According to the Environmental Protection Agency (EPA), the average person breathes in 3,400 gallons of air each day. Because air is so essential to life, it is important that it be free of pollutants. Throughout the world poor air quality contributes to hundreds of thousands of deaths and diseases each year, not to mention dying forests and lakes and the corrosion of stone buildings and monuments. Air quality is also important to the quality of life and recreation because air pollution causes haze that decreases visibility during outdoor activities.

Fossil fuels and chemicals have played a major role in society's pursuit of economic growth and higher standards of living. However, burning fossil fuels and releasing toxic chemicals into the air alter Earth's chemistry and can threaten the very air on which life depends.

## WHAT ARE THE MAJOR AIR POLLUTANTS?

Air quality standards established under the Clean Air Act of 1970 (CAA; PL 91-604) are designed to protect public health and welfare. The act was amended by the CAA Amendments of 1990 (CAAA; PL 101-549) with the goal of cleaning up urban air. It established air quality standards called the National Ambient Air Quality Standards (NAAQS) for six major air pollutants:

- carbon monoxide (CO)

- lead (Pb)

- nitrogen oxides ($NO_5$)

- ozone ($O_3$)

- particulate matter (PM)

- sulfur oxides ($SO_5$)

These are called the priority or criteria pollutants and are identified as serious threats to human health. The

CAA also required states to develop plans to implement and maintain the NAAQS. The states can have stricter rules than the federal program but not more lenient ones.

The EPA has documented air pollution trends in the United States annually since 1973. Its *Latest Findings on National Air Quality: 2002 Status and Trends* (2003) reports two kinds of trends for priority pollutants: emissions and air quality concentrations. Emissions are calculated estimates of the total tonnage of these pollutants released into the air annually. Air quality concentrations are based on data collected at thousands of monitoring sites around the country. The EPA maintains a database called the National Emission Inventory that characterizes the emissions of air pollutants in the United States based on data input from state and local agencies.

From 1970 to 2002 total emissions of the six priority pollutants decreased by 48 percent. This occurred even as the United States experienced massive increases in gross domestic product and vehicle miles traveled and moderate increases in overall energy consumption and population. Figure 5.1 compares emissions in 1970 and 2001 for lead and between 1970 and 2002 for all other principal air pollutants. The figure shows that emissions have declined for each pollutant. However, there is still much work to be done to clear the air. The EPA estimates that during 2002 approximately 160 million tons of air pollutants were emitted in the United States.

### Carbon Monoxide

Carbon monoxide (CO) is a colorless, odorless gas created when the carbon in certain fuels is not burned completely. These fuels include coal, natural gas, oil, gasoline, and wood.

EMISSIONS AND SOURCES. In 2002 approximately 100 million tons of CO were emitted into the air. As shown in Figure 5.1, CO emissions have decreased by 48

FIGURE 5.1

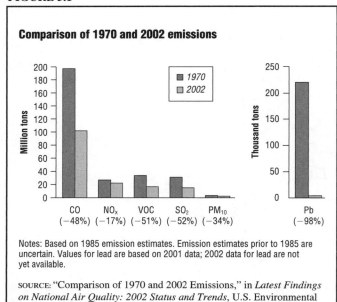

**Comparison of 1970 and 2002 emissions**

Notes: Based on 1985 emission estimates. Emission estimates prior to 1985 are uncertain. Values for lead are based on 2001 data; 2002 data for lead are not yet available.

SOURCE: "Comparison of 1970 and 2002 Emissions," in *Latest Findings on National Air Quality: 2002 Status and Trends*, U.S. Environmental Protection Agency, Office of Air Quality Planning and Standards, Washington, DC, August 2003

FIGURE 5.2

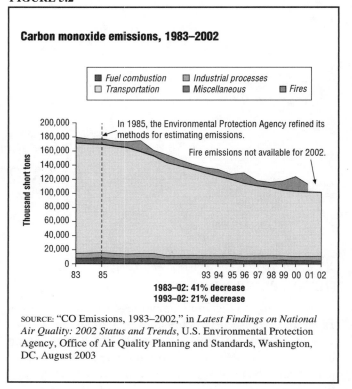

**Carbon monoxide emissions, 1983–2002**

SOURCE: "CO Emissions, 1983–2002," in *Latest Findings on National Air Quality: 2002 Status and Trends*, U.S. Environmental Protection Agency, Office of Air Quality Planning and Standards, Washington, DC, August 2003

percent since 1970. Figure 5.2 shows the primary sources of CO emissions for 1983 to 2002. Transportation has historically been the largest source. In 2002 transportation accounted for 82 percent of known CO emissions. In recent years forest fires (planned and unplanned) have contributed 5–15 percent of the nation's carbon monoxide emissions. Non-transportation fuel combustion, industrial processes, and miscellaneous other sources are minor contributors of CO emissions, each contributing less than 5 percent of the total.

Non-transportation fuel combustion occurs at power plants, homes, and businesses that burn fuel to produce power or heat. In 1940 cars and trucks created about 28 percent of all CO emissions, while homes burning coal and oil made up about 50 percent. From 1940 through 1970 emissions from cars and trucks nearly tripled. By 1970 they accounted for 71 percent of all CO emissions and continued to increase into the 1980s. Their contribution to the total began to fall as vehicle fuel efficiency improved.

The EPA estimates that in 2002 motor vehicle exhaust from road vehicles accounted on average for 60 percent of all CO emissions. The value was as much as 95 percent in cities with heavy traffic congestion. Non-road vehicles account for the remainder of transportation-related CO emissions. Non-road vehicles include tractors, lawn mowers, construction equipment (such as bulldozers), industrial equipment (such as forklifts), snowmobiles, all-terrain vehicles, boats, trains, and airplanes.

**AIR QUALITY.** Air quality concentrations of CO from 1983 to 2002 are shown in Figure 5.3 based on monitor-

ing data from 205 sites around the country. Over this time period CO concentrations decreased by 65 percent. In 2002 the average measured CO concentration at the monitoring sites was just under three parts per million (three parts CO per million parts of air).

Despite this data the EPA reports that in 2002 approximately 700,000 Americans lived in counties with air quality concentrations of CO above the NAAQS. These nonattainment areas, as they are called, are designated as having "serious" or "moderate" CO pollution depending on the air quality concentrations. As of January 2004 serious CO nonattainment areas include the following:

• Anchorage and Fairbanks, Alaska

• Las Vegas, Nevada

• Los Angeles, California

• Phoenix, Arizona

• Spokane, Washington

Moderate CO nonattainment areas include Provo, Utah; El Paso, Texas; Missoula, Montana; and Reno, Nevada.

**ADVERSE HEALTH EFFECTS.** CO is a dangerous gas that enters a person's bloodstream through the lungs. It reduces the ability of blood to carry oxygen to the body's cells, organs, and tissues. The health danger is highest for people suffering from cardiovascular diseases.

**FIGURE 5.3**

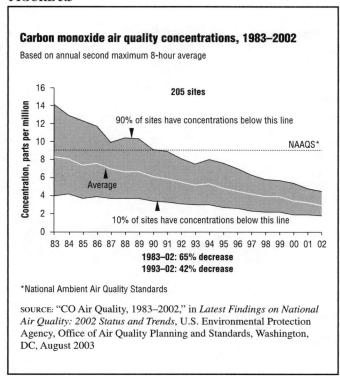

**Carbon monoxide air quality concentrations, 1983–2002**

Based on annual second maximum 8-hour average

205 sites

90% of sites have concentrations below this line

NAAQS*

Average

10% of sites have concentrations below this line

**1983–02: 65% decrease**
**1993–02: 42% decrease**

*National Ambient Air Quality Standards

SOURCE: "CO Air Quality, 1983–2002," in *Latest Findings on National Air Quality: 2002 Status and Trends*, U.S. Environmental Protection Agency, Office of Air Quality Planning and Standards, Washington, DC, August 2003

**FIGURE 5.4**

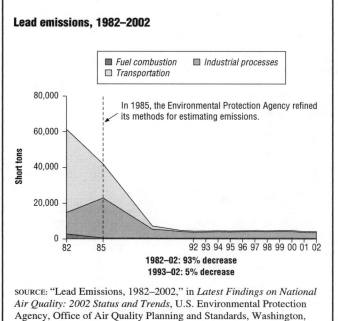

**Lead emissions, 1982–2002**

■ Fuel combustion   ■ Industrial processes
□ Transportation

In 1985, the Environmental Protection Agency refined its methods for estimating emissions.

**1982–02: 93% decrease**
**1993–02: 5% decrease**

SOURCE: "Lead Emissions, 1982–2002," in *Latest Findings on National Air Quality: 2002 Status and Trends*, U.S. Environmental Protection Agency, Office of Air Quality Planning and Standards, Washington, DC, August 2003

## Lead

Lead is a metal that can enter the atmosphere via combustion or industrial processing of lead-containing materials.

**EMISSIONS AND SOURCES.** Prior to 1985 the major source of lead emissions in the United States was the leaded gasoline used in automobiles. Conversion to unleaded gasoline produced a dramatic reduction in lead emissions. As shown in Figure 5.4, transportation has virtually been eliminated as a source of lead emissions. Nontransportation fuel combustion is also a minor contributor. Industrial processes (chiefly metals smelting and battery manufacturing) are responsible for the bulk of lead emissions. The EPA reports that lead emissions declined by 93 percent between 1982 and 2002.

**AIR QUALITY.** Air quality concentrations of lead based on monitoring data from 42 sites from 1983 to 2002 are shown in Figure 5.5. Over this time period lead concentrations decreased by 94 percent.

Despite great progress in lead reduction, the EPA reports that in 2002 approximately 200,000 Americans lived in counties with air quality concentrations of lead above the NAAQS. As of January 2004 there are only three lead nonattainment areas in the country: East Helena, Montana, and portions of both Iron County and Jefferson County in Missouri.

**ADVERSE HEALTH EFFECTS.** Lead is a particularly dangerous pollutant because it accumulates in the blood, bones, and soft tissues of the body. It can adversely affect the nervous system, kidneys, liver, and other organs. Excessive concentrations are associated with neurological impairments, mental retardation, and behavioral disorders. Even low doses of lead can damage the brains and nervous systems of fetuses and young children. Atmospheric lead that falls onto vegetation poses an ingestion hazard to humans and animals.

### Nitrogen Dioxide

Nitrogen dioxide ($NO_2$) is a reddish-brown gas that forms in the atmosphere when nitrogen oxide (NO) is oxidized. The chemical formula $NO_5$ is used collectively to describe NO, $NO_2$, and other nitrogen oxides.

**EMISSIONS AND SOURCES.** $NO_2$ primarily comes from burning fuels such as gasoline, natural gas, coal, and oil. The exhaust from transportation vehicles is the major source of $NO_2$, accounting for 56 percent of emissions during 2002. Fuel combustion in power plants, homes, and businesses accounted for 37 percent of $NO_5$ emissions. Industrial processes and miscellaneous sources were minor contributors.

Overall, emissions of $NO_5$ decreased by 15 percent between 1983 and 2002. (See Figure 5.6.) Most of this improvement occurred during the late 1990s and early 2000s.

The EPA notes that despite the overall decrease in $NO_5$ emissions in this country, emissions from non-road

FIGURE 5.5

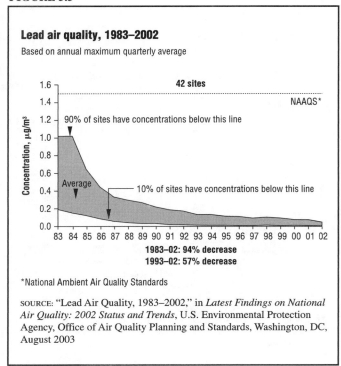

Lead air quality, 1983–2002

Based on annual maximum quarterly average

1983–02: 94% decrease
1993–02: 57% decrease

*National Ambient Air Quality Standards

SOURCE: "Lead Air Quality, 1983–2002," in *Latest Findings on National Air Quality: 2002 Status and Trends*, U.S. Environmental Protection Agency, Office of Air Quality Planning and Standards, Washington, DC, August 2003

FIGURE 5.6

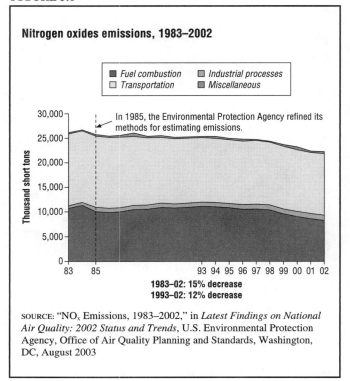

Nitrogen oxides emissions, 1983–2002

1983–02: 15% decrease
1993–02: 12% decrease

SOURCE: "NO$_x$ Emissions, 1983–2002," in *Latest Findings on National Air Quality: 2002 Status and Trends*, U.S. Environmental Protection Agency, Office of Air Quality Planning and Standards, Washington, DC, August 2003

engines increased dramatically between 1983 and 2002. Non-road engines are those in sources such as tractors, lawn mowers, construction and industrial equipment, off-road recreational vehicles, boats, trains, and airplanes. As of April 2004 the EPA has proposed new regulations to control NO$_5$ emissions from non-road diesel engines.

**AIR QUALITY.** NO$_2$ is a major precursor of smog and also contributes to acid rain and haze. It can also undergo reactions in the air that lead to the formation of particulate matter and ozone.

Air quality concentrations of NO$_2$ based on monitoring data from 125 sites around the country from 1983 to 2002 are shown in Figure 5.7. Over this time period NO$_2$ concentrations decreased by 21 percent. In 2002 the average measured NO$_2$ concentration at the monitoring sites was approximately 0.02 parts per million. That year all U.S. counties attained the EPA standards for NO$_2$ air quality.

**ADVERSE HEALTH EFFECTS.** Inhalation of even low concentrations of NO$_2$ for short time periods can be harmful to the human body's breathing functions. Longer exposures are considered damaging to the lungs and may cause people to be more susceptible to certain respiratory problems, such as infections.

## Ozone

Ozone is a gas naturally present in Earth's upper atmosphere. Approximately 90 percent of Earth's ozone lies in the stratosphere at altitudes greater than about 20 miles. Ozone molecules at this level absorb ultraviolet

(UV) radiation from the sun and prevent it from reaching the ground. Thus, stratospheric ozone (the "ozone layer") is good for the environment. Tropospheric (or ground-level) ozone, on the other hand, is a potent air pollutant with serious health consequences.

Ground-level ozone has become a persistent problem in many parts of the world. Ground-level ozone is the most complex, pervasive, and difficult to control of the six priority pollutants.

**EMISSIONS AND SOURCES.** Unlike other pollutants, ground-level ozone is not emitted directly into the air. It forms mostly on sunny, hot days due to complex chemical reactions that take place when the atmosphere contains other pollutants, primarily volatile organic compounds (VOCs) and nitrogen oxides (NO$_5$) emitted from industrial chemical processes and fossil fuel combustion. Such pollutants are called ozone precursors because their presence in the atmosphere leads to ozone creation.

VOCs are carbon-containing chemicals that easily become vapors or gases. Paint thinners, degreasers, and other solvents contain a great number of VOCs, which are also released from burning fuels such as coal, natural gas, gasoline, and wood. Cars are a major source of VOCs. From 1940 through 1970 VOC emissions increased about 77 percent, mainly because of the increase in car and truck traffic and industrial production. However, since 1970 national VOC emissions have decreased as a result of emission controls placed on cars and trucks and less open burning of solid waste.

FIGURE 5.7

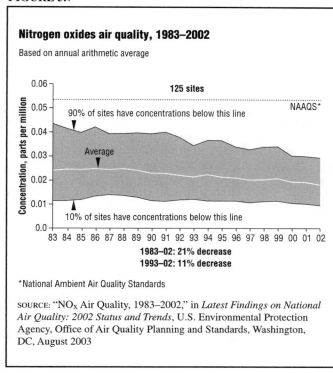

**Nitrogen oxides air quality, 1983–2002**

Based on annual arithmetic average

SOURCE: "NO$_X$ Air Quality, 1983–2002," in *Latest Findings on National Air Quality: 2002 Status and Trends*, U.S. Environmental Protection Agency, Office of Air Quality Planning and Standards, Washington, DC, August 2003

FIGURE 5.8

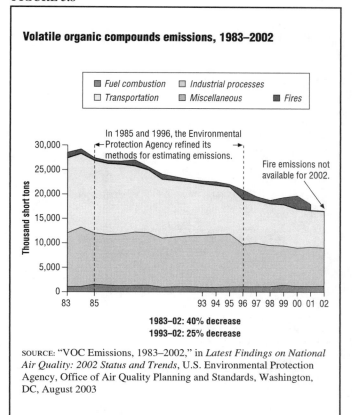

**Volatile organic compounds emissions, 1983–2002**

SOURCE: "VOC Emissions, 1983–2002," in *Latest Findings on National Air Quality: 2002 Status and Trends*, U.S. Environmental Protection Agency, Office of Air Quality Planning and Standards, Washington, DC, August 2003

VOC emissions dropped by 40 percent from 1983 to 2002. (See Figure 5.8.) Most of the decline occurred in industrial processing and transportation sources. However, these two sources still account for the vast majority of VOC emissions. In 2002 industrial processes and transportation each accounted for approximately 40 percent of the emissions. In recent years forest fires have contributed 5–10 percent of VOC emissions. Fuel combustion and miscellaneous sources are also minor contributors.

VOCs and NO$_5$ are the primary ozone precursors. Although emissions of both decreased during the 1980s and 1990s, the EPA notes that additional reduction in NO$_5$ emissions must take place to achieve any substantial improvements in ozone air quality.

**AIR QUALITY.** Ozone has different health and environmental effects depending on the time of exposure. The EPA monitors average eight-hour and one-hour ozone levels and sets different standards for each. Ozone concentrations can vary greatly from year to year depending on the emissions of ozone precursors and weather conditions.

As shown in Figure 5.9, the average national ozone concentration, based on an eight-hour average, decreased by 14 percent between 1983 and 2002 but increased by 4 percent between 1993 and 2002. The average one-hour concentration decreased by 22 percent between 1983 and 2002. Between 1993 and 2002, however, the decrease was only 2 percent. (See Figure 5.10.)

In 2002 both the one-hour and eight-hour averages for ozone air quality exceeded the NAAQS. The one-hour

exceedance affected 68 million people, while the eight-hour exceedance affected 136.4 million people. This means that ozone has, by far, the largest nonattainment area of any of the six priority pollutants. As of April 15, 2004, the nonattainment area for the eight-hour ozone standard includes more than 400 counties, as shown in Figure 5.11. Most are clustered around major cities throughout the Northeast, Southeast, and Midwest, and in Texas, Colorado, Arizona, Nevada, and California.

Ground-level ozone is the primary component in smog. Smog, a word made up by combining "smoke" and "fog," is probably the most well-known form of air pollution. It retards crop and tree growth, impairs health, and limits visibility. When temperature inversions occur (the warm air stays near the ground instead of rising) and winds are calm, such as during the summer, smog may hang over a huge area for days at a time. As traffic and other pollution sources add more pollutants to the air, the smog gets worse. Wind often blows smog-forming pollutants away from their sources; this is why smog frequently can be more serious miles away from where the pollutants were created.

Most people associate dirty air with cities and the areas around them. There is good reason for this, because some of the worst smog in the country occurs in such urban areas as Los Angeles, California—a city known for its air quality problems. In a major industrial nation such

**FIGURE 5.9**

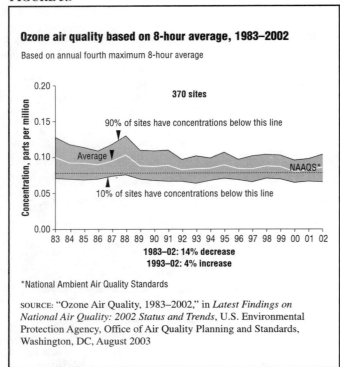

Ozone air quality based on 8-hour average, 1983–2002

Based on annual fourth maximum 8-hour average

370 sites

90% of sites have concentrations below this line

Average

10% of sites have concentrations below this line

NAAQS*

83 84 85 86 87 88 89 90 91 92 93 94 95 96 97 98 99 00 01 02

**1983–02: 14% decrease**
**1993–02: 4% increase**

*National Ambient Air Quality Standards

SOURCE: "Ozone Air Quality, 1983–2002," in *Latest Findings on National Air Quality: 2002 Status and Trends*, U.S. Environmental Protection Agency, Office of Air Quality Planning and Standards, Washington, DC, August 2003

**FIGURE 5.10**

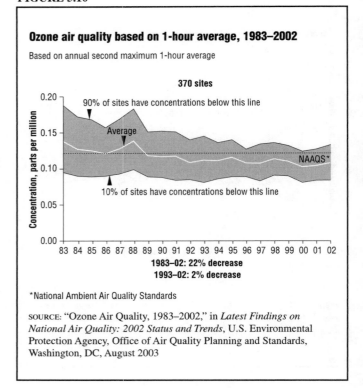

Ozone air quality based on 1-hour average, 1983–2002

Based on annual second maximum 1-hour average

370 sites

90% of sites have concentrations below this line

Average

10% of sites have concentrations below this line

NAAQS*

83 84 85 86 87 88 89 90 91 92 93 94 95 96 97 98 99 00 01 02

**1983–02: 22% decrease**
**1993–02: 2% decrease**

*National Ambient Air Quality Standards

SOURCE: "Ozone Air Quality, 1983–2002," in *Latest Findings on National Air Quality: 2002 Status and Trends*, U.S. Environmental Protection Agency, Office of Air Quality Planning and Standards, Washington, DC, August 2003

as the United States, however, smog is not limited just to cities. The Great Smoky Mountains, located in western North Carolina and eastern Tennessee, are seeing more air pollution. Harmful emissions from various coal-burning facilities located outside the mountain range, as well as pollution from motor vehicles, are damaging the mountain's environment.

Ground-level ozone is harmful to ecosystems, particularly vegetation. Ozone exposure reduces forest yields by stunting the growth of seedlings and increasing stresses on trees. Such damage can take years to become evident. Between 1993 and 2002 the EPA monitored ozone levels based on eight-hour average concentrations at 28 national parks around the country. The results indicated that ozone levels increased at 18 of the parks, remained unchanged at four other parks, and decreased at six parks. Of the eighteen parks showing increased ozone levels, five of them experienced what the EPA calls "statistically significant" upward trends. These parks are as follows:

- Great Smoky Mountains (Tennessee)

- Craters of the Moon (Idaho)

- Mesa Verde (Colorado)

- Denali (Alaska)

- Acadia (Maine)

Only one national park—Saguaro National Park in Arizona—showed significant improvements in ozone levels.

**ADVERSE HEALTH EFFECTS.** Even the smallest amounts of ozone can cause breathing difficulties. Ozone exposure can cause serious problems with lung functions, leading to infections, chest pain, and coughing.

According to the EPA, ozone exposure is linked with increased emergency room visits and hospital admissions due to such respiratory problems as lung inflammation and asthma. Ozone causes or aggravates these problems, particularly in people working outdoors, the elderly, and children. Children are especially susceptible to the harmful effects of ozone because they spend a great deal of time outside and because their lungs are still developing.

According to the Centers for Disease Control and Prevention, the percentage of American children with asthma more than doubled between 1980 and 2001. In 1980 approximately 3.7 percent of all children age 17 and younger suffered from asthma. By 2001 this figure had climbed to 8.7 percent. In general asthma levels are greater among children that live in inner cities, areas also prone to higher concentrations of ozone, smog, and other air pollutants.

Long-term exposure of any age group to moderate levels of ozone is thought to cause irreversible lung damage due to premature aging of the tissues.

The EPA maintains an Air Quality Index (AQI) as a means for warning the public when air pollutants exceed unhealthy levels. AQI values range from 0 to 500. Higher values correspond to greater levels of air pollution and increased risk to human health. An AQI value of 100 is

FIGURE 5.11

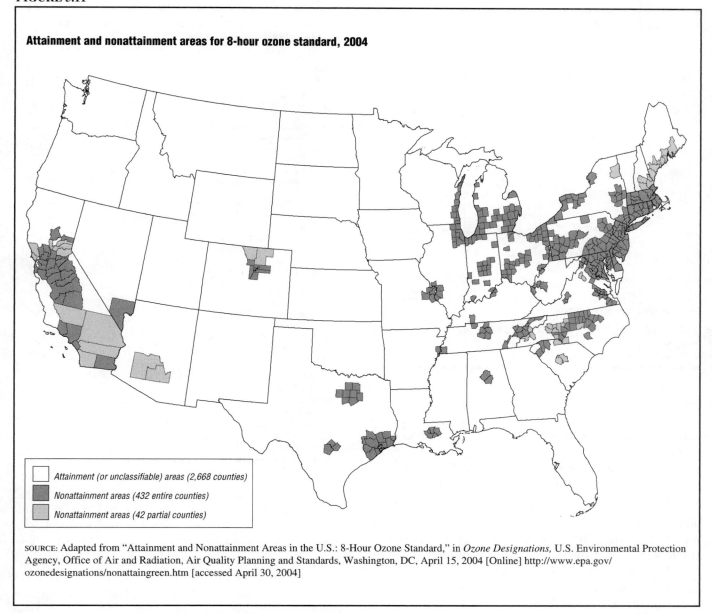

**Attainment and nonattainment areas for 8-hour ozone standard, 2004**

Attainment (or unclassifiable) areas (2,668 counties)

Nonattainment areas (432 entire counties)

Nonattainment areas (42 partial counties)

SOURCE: Adapted from "Attainment and Nonattainment Areas in the U.S.: 8-Hour Ozone Standard," in *Ozone Designations,* U.S. Environmental Protection Agency, Office of Air and Radiation, Air Quality Planning and Standards, Washington, DC, April 15, 2004 [Online] http://www.epa.gov/ozonedesignations/nonattaingreen.htm [accessed April 30, 2004]

assigned to the concentration of air pollutant equal to its NAAQS. For example, the average eight-hour ozone level considered unhealthy is 0.08 parts per million (ppm). Table 5.1 shows the ozone AQI. Index values are commonly reported during summertime radio and television newscasts to warn people about the dangers of ozone exposure.

In *State of the Air: 2003* (May 2003), the American Lung Association assessed the quality of air in U.S. communities, giving them grades ranging from "A" through "F" based on how often their air quality exceeds the "unhealthy" limits of the EPA's Air Quality Index. The 2003 report found that 137 million Americans live in areas that received an "F." That is about 69 percent of the nation's population who live in counties where there are ozone monitors.

Living in the monitored counties that received a failing grade are more than seven million adult asthmatics and nearly two million children who had an asthma attack within the previous year, 4.7 million people with chronic bronchitis, and 1.5 million people suffering from emphysema. Just over 16 million older adults (age 65 and up) lived in monitored counties that received a failing grade. The organization found that, although cities in southern California generally scored poorly, such Sunbelt states as Texas, Tennessee, Georgia, North Carolina, and South Carolina also scored low. The American Lung Association's list of the 25 metropolitan areas with the worst ozone pollution in 2003 is shown in Table 5.2.

**Particulate Matter**

Particulate matter (PM) is the general term for the mixture of solid particles and/or liquid droplets found in the air. Primary particles are those emitted directly to the atmosphere—for example, dust, dirt, and soot (black carbon).

## TABLE 5.1

### Air Quality Index (AQI): Ozone

| Index values | Levels of health concern | Cautionary statements |
|---|---|---|
| 0–50 | Good | None |
| 51–100* | Moderate | Unusually sensitive people should consider limiting prolonged outdoor exertion. |
| 101–150 | Unhealthy for sensitive groups | Active children and adults, and people with respiratory disease, such as asthma, should limit prolonged outdoor exertion. |
| 151–200 | Unhealthy | Active children and adults, and people with respiratory disease, such as asthma, should avoid prolonged outdoor exertion; everyone else, especially children, should limit prolonged outdoor exertion. |
| 201–300 | Very unhealthy | Active children and adults, and people with respiratory disease, such as asthma, should avoid all outdoor exertion; everyone else, especially children, should limit outdoor exertion. |
| 301–500 | Hazardous | Everyone should avoid all outdoor exertion. |

*Generally, an AQI of 100 for ozone corresponds to an ozone level of 0.08 parts per million (averaged over 8 hours).

SOURCE: "Air Quality Index (AQI): Ozone," in *Air Quality Index—A Guide to Air Quality and Your Health*, U.S. Environmental Protection Agency, Office of Air and Radiation, Washington, DC, June 2000

## TABLE 5.2

### The American Lung Association ranks America's 25 most ozone-polluted cities in 2004

| 2004 rank | Metropolitan statistical areas | Total population |
|---|---|---|
| 1 | Los Angeles-Long Beach-Riverside, CA | 17,044,188 |
| 2 | Fresno-Madera, CA | 964,897 |
| 3 | Bakersfield, CA | 694,059 |
| 4 | Visalia-Porterville, CA | 381,772 |
| 5 | Houston-Baytown-Huntsville, TX | 5,086,741 |
| 6 | Merced, CA | 225,398 |
| 7 | Sacramento-Arden-Arcade-Truckee, CA-NV | 2,068,427 |
| 8 | Hanford-Corcoran, CA | 135,043 |
| 9 | Knoxville-Sevierville-La Follette, TN | 796,760 |
| 10 | Dallas-Fort Worth, TX | 5,676,171 |
| 11 | Washington-Baltimore-N. Virginia, DC-MD-VA-WV | 7,826,485 |
| 12 | Philadelphia-Camden-Vineland, PA-NJ-DE-MD | 5,899,571 |
| 13 | New York-Newark-Bridgeport, NY-NJ-CT-PA | 21,705,461 |
| 14 | Charlotte-Gastonia-Salisbury, NC-SC | 1,993,372 |
| 15 | Cleveland-Akron-Elyria, OH | 2,950,615 |
| 16 | Greensboro-Winston-Salem-High Point, NC | 1,315,361 |
| 17 | Pittsburgh-New Castle, PA | 2,512,302 |
| 18 | San Diego-Carlsbad-San Marcos, CA | 2,906,660 |
| 18 | Phoenix-Mesa-Scottsdale, AZ | 3,500,151 |
| 20 | Modesto, CA | 482,440 |
| 21 | Atlanta-Sandy Springs-Gainesville, GA | 4,844,726 |
| 22 | Morristown-Newport, TN | 159,648 |
| 23 | Raleigh-Durham-Cary, NC | 1,401,331 |
| 23 | Lancaster, PA | 478,561 |
| 25 | Sheboygan, WI | 112,480 |

SOURCE: "Table 2b: People at Risk in 25 Most Ozone-Polluted Cities," in *State of the Air: 2004*, American Lung Association, New York, NY, April 29, 2004. Reprinted with permission © 2004 American Lung Association. For more information on how you can fight lung disease, the third-leading cause of death in the United States, please contact The American Lung Association at 1-800-LUNG-USA (1-800-586-4872) or visit the Web site at http://www.lungusa.org.

Secondary particles form in the atmosphere due to complex chemical reactions among gaseous emissions and include sulfates, nitrates, ammoniums, and organic carbon compounds. For example, sulfate particulates can form when sulfur dioxide emissions from industrial facilities and power plants undergo chemical reactions in the atmosphere.

The EPA tracks two sizes of particulate matter: PM10 and PM2.5. PM10 are all particles less than or equal to 10 micrometers ($\mu$) in diameter. This is roughly one-seventh the diameter of a human hair and small enough to be breathed into the lungs. PM2.5 are the smallest of these particles (less than or equal to 2.5 $\mu$m in diameter). PM2.5 is also called fine PM. The particles ranging in size between 2.5 and 10 $\mu$m in diameter are known as coarse PM. Most coarse PM is primary particles, while most fine PM is secondary particles.

**EMISSIONS AND SOURCES.** Sources of particulate matter include unpaved roads, agriculture and forestry, residential wood stoves and fireplaces, and fuel combustion in vehicles, power plants, and industry.

The EPA tracks trends in direct PM emissions from anthropogenic (human-caused) sources. As shown in Figure 5.12, direct PM10 emissions declined by 22 percent between 1993 and 2002. The major sources have historically been fuel combustion, industrial processes, and transportation. In 2002 industrial processes accounted for 46 percent of direct PM10 emissions. Fuel combustion at power plants and in homes and businesses contributed another 33 percent. The exhaust from transportation vehicles contributed 21 percent.

These sources are called "traditionally inventoried" PM sources. As shown in Figure 5.13, they actually contribute only a small portion of total direct emission of PM10. In 2002 the EPA estimates that fugitive dust was the primary culprit, accounting for 63 percent of all emissions. This is dust thrown up into the air when vehicles travel over unpaved roads and during land-disturbing construction activities such as bulldozing. Agriculture and forestry operations also stir up soil; in 2002 they accounted for 22 percent of direct PM10 emissions.

Fugitive dust, agriculture, and forestry combine to contribute 85 percent of all direct PM10 emissions. However, these sources are not as great a concern to air quality as the traditionally inventoried sources. This is because soil, dust, and dirt thrown up into the air does not typically travel far from its original location or climb very far into the atmosphere. The final source category shown in Figure 5.13 is called other combustion. This includes emissions associated with wildfires and managed burning on forests and grasslands. In 2002 these sources contributed 5 percent of all direct PM10 emissions.

Direct emissions of PM2.5 also declined between 1993 and 2002. They dropped from approximately 2,230 tons per year in 1993 to 1,850 tons per year in 2003, a

FIGURE 5.12

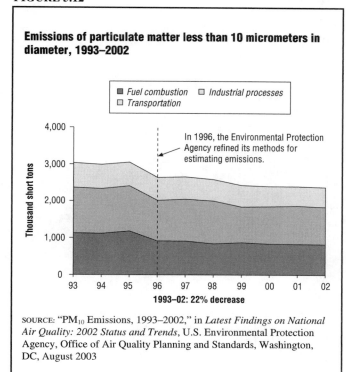

**Emissions of particulate matter less than 10 micrometers in diameter, 1993–2002**

In 1996, the Environmental Protection Agency refined its methods for estimating emissions.

1993–02: 22% decrease

SOURCE: "PM₁₀ Emissions, 1993–2002," in *Latest Findings on National Air Quality: 2002 Status and Trends*, U.S. Environmental Protection Agency, Office of Air Quality Planning and Standards, Washington, DC, August 2003

FIGURE 5.13

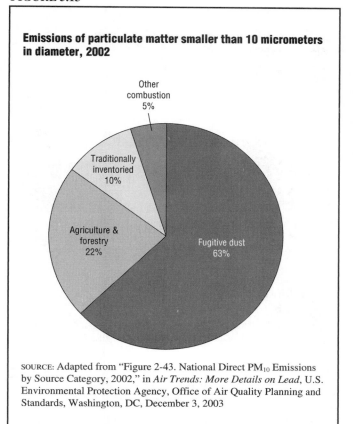

**Emissions of particulate matter smaller than 10 micrometers in diameter, 2002**

SOURCE: Adapted from "Figure 2-43. National Direct PM₁₀ Emissions by Source Category, 2002," in *Air Trends: More Details on Lead*, U.S. Environmental Protection Agency, Office of Air Quality Planning and Standards, Washington, DC, December 3, 2003

decrease of 17 percent. The major sources of direct PM2.5 are the same as those shown for PM10 in Figure 5.12. The historical percentage breakdown of these sources is also very similar.

Most PM2.5 is not comprised of primary particles from direct emissions but of secondary particles that form in the atmosphere. The EPA tracks secondary PM2.5 particle types at monitoring sites around the country. Data collected in 2001 and 2002 indicate that sulfates, ammonium, and carbon are the principal secondary particles found in the eastern part of the nation. These pollutants are largely associated with coal-fired power plants. In western states (particularly California), carbon and nitrates comprise most of the secondary particles. On a national level, secondary PM2.5 concentrations are generally higher in urban areas than in rural areas.

**AIR QUALITY.** When PM hangs in the air, it creates a haze, limiting visibility. PM is one of the major components of smog and can have adverse effects on vegetation and sensitive ecosystems. Long-term exposure to PM can damage painted surfaces, buildings, and monuments.

From 1940 to 1971, PM in the air generally increased. Pollution-control laws, however, led to a drop in PM, beginning in the 1970s. Figure 5.14 shows the historical trend in PM10 air quality based on data collected by the EPA from 804 monitoring sites. Between 1993 and 2002 PM10 concentrations decreased by 13 percent. The EPA reports that in 2002 there were 15 million people living in

counties with PM10 concentrations greater than the NAAQS. The vast majority of these people lived in southern California. PM10 nonattainment areas were also designated in parts of all other western states, particularly Arizona and Nevada.

In 1999 the EPA began nationwide tracking of PM2.5 air quality concentrations. Between 1999 and 2002 these concentrations decreased by 8 percent based on data collected from 858 monitoring sites. In February 2004 the EPA asked states and tribes to recommend PM2.5 designations for counties within their jurisdictions. Final designations for PM2.5 attainment and nonattainment areas are expected to be published in late 2004 or early 2005. State and local governments with nonattainment areas will be given three years to develop plans for reducing levels of fine particulates.

**ADVERSE HEALTH EFFECTS.** PM can irritate the nostrils, throat, and lungs and aggravate respiratory conditions such as bronchitis and asthma. PM exposure can also endanger the circulatory system and is linked with cardiac arrhythmias (episodes of irregular heartbeats) and heart attacks. PM2.5 are the most damaging, because their small size allows them access to deeper regions of the lungs. These small particles (less than or equal to 2.5 $\mu$m in diameter) have been linked with the most serious health effects in humans. Particulates pose the greatest health

FIGURE 5.14

**Air quality for particulate matter less than 10 micrometers in diameter, 1993–2002**

Based on seasonally weighted annual average

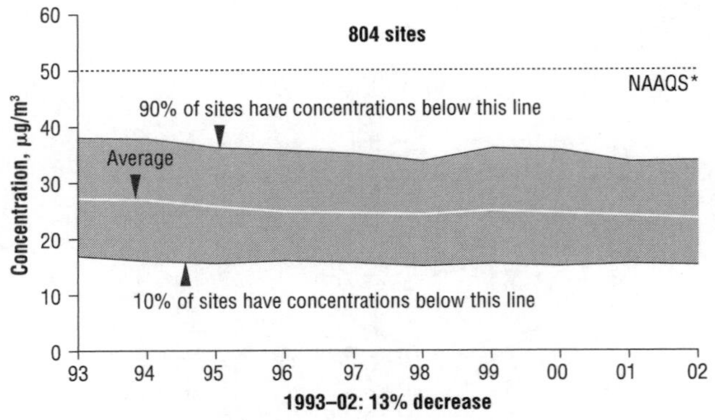


*National Ambient Air Quality Standards

SOURCE: "$PM_{10}$ Air Quality, 1993–2002," in *Latest Findings on National Air Quality: 2002 Status and Trends*, U.S. Environmental Protection Agency, Office of Air Quality Planning and Standards, Washington, DC, August 2003

FIGURE 5.15

**Sulfur dioxide ($SO_2$) emissions, 1983–2002**

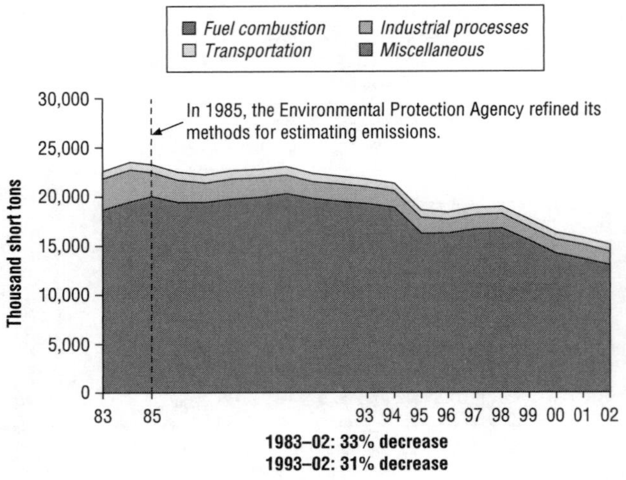


SOURCE: "$SO_2$ Emissions, 1983–2002," in *Latest Findings on National Air Quality: 2002 Status and Trends*, U.S. Environmental Protection Agency, Office of Air Quality Planning and Standards, Washington, DC, August 2003

risk to those with heart or lung problems, the elderly, and especially children, who are particularly susceptible due to the greater amount of time they spend outside and the fact that their lungs are not fully developed.

## Sulfur Dioxide

Sulfur dioxide ($SO_2$) is a gas composed of sulfur and oxygen. The chemical formula $SO_5$ is used collectively to describe sulfur oxide, $SO_2$, and other sulfur oxides.

**EMISSIONS AND SOURCES.** One of the primary sources of sulfur dioxide is the combustion of fossil fuels containing sulfur. Coal (particularly high-sulfur coal common to the eastern United States) and oil are the major fuel sources associated with $SO_2$. Power plants have historically been the main source of $SO_2$ emissions. Some industrial processes and metal smelting also cause $SO_2$ to form.

From 1940 to 1970 $SO_2$ emissions increased as a result of the growing use of fossil fuels, especially coal, in industry and power plants. Since 1970 total $SO_2$ emissions have dropped because of greater reliance on cleaner fuels with lower sulfur content and the increased use of pollution control devices, such as scrubbers, to clean emissions. $SO_2$ emissions decreased by 33 percent between 1983 and 2002. (See Figure 5.15.)

In 2002 fuel combustion in power plants accounted for 85 percent of $SO_2$ emissions. The remainder was attributed to industrial processes (9 percent), transportation (5 percent), and miscellaneous sources (1 percent).

**AIR QUALITY.** Trends in air quality concentrations of $SO_2$ are shown in Figure 5.16. The average concentration fell by 54 percent between 1983 and 2002. In 2002 the EPA reported that all counties in the United States met NAAQS for sulfur dioxide.

$SO_2$ is a major contributor to acid rain, haze, and particulate matter. Acid rain is of particular concern because acid deposition harms aquatic life by lowering the pH (level of acidity; a lower value indicates more acid) of surface waters, impairs the growth of forests, causes depletion of natural soil nutrients, and corrodes buildings, cars, and monuments. Acid rain is largely associated with the eastern United States because eastern coal tends to be higher in sulfur content than coal mined in the western United States.

In 1990 the U.S. Congress established the Acid Rain Program under Title IV of the 1990 Clean Air Act Amendments. The goal of the program was (and is) to reduce annual emission of $SO_2$ by 10 million tons and of $NO_5$ by two million tons between 1980 and 2010. A permanent national cap of 8.95 million tons of $SO_2$ per year is to be in effect for electric utilities by 2010. The program expects to meets its goals by tightening annual emission limits on thousands of power plants around the country.

**ADVERSE HEALTH EFFECTS.** Inhaling sulfur dioxide in polluted air can impair breathing in those with asthma or even in healthy adults who are active outdoors. As with other air pollutants, children, the elderly, and those with

FIGURE 5.16

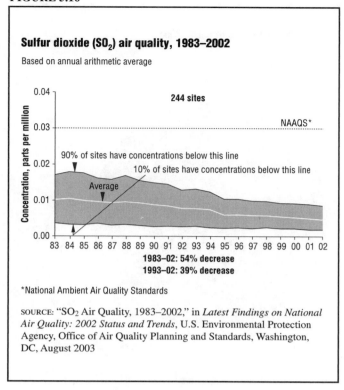

**Sulfur dioxide (SO₂) air quality, 1983–2002**

Based on annual arithmetic average

1983–02: 54% decrease
1993–02: 39% decrease

*National Ambient Air Quality Standards

SOURCE: "SO₂ Air Quality, 1983–2002," in *Latest Findings on National Air Quality: 2002 Status and Trends*, U.S. Environmental Protection Agency, Office of Air Quality Planning and Standards, Washington, DC, August 2003

FIGURE 5.17

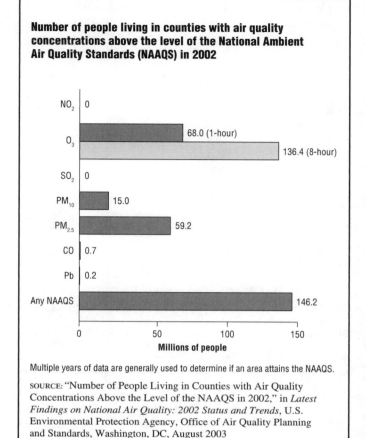

**Number of people living in counties with air quality concentrations above the level of the National Ambient Air Quality Standards (NAAQS) in 2002**

Multiple years of data are generally used to determine if an area attains the NAAQS.

SOURCE: "Number of People Living in Counties with Air Quality Concentrations Above the Level of the NAAQS in 2002," in *Latest Findings on National Air Quality: 2002 Status and Trends*, U.S. Environmental Protection Agency, Office of Air Quality Planning and Standards, Washington, DC, August 2003

preexisting respiratory and cardiovascular diseases and conditions are most susceptible to adverse effects from breathing this gas.

## Summary for Primary Pollutants

In 2002 146.2 million people lived in counties with air quality concentrations greater than the level of the NAAQS. (See Figure 5.17.) Ozone (O₃) was, by far, the largest problem, followed by particulate matter (PM), carbon monoxide (CO), and lead (Pb). Note that adding the values for individual pollutants does not equal the nationwide total. This is because some people live in counties that are nonattainment for multiple pollutants. Table 5.3 summarizes the main sources and health risks of the priority pollutants.

## Air Toxics

Hazardous air pollutants (HAPs), also referred to as air toxics, are pollutants that may cause severe health effects or ecosystem damage. Serious health risks linked to HAPs include cancer, immune system disorders, neurological problems, reproductive effects, and birth defects. The Clean Air Act (CAA) lists 188 substances as HAPs and targets them for regulation in section 112 (b) (1). The air toxics program complements the NAAQS program. Examples of HAPs are benzene, dioxins, arsenic, beryllium, mercury, and vinyl chloride.

Major sources of HAP emissions include transportation vehicles, construction equipment, power plants, factories, and refineries. Some air toxics come from common

sources. For example, benzene emissions are associated with gasoline. Air toxics are not subject to intensive national monitoring; the EPA and state environmental agencies monitor air toxic levels at approximately 300 sites nationwide. However, the EPA is working to build a more extensive monitoring network. In 2003 the agency launched the National Air Toxic Trend Site (NATTS) network. It is designed to follow trends in high-risk air toxics such as benzene, chromium, and formaldehyde.

Air pollutant emissions have been linked with increased risk of severe health problems and even death. The results of a study conducted between 1982 and 1998 on 500,000 adults across the United States were reported by C. Pope et al. in "Lung Cancer, Cardiopulmonary Mortality, and Long-Term Exposure to Fine Particulate Air Pollution," (*Journal of the American Medical Association*, March 6, 2002). The researchers found that high levels of fine particulates and sulfur oxides in the air were linked with higher mortality rates from cancer and cardiopulmonary illnesses. Each increase of 10 micrograms of these pollutants in a cubic meter of air was associated with an 8 percent increase in lung cancer mortality, a 6 percent increase in deaths due to cardiopulmonary diseases, and a 4 percent increase in all cancer deaths.

**TABLE 5.3**

## Air pollutants, health risks, and contributing sources

| Pollutants | Health risks | Contributing sources |
|---|---|---|
| Ozone* (O$_3$) | Asthma, reduced respiratory function, eye irritation | Cars, refineries, dry cleaners |
| Particulate matter (PM-IO) | Bronchitis, cancer, lung damage | Dust, pesticides |
| Carbon monoxide (CO) | Blood oxygen carrying capacity reduction, cardiovascular and nervous system impairments | Cars, power plants, wood stoves |
| Sulphur dioxide (SO$_2$) | Respiratory tract impairment, destruction of lung tissue | Power plants, paper mills |
| Lead (Pb) | Retardation and brain damage, esp. children | Cars, nonferrous smelters, battery plants |
| Nitrogen dioxide (NO$_2$) | Lung damage and respiratory illness | Power plants, cars, trucks |

*Ozone refers to tropospheric ozone, which is hazardous to human health.

SOURCE: Fred Seitz and Christine Plepys, "Table 1. Criteria Air Pollutants, Health Risks and Sources," in *Monitoring Air Quality in Healthy People 2000, Healthy People 2000—Statistical Notes,* Centers for Disease Control and Prevention, National Center for Health Statistics, Hyattsville, MD, September 1995

**FIGURE 5.18**

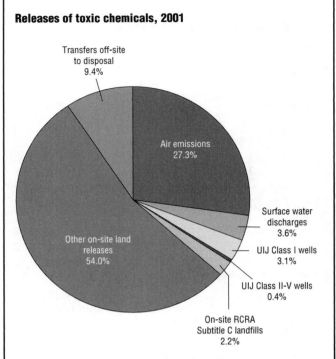

Releases of toxic chemicals, 2001

Note: UIJ=underground injection; RCRA=Resource Conservation and Recovery Act

SOURCE: "Figure ES-1: Distribution of TRI On-Site and Off-Site Releases, 2001," in *2001 Toxics Release Inventory Public Data Release,* U.S. Environmental Protection Agency, Office of Waste, Washington, DC, 2003

**NATIONAL AIR TOXIC ASSESSMENT.** In May 2002 the EPA released the latest findings from its National Air Toxic Assessment. The data, which are for 1996, show that approximately 4.6 million tons of air toxics were released into the air, down from a baseline value of six million tons for 1990–1993. Air toxics were emitted from many sources, including industrial and mobile (vehicles and non-road equipment) sources. The known carcinogens posing the greatest risks to human health were benzene and chromium. The suspected carcinogen showing the greatest risk was formaldehyde.

**THE TOXICS RELEASE INVENTORY.** The Toxics Release Inventory (TRI) was established under the Emergency Planning and Community Right-to-Know Act of 1986. The TRI program requires annual reports on the waste management activities and toxic chemical releases of certain industrial facilities using specific toxic chemicals. The TRI list includes more than 650 toxic chemicals.

TRI reports have been required since 1987 from manufacturing facilities (called the "original" industries). In 1998 the reporting requirements were also applied to a second group of industries called the "new" industries. These include metal and coal mining, electric utilities burning coal or oil, chemical wholesale distributors, petroleum terminals, bulk storage facilities, Resource Conservation and Recovery Act subtitle C hazardous water treatment and disposal facilities, solvent recovery services, and federal facilities. However, the requirements

only apply to facilities with 10 or more full-time employees that use certain thresholds of toxic chemicals.

In July 2003 the *2001 Toxics Release Inventory (TRI) Public Data Release Report* was published. There were 6.16 billion pounds of chemical releases reported by covered facilities during 2001. The vast majority of the releases (91 percent) were on-site releases to air, land, and water. The remainder were off-site releases (when a facility sends toxic chemicals to another facility where they are then released). Although most (56.2 percent) of the releases were to landfills or surface impoundments, air emissions did account for just over 27 percent of the total. (See Figure 5.18.)

## THE AUTOMOBILE'S CONTRIBUTION TO AIR POLLUTION

Automobiles dominate the transportation sector's share of energy-related carbon emissions. The transportation sector accounts for more than 65 percent of U.S. petroleum consumption and more than 75 percent of carbon monoxide (CO) emissions.

As shown in Figure 5.19, the United States leads the world in car ownership. In 2001 there were nearly 850 vehicles per 1000 people in the United States. This rate has more than doubled since 1960, when there were approximately 400 vehicles per 1000 people. The symbols along the U.S. trendline indicate vehicle ownership in 2001 in other countries. For example, in 2001 in western Europe ownership was just over 500 cars per 1000 people, a level experienced in the United States back in 1969. Vehicle ownership in 2001 was even lower in other parts of the world. With industrialization occurring in many developing countries, an increase in global automobile use and the emissions that accompany them is inevitable.

American states that do not meet CAA standards must do something to bring emissions into compliance with national standards. Because of California's extreme air pollution problems, the Clean Air Act Amendments (CAAA) allowed states to set stricter emission standards than those required by the amendments, which California did. These included strict new laws on automobile pollution. The remaining 49 states were given the option of choosing either the standards of California or the federal CAAA.

States have the freedom to cut their emissions in whatever manner they choose. Some states are phasing in tougher tests for auto emissions. In major metropolitan areas, particularly in the Northeast, owners of cars and light trucks are required to take their vehicles to centralized, high-technology inspection stations. The EPA estimates that approximately three-quarters of the vehicles pass the inspections on the first try. For those that do not pass, repairs must be made. Public protest over such inspections caused several states to temporarily rescind vehicle checks. These states must find other ways to cut their emissions. If they cannot bring their emissions down to comply with EPA standards, they may have to reinstate vehicle inspections. Politicians are finding such measures very unpopular with their constituents.

One of the major failings in reducing auto-induced smog is that efforts have focused on reducing tailpipe emissions instead of eliminating their formation in the first place. Automakers have shown that they can adapt to tighter emission standards by introducing lighter engines, fuel injection, catalytic converters, and other technological improvements. Some experts believe, however, that efforts could be better spent by drastically reducing emissions and promoting alternative energy as well as alternative transportation such as mass transit systems, carpools, and bicycles.

**Reformulated Gasoline**

Reformulated gasoline (RFG) is gasoline that contains added oxygen. Oxygenation of fuel makes combustion more complete. Incomplete fuel combustion is a major cause of carbon monoxide (CO) emissions. RFG is specially blended to have lower concentrations of certain volatile organic compounds (VOCs) in order to reduce

FIGURE 5.19

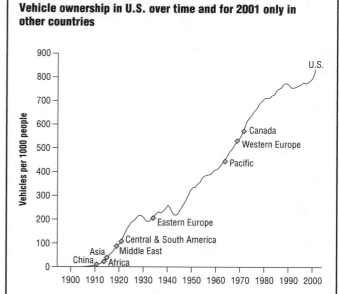

**Vehicle ownership in U.S. over time and for 2001 only in other countries**

SOURCE: Stacy C. Davis and Susan W. Diegel, "Figure 3.1: Vehicles per Thousand People: U.S. Compared to Other Countries," in *Transportation Energy Data Book: Edition 23,* U.S. Department of Energy, Office of Planning, Budget Formulation and Analysis, Energy Efficiency and Renewable Energy, Washington, DC, October 2003

ozone formation and emissions of air toxins. Although RFG combustion results in lower CO emissions, higher carbon dioxide ($CO_2$) emissions result due to the presence of additional oxygen. The most frequently used oxygenates in RFG are methyl tertiary-butyl ether (MTBE) and ethanol. Fuel ethanol is derived from fermented agricultural products such as corn. RFG results in a lower-octane fuel and usually an increase in price.

The CAA standards that went into effect in 1995 required those areas with the worst polluted air to sell RFG. Denver, Colorado, known for its "brown cloud" of pollution, enacted the nation's first oxygenated fuels program—an entire winter period during which all fuels sold at gas stations were required to have a 3 percent oxygen content. Other areas voluntarily chose to participate in RFG programs. By the early 2000s RFG accounted for more than one-third of all gasoline sold.

Figure 5.20 shows U.S. production and imports of MTBE and fuel ethanol from 1985 to 2002. More than 99 percent of fuel ethanol is produced in the United States, primarily in the corn-growing regions of the Midwest. MTBE sources are approximately 77 percent domestic and 23 percent foreign.

MTBE has historically dominated as the RFG oxygenator of choice. In 2002 it accounted for nearly 63 percent of oxygenate consumption. This value is down from

FIGURE 5.20

**Oxygenate production and imports (million of gallons), 1985–2003**

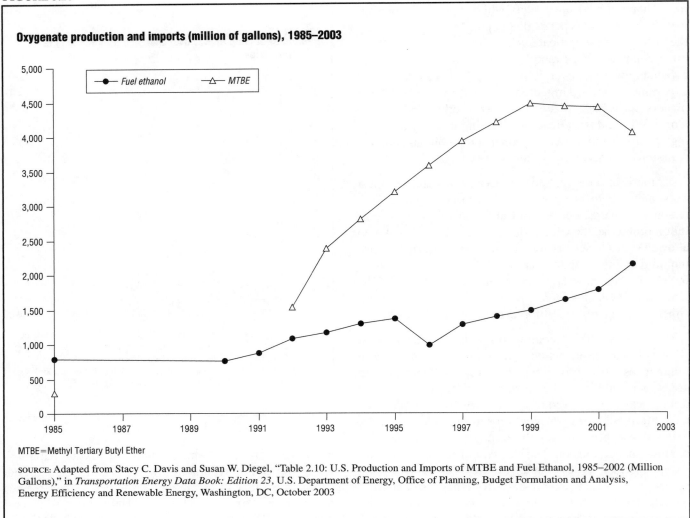

MTBE=Methyl Tertiary Butyl Ether

SOURCE: Adapted from Stacy C. Davis and Susan W. Diegel, "Table 2.10: U.S. Production and Imports of MTBE and Fuel Ethanol, 1985–2002 (Million Gallons)," in *Transportation Energy Data Book: Edition 23*, U.S. Department of Energy, Office of Planning, Budget Formulation and Analysis, Energy Efficiency and Renewable Energy, Washington, DC, October 2003

71 percent in 1998. MTBE demand has fallen, primarily due to environmental concerns. It is very soluble in water and therefore tends to migrate to water supplies. As of August 2003 the U.S. Geological Survey reported that MTBE was found in source water four to five times more often in RFG areas than in non-RFG areas. Several states took steps to decrease or ban the use of MTBE in RFG. (See Table 5.4.) The ban in California is expected to significantly affect MTBE markets as the state accounted for almost one-third of all U.S. consumption during 2003. MTBE bans are expected to greatly increase use of fuel ethanol in RFG.

Beginning in the late 1990s, customer complaints about gasoline prices caused many areas that had voluntarily opted to sell RFG to back out. In the summer of 2000 RFG prices in Milwaukee and Chicago exceeded $2.75 per gallon. These areas used ethanol exclusively in RFG. Even before these price spikes occurred, many states were concerned about the effects of RFG requirements on gasoline prices. In 1999 California asked the federal government for a waiver from the oxygenate requirement. Such a waiver would allow the state to temporarily use non-RFG in case of RFG supply and distribution problems. In 2001 the waiver was denied.

Climbing gasoline prices over the next few years reignited interest in RFG waivers. In April 2004 California resubmitted its waiver request after the Bush administration granted oxygenate waivers to New Hampshire and Arizona. California lawmakers pointed out that the waiver could reduce record high gasoline prices in the state.

### Catalytic Converters

Tailpipe catalytic converters are one of the most successful technologies in the history of smog control, eliminating 96 to 98 percent of carbon monoxide (CO) and unburned hydrocarbons and approximately 75 percent of nitrogen oxides.

A catalytic converter is a stainless steel vessel that contains certain chemicals (catalysts) known to affect emission gas reactions. Striving to eliminate pollutants entirely, researchers are perfecting new "preheating" con-

TABLE 5.4

**Overview of state bans on MTBE***

| State | MTBE ban schedule | MTBE consumption* (% of U.S. total) |
|---|---|---|
| California | MTBE ban starting January 1, 2004 | 31.7 |
| Colorado | MTBE ban started April 30, 2002 | 0 |
| Connecticut | MTBE ban starting October 1, 2003 | 3.1 |
| Illinois | MTBE prohibited by July 2004 | 0 |
| Indiana | MTBE limited to 0.5% by volume, starting July 23, 2004 | 0 |
| Iowa | 0.5% MTBE by volume cap, already in effect | 0 |
| Kansas | MTBE limited to 0.5% by volume, starting July 1, 2004 | 0 |
| Kentucky | MTBE ban starting January 1, 2006; beginning in January 1, 2004, ethanol encouraged to be used in place of MTBE | 0.8 |
| Maine | Law merely expresses state's "goal" to ban MTBE; it's not an actual ban. The "goal" is to phase out gasoline or fuel products treated with MTBE by January 1, 2003 | 0 |
| Michigan | MTBE prohibited by June 1, 2003 | 0 |
| Minnesota | All ethers (MTBE, ETBE, TAME) limited to 1/3 of 1.0% by weight after July 1, 2000; after July 1, 2005, total ether ban | 0 |
| Missouri | MTBE limited to 0.5% by volume, starting July 1, 2005 | 1.1 |
| Nebraska | MTBE limited to 1.0% by volume, starting July 13, 2000 | 0 |
| New York | MTBE ban starting January 1, 2004 | 7.5 |
| Ohio | MTBE ban starting July 1, 2005 | 0 |
| S. Dakota | 0.5% MTBE by volume cap, already in effect | 0 |
| Washington | MTBE ban starting December 31, 2003 | 0 |

*Methyl Tertiary Butyl Ether. Data include MTBE blended into RFG and oxygenated gasoline only. MTBE may also be found in conventional gasoline, but not in significant amounts.

SOURCE: "Table 1. Overview of State MTBE Bans," in *Status and Impact of State MTBE Bans*, U.S. Department of Energy, Energy Information Administration, Washington, DC, March 27, 2003 [Online] http://www.eia.doe.gov/oiaf/servicerpt/mtbeban/table1.html [accessed April 28, 2004]

verters. Converters typically cannot function properly until the car has warmed to a specific temperature, causing emissions to be greatest during the first few miles. The new converters would function as soon as a car is started.

The implementation of catalytic converters in automobiles during the early 1970s was a major reason for conversion from leaded to unleaded gasoline in the United States because the catalysts within the converter were not compatible with lead. Thus the catalytic converter has been directly and indirectly responsible for dramatic reductions in several major air pollutants. In 1998, however, the EPA expressed concerns that catalytic converters were contributing to global warming because they convert many nitrogen-oxygen compounds to nitrous oxide (NO), a common gas associated with global warming. NO levels may increase, at least in part, from the growth in the number of vehicle miles traveled by cars that have catalytic converters.

## Government Regulation—Corporate Average Fuel Economy Standards

In 1973 the Organization of Petroleum Exporting Countries (OPEC) imposed an oil embargo that provided a painful reminder to Americans of how dependent the country had become on foreign sources of fuel. Although the United States makes up only 5 percent of the world's population, it consumes approximately one-quarter of the world's supply of oil, much of which is imported from the Middle East. The 1973 oil embargo prompted Congress to pass the 1975 Automobile Fuel Efficiency Act (PL 96-426), which set the initial Corporate Average Fuel Economy (CAFE) standards.

CAFE standards required each domestic automaker to increase the average mileage of the new cars sold to 27.5 miles per gallon (mpg) by 1985. Under CAFE rules automakers could still sell the big, less efficient cars with powerful eight-cylinder engines, but to meet average fuel efficiency rates they also had to sell smaller, more efficient cars. Automakers that failed to meet each year's CAFE standards were required to pay fines. Those who managed to surpass the rates earned credits that they could use in years when they fell below CAFE requirements.

Opponents to CAFE standards complained that the congressional fuel economy campaign would saddle American motorists with car features they would not like and would not buy. They also pointed out that the standards would increase demand for smaller cars, which would in turn raise the number of highway deaths and injuries, limit consumer choice of larger and family-sized vehicles, and place thousands of auto-related jobs at risk.

During the early 1980s automakers became more inventive and managed to keep their cars relatively large and roomy with such innovations as electronic fuel injection and front-wheel drive. Ford's prestigious Lincoln Town Car managed to achieve better mileage in 1985 than its small Pinto did in 1974.

The CAFE standard was lowered during the mid-to-late 1980s and then raised back to 27.5 mpg for 1990 model automobiles. In 2003 the standard remained at 27.5 mpg for automobiles and 20.7 mpg for light trucks, including pickups, minivans, sport utility vehicles (SUVs), and vans. According to the U.S. Department of Transportation (DOT), automobile manufacturers

FIGURE 5.21

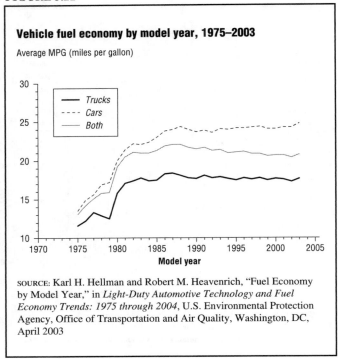

**Vehicle fuel economy by model year, 1975–2003**

Average MPG (miles per gallon)

SOURCE: Karl H. Hellman and Robert M. Heavenrich, "Fuel Economy by Model Year," in *Light-Duty Automotive Technology and Fuel Economy Trends: 1975 through 2004*, U.S. Environmental Protection Agency, Office of Transportation and Air Quality, Washington, DC, April 2003

FIGURE 5.22

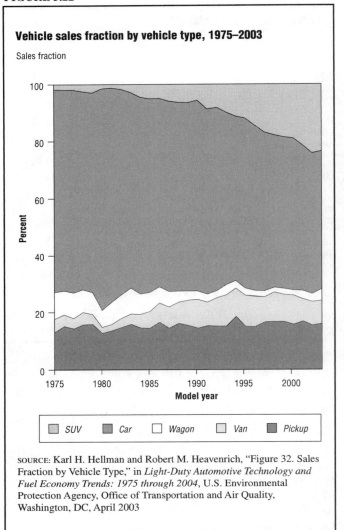

**Vehicle sales fraction by vehicle type, 1975–2003**

Sales fraction

SOURCE: Karl H. Hellman and Robert M. Heavenrich, "Figure 32. Sales Fraction by Vehicle Type," in *Light-Duty Automotive Technology and Fuel Economy Trends: 1975 through 2004*, U.S. Environmental Protection Agency, Office of Transportation and Air Quality, Washington, DC, April 2003

achieved an average 29.5 mpg fuel economy for model year 2003 cars, thus meeting the standard. Likewise, the average light truck fuel economy was 21.8 mpg, also above the standard.

### "Real World" Estimates of Fuel Economy

Fuel economy rates calculated by DOT for comparison to CAFE standards are not the same as rates reported to consumers on new vehicle labels or commonly published by the EPA or U.S. Department of Energy (DOE) in fuel economy guides. DOT CAFE rates are calculated based on laboratory data and take into account credits issued to manufacturers for alternative fuel capabilities and other factors. Rates published by the EPA and DOE are approximately 15 percent lower, as they reflect only actual "real world" experience.

According to the 2003 EPA report *Light-Duty Automotive Technology and Fuel Economy Trends: 1975 Through 2003,* "real world" estimates of fuel economy for model year 2003 cars and light trucks were 24.8 mpg and 17.7 mpg, respectively. Figure 5.21 shows average "real world" fuel economy rates for model years 1975 through 2003. The combined rate for both cars and trucks has actually decreased since the mid-1980s because of increased demand for light trucks, particularly SUVs. (See Figure 5.22.)

### Market Factors

In 1985 cars accounted for nearly 80 percent of new sales. By 2003 their market share had fallen to around 50 percent. Sales of SUVs and minivans increased dramatically over the same time period. These light trucks have lower fuel economy rates than cars. Also, many states have increased interstate speed limits since the 1980s. This has also lowered overall fuel efficiency.

SUVs and minivans fall under less stringent emissions standards than automobiles because they originated as modifications of light-truck bodies and are classified as trucks. Automakers and buyers of trucks and SUVs oppose tightening restrictions on emissions of these vehicles, although critics contend that new SUVs are more like cars than trucks in design.

U.S. industry officials claim that increasing fuel efficiency is not cost effective. With each mile added in efficiency, the costs to obtain that improvement increase to the point that it is no longer cost effective. This is the same objection that is made, in general, about cleaning up many environmental hazards—that the first and most drastic improvements are the least expensive and thereafter cleanup becomes more costly.

TABLE 5.5

## Characteristics of major alternative fuels

| | Compressed natural gas (CNG) | Biodiesel (B20) | Electricity | Ethanol (E85) | Hydrogen | Liquified natural gas (LNG) | Liquified petroleum gas (LPG) | Methanol (M85) |
|---|---|---|---|---|---|---|---|---|
| Main fuel source | Underground reserves | Soy bean oil, waste cooking oil, animal fats, and rapeseed oil | Coal; however, nuclear, natural gas, hydroelectric, and renewable resources can also be used. | Corn, grains, or agricultural waste | Natural gas, methanol, and other energy sources. | Underground reserves | A by-product of petroleum refining or natural gas processing | Natural gas, coal, or, woody biomass |
| Physical state | Compressed gas | Liquid | Electricity | Liquid | Compressed gas or liquid | Liquid | Liquid | Liquid |
| Environmental impacts of burning fuel | CNG vehicles can demonstrate a reduction in ozone-forming emissions (CO and $NO_x$) compared to some conventional fuels; however, HC emissions may be increased. | Reduces particulate matter and global warming gas emissions compared to conventional diesel; however, $NO_x$ emissions may be increased. | Electric vehicles have zero tailpipe emissions; however, some amount of emissions can be contributed to power generation. | E85 vehicles can demonstrate a 25% reduction in ozone-forming emissions (CO and $NO_x$) compared to reformulated gasoline. | Zero regulated emissions for fuel cell-powered vehicles, and only NOx emissions possible for internal combustion engines operating on hydrogen. | LNG vehicles can demonstrate a reduction in ozone-forming emissions (CO and $NO_x$) compared to some conventional fuels; however, HC emissions may be increased. | LPG vehicles can demonstrate a 60% reduction in ozone-forming emissions (CO and $NO_x$) compared to reformulated gasoline. | M85 vehicles can demonstrate a 40% reduction in ozone-forming emissions (CO and $NO_x$) compared to reformulated gasoline. |
| Energy security impacts | CNG is domestically produced. The United States has vast natural gas reserves. | Biodiesel is domestically produced and has a fossil energy ratio of 3.3 to 1, which means that its fossil energy inputs are similar to those of petroleum. | Electricity is generated mainly through coal fired power plants. Coal is the country's most plentiful, economical, and price-stable fossil fuel. | Ethanol is produced domestically and it is renewable. | Hydrogen can help reduce U.S. dependence on foreign oil by being produced by renewable resources. | LNG is domestically produced and it typically costs less than gasoline and diesel fuels. | LPG is the most widely available alternative fuel. Its disadvantage is that 45% of the fuel in the U.S. is derived from oil. | Methanol can be domestically produced from renewable resources. |

SOURCE: Adapted from data in *Custom Alternative Fuels Comparison Chart*, U.S. Department of Energy, National Renewable Energy Laboratory, Alternative Fuels Data Center, Golden, CO, May 2004 [Online] http://www.eere.energy.gov/cleancities/afdc/altfuel/fuel_comp.html [accessed May 1, 2004]

In contrast, for 2005 the European Commission has proposed an ambitious target of 47 mpg for gasoline-driven cars (compared to the current average of 29 mpg) and 52 mpg for diesel-powered cars. It must be noted that, while European countries do not generally legislate fuel efficiency, gasoline in Europe tends to cost more than twice what it does in the United States. That serves as a powerful incentive to European drivers to buy fuel-efficient vehicles.

Although some manufacturers in the United States claim to be able to produce more efficient cars, they also contend that American consumers are not interested in buying them. They believe consumers instead seek features that raise both the size and the price of a car. In essence, manufacturers claim, there is little market for a more efficient vehicle.

### Alternative Fuels

The DOE defines alternative fuels as those that are "substantially non-petroleum and yield energy security and environmental benefits." The DOE recognizes all of the following as alternative fuels:

- compressed or liquefied natural gas
- coal-derived liquid fuels
- liquefied petroleum gas
- alcohol fuels (mixtures that contain at least 70 percent alcohol)
- bio fuels (fuels derived from biological materials)
- electricity
- solar energy
- hydrogen

Table 5.5 summarizes information for the major alternative fuels related to their physical state, sources, environmental impacts, and availability.

Although these fuels offer advantages, their use may substitute one problem for another. For example, the alcohol fuel methanol reduces ozone formation but increases formaldehyde, a human carcinogen, and is twice as toxic as gasoline if it comes in contact with the skin. Engines require twice as much methanol as gasoline to travel a similar distance. Natural gas reduces hydrocarbons and CO but increases $NO_5$.

The DOE reports that 518,919 alternative fuel vehicles (AFVs) were in use in the United States in 2002, up from 251,352 AFVs in 1992. (See Table 5.6.) Figure 5.23 provides a breakdown of vehicles in use in 2002 by fuel

**TABLE 5.6**

## Alternative-fueled vehicles by type, 1992–2002

| Year | Liquefied petroleum gases[1] | Compressed natural gas | Liquefied natural gas | Methanol, 85 percent[2] | Methanol, heat | Ethanol, 85 percent[2] | Ethanol, 95 percent[2] | Electricity | Total |
|------|------|------|------|------|------|------|------|------|------|
| | | | | Number of vehicles in use | | | | | |
| 1992 | 221,000 | 23,191 | 90 | 4,850 | 404 | 172 | 38 | 1,607 | 251,352 |
| 1993 | 269,000 | 32,714 | 299 | 10,263 | 414 | 441 | 27 | 1,690 | 314,848 |
| 1994 | 264,000 | 41,227 | 484 | 15,484 | 415 | 605 | 33 | 2,224 | 324,472 |
| 1995 | 259,000 | 50,218 | 603 | 18,319 | 386 | 1,527 | 136 | 2,860 | 333,049 |
| 1996 | 263,000 | 60,144 | 663 | 20,265 | 172 | 4,536 | 361 | 3,280 | 352,421 |
| 1997 | 263,000 | 68,571 | 813 | 21,040 | 172 | 9,130 | 347 | 4,453 | 367,526 |
| 1998 | 266,000 | 78,782 | 1,172 | 19,648 | 200 | 12,788 | 14 | 5,243 | 383,847 |
| 1999 | R267,833 | R91,267 | 1,681 | 18,964 | 198 | R24,604 | 14 | 6,964 | R411,525 |
| 2000 | R272,193 | R100,738 | R2,090 | R10,426 | R0 | R58,621 | R4 | R11,834 | R455,906 |
| 2001 | R276,597 | R113,835 | R2,576 | R7,827 | R0 | R71,336 | R0 | R17,848 | R490,019 |
| 2002P | 281,286 | 126,341 | 3,187 | 5,873 | 0 | 82,477 | 0 | 19,755 | 518,919 |

[1]Vehicles in use represent lower bound estimates, rounded to the nearest thousand.
[2]Remaining portion is motor gasoline.
R=Revised. P=Preliminary.
Note: Totals may not equal sum of components due to independent rounding.

SOURCE: "Table 10.7. Estimated Alternative-Fueled Vehicles and Fuel Consumption by Type, 1992–2002," in *Annual Energy Review 2002*, U.S. Department of Energy, Energy Information Administration, Washington, DC, October 2003

type. Most relied on liquefied petroleum gas (55 percent) followed by compressed natural gas (24 percent) and alcohol fuel containing at least 85 percent ethanol (16 percent). A breakdown by ownership is provided in Figure 5.24. A majority (63 percent) of AFVs are in private hands. The remainder are used by federal, state, and local governments.

**ALTERNATIVE FUELS AND THE MARKETPLACE.** AFVs cannot become a viable transportation option unless a fuel supply is readily available. Ideally, the infrastructure for supplying alternative fuels will be developed simultaneously with the vehicles. Table 5.7 lists the number of alternative fuel stations in each state as of April 30, 2004. There are 6,230 of these stations around the country. Most of them (64 percent) provide liquefied petroleum gas, followed by compressed natural gas (17 percent) and electricity (13 percent). Together California and Texas account for slightly more than one-third of all the stations.

Many state policies and programs encourage the use of alternative fuels. California, for example, requires the sale of electric vehicles (EVs). This has caused vehicle manufacturers to expedite vehicle research and development. In fact, EVs are already selling in California, and some rental car agencies now offer EVs to customers at prices only slightly higher than gasoline-powered cars.

Market success of alternative fuels and AFVs depends on public acceptance. People are accustomed to using gasoline as their main transportation fuel and it is readily available. As federal and state requirements for alternative fuels increase, so should the availability of such fuels as well as their acceptance by the general public. In the long run, electricity and hydrogen seem the most promising of the alternative fuels for vehicles.

**ELECTRIC VEHICLES—PROMISE AND REALITY.** The electric vehicle (EV) is not a new invention. Popular during the 1890s, the quiet, clean, and simple vehicle was expected to dominate the automotive market of the twentieth century. Instead, it quietly disappeared as automakers chose to invest billions of dollars in the internal combustion engine. It has taken a century, but the EV has returned.

The primary difficulty with EVs lies in inadequate battery power. The cars must be recharged often. These vehicles also use lead-acid or nickel-cadmium batteries and have a range of 70 to 100 miles on a single charge. The range is reduced by factors such as cold temperatures, the use of air conditioning, vehicle load, and steep terrain.

In addition, EVs are expensive. Despite their high price, EVs have many advantages. They are relatively noiseless and simple in design and operation. They cost less to refuel and service and have fewer parts to break down. Their owners are likely to spend less time on maintenance and, if they recharge at home, will rarely have to go to the service station. These time-saving features have real value in the busy world of the twenty-first century. Over time the cost gap between cars that pollute and EVs that do not will narrow. With advances in battery development, the gap could close entirely.

Automobile industry experts believe EVs will assume a "second car" role for commuters and for short trips, much like the microwave oven has become an addition to, rather than a replacement for, conventional ovens for cooking.

**HYDROGEN-FUELED VEHICLES ON THE HORIZON.** Hydrogen is the most simple naturally occurring element and can be found in materials such as water, natural gas, and coal. For decades advocates of hydrogen have pro-

FIGURE 5.23

**Alternative fuel vehicles in use by fuel type, 2002**

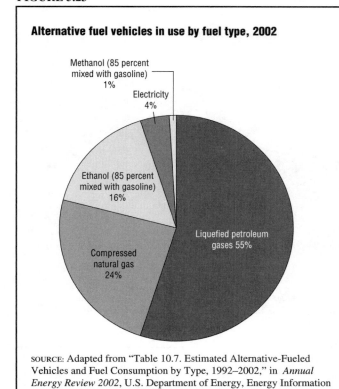

SOURCE: Adapted from "Table 10.7. Estimated Alternative-Fueled Vehicles and Fuel Consumption by Type, 1992–2002," in *Annual Energy Review 2002*, U.S. Department of Energy, Energy Information Administration, Washington, DC, October 2003

FIGURE 5.24

**Ownership of alternative fuel vehicles, 2002**

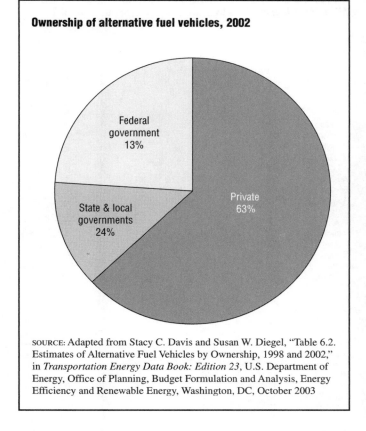

SOURCE: Adapted from Stacy C. Davis and Susan W. Diegel, "Table 6.2. Estimates of Alternative Fuel Vehicles by Ownership, 1998 and 2002," in *Transportation Energy Data Book: Edition 23*, U.S. Department of Energy, Office of Planning, Budget Formulation and Analysis, Energy Efficiency and Renewable Energy, Washington, DC, October 2003

moted it as the fuel of the future—abundant, clean, and cheap. Hydrogen researchers from universities, laboratories, and private companies claim that their industry has already produced vehicles that could be ready for consumers if problems of fuel supply and distribution could be solved. Other experts contend that economics and safety concerns will limit hydrogen's wider use for decades.

In 2002 the DOE formed a government-industry partnership called Freedom Cooperative Automotive Research (FreedomCAR). The goal of FreedomCAR is to develop highly fuel-efficient vehicles that operate using hydrogen produced from renewable energy sources. Industrial partners in the venture include Ford, General Motors, and Daimler-Chrysler. FreedomCAR research takes place at facilities operated by the DOE's National Renewable Energy Laboratory in Golden, Colorado.

In his 2003 State of the Union Address, President George W. Bush announced the creation of the Hydrogen Fuel Initiative (HFI). This $1.2 billion program is designed to develop the technology needed for commercially viable hydrogen-powered fuel cells by the year 2020. Fuel cells designed for transportation vehicles and home/business use are to be developed. The HFI has three primary missions as part of its goals:

- Lower the cost of hydrogen production to make it cost effective with gasoline production by the year 2010

- Develop hydrogen fuel cells that provide the same vehicle range (at least 300 miles of travel) as conventional gasoline fuel tanks

- Lower the cost of hydrogen fuel cells to be comparable in cost with internal combustion engines

**HYBRID VEHICLES.** Many experts believe that the most feasible solution in the near future is to produce vehicles that use a combination of gasoline and one of the alternative fuel sources. These are called hybrid vehicles. Figure 5.25 depicts a hybrid automobile that relies on a small internal combustion engine and electricity (from batteries).

As of early 2004 there are three hybrid automobiles for sale in the United States. They are the Honda Insight, Toyota Prius, and Honda Civic Hybrid. The DOE calls them Advanced Technology Vehicles. Table 5.8 compares sales and technical specifications for the three models. With fuel efficiencies in the range of 47–57 mpg, these cars provide roughly twice as many miles per gallon of gasoline than conventional cars on the market. Manufacturers continue research on hybrid cars, which they hope will eventually satisfy American tastes and pocketbooks and provide even greater fuel efficiency.

**MANDATING OF ALTERNATIVE FUEL VEHICLES.** The Energy Policy Act of 1992 (PL 102-486), passed in the wake of the 1991 Persian Gulf War, required that federal

TABLE 5.7

**Alternative fuel stations by state and fuel type, as of April 2004**

| State | CNG | E85 | LPG | ELEC | BD | HY | LNG | Totals by state |
|---|---|---|---|---|---|---|---|---|
| Alabama | 9 | 0 | 77 | 34 | 0 | 0 | 1 | 121 |
| Alaska | 0 | 0 | 9 | 0 | 0 | 0 | 0 | 9 |
| Arizona | 27 | 1 | 112 | 54 | 3 | 1 | 8 | 206 |
| Arkansas | 4 | 0 | 85 | 0 | 0 | 0 | 0 | 89 |
| California | 194 | 1 | 352 | 514 | 17 | 5 | 35 | 1,118 |
| Colorado | 35 | 9 | 85 | 6 | 8 | 0 | 0 | 143 |
| Connecticut | 25 | 0 | 29 | 5 | 1 | 0 | 0 | 60 |
| Delaware | 3 | 0 | 4 | 0 | 4 | 0 | 0 | 11 |
| Dist. of Columbia | 2 | 0 | 0 | 0 | 0 | 0 | 0 | 2 |
| Florida | 44 | 0 | 154 | 3 | 1 | 0 | 0 | 202 |
| Georgia | 65 | 0 | 54 | 87 | 2 | 0 | 0 | 208 |
| Hawaii | 0 | 0 | 7 | 11 | 3 | 0 | 0 | 21 |
| Idaho | 9 | 1 | 33 | 0 | 2 | 0 | 2 | 47 |
| Illinois | 22 | 13 | 92 | 0 | 3 | 0 | 0 | 130 |
| Indiana | 23 | 0 | 54 | 0 | 1 | 0 | 3 | 81 |
| Iowa | 0 | 11 | 44 | 0 | 1 | 0 | 0 | 56 |
| Kansas | 6 | 2 | 67 | 0 | 6 | 0 | 1 | 82 |
| Kentucky | 1 | 7 | 27 | 0 | 0 | 0 | 0 | 35 |
| Louisiana | 11 | 0 | 45 | 0 | 0 | 0 | 0 | 56 |
| Maine | 0 | 0 | 15 | 0 | 2 | 0 | 0 | 17 |
| Maryland | 20 | 3 | 28 | 1 | 7 | 0 | 0 | 59 |
| Massachusetts | 13 | 0 | 44 | 41 | 2 | 0 | 0 | 100 |
| Michigan | 20 | 4 | 138 | 5 | 9 | 0 | 0 | 176 |
| Minnesota | 5 | 87 | 58 | 0 | 1 | 0 | 0 | 151 |
| Mississippi | 0 | 0 | 34 | 0 | 0 | 0 | 0 | 34 |
| Missouri | 8 | 7 | 151 | 0 | 1 | 0 | 0 | 167 |
| Montana | 5 | 2 | 41 | 0 | 1 | 0 | 1 | 50 |
| Nebraska | 1 | 5 | 27 | 0 | 1 | 0 | 0 | 34 |
| Nevada | 17 | 0 | 34 | 0 | 6 | 1 | 0 | 58 |
| New Hampshire | 2 | 0 | 30 | 12 | 3 | 0 | 0 | 47 |
| New Jersey | 18 | 0 | 25 | 0 | 0 | 0 | 0 | 43 |
| New Mexico | 13 | 2 | 81 | 0 | 1 | 0 | 0 | 97 |
| New York | 55 | 0 | 67 | 11 | 0 | 0 | 0 | 133 |
| North Carolina | 9 | 2 | 75 | 6 | 22 | 0 | 0 | 114 |
| North Dakota | 4 | 2 | 18 | 0 | 0 | 0 | 0 | 24 |
| Ohio | 38 | 2 | 77 | 0 | 2 | 0 | 0 | 119 |
| Oklahoma | 57 | 1 | 99 | 0 | 0 | 0 | 0 | 157 |
| Oregon | 16 | 0 | 49 | 4 | 5 | 0 | 1 | 75 |
| Pennsylvania | 51 | 0 | 105 | 0 | 1 | 0 | 1 | 158 |
| Rhode Island | 6 | 0 | 7 | 2 | 0 | 0 | 0 | 15 |
| South Carolina | 4 | 1 | 62 | 0 | 2 | 0 | 0 | 69 |
| South Dakota | 0 | 9 | 26 | 0 | 0 | 0 | 0 | 35 |
| Tennessee | 2 | 1 | 60 | 0 | 0 | 0 | 0 | 63 |
| Texas | 40 | 0 | 969 | 6 | 1 | 0 | 6 | 1,022 |
| Utah | 65 | 2 | 38 | 0 | 0 | 0 | 1 | 106 |
| Vermont | 1 | 0 | 15 | 11 | 0 | 0 | 0 | 27 |
| Virginia | 22 | 1 | 58 | 11 | 6 | 0 | 2 | 100 |
| Washington | 22 | 1 | 83 | 6 | 14 | 0 | 0 | 126 |
| West Virginia | 5 | 0 | 9 | 0 | 0 | 0 | 0 | 14 |
| Wisconsin | 22 | 10 | 77 | 0 | 1 | 0 | 0 | 110 |
| Wyoming | 14 | 1 | 36 | 0 | 2 | 0 | 0 | 53 |
| Totals by fuel: | 1,035 | 188 | 3,966 | 830 | 142 | 7 | 62 | 6,230 |

Notes: CNG=Compressed natural gas; E85=85% ethanol; LPG=propane; ELEC=electric; BD=biodiesel; HY=hydrogen; LNG=Liquefied natural gas.

SOURCE: "Alternative Fuel Station Counts Listed by State and Fuel Type," in *Alternative Fuels Data Center*, U.S. Department of Energy, National Renewable Energy Laboratory, Alternative Fuels Data Center, Golden, CO, April 30, 2004 [Online] http://www.afdc.nrel.gov/refuel/state_tot.shtml [accessed April 30, 2004]

---

and state governments and fuel provider fleet owners increase the percentages of vehicles powered by alternative fuels. The fleet requirements affect those who own or control at least 50 vehicles in the United States and fleets of at least 20 vehicles that are centrally fueled (or capable of being centrally fueled) within a metropolitan area of 250,000 or more.

FIGURE 5.25

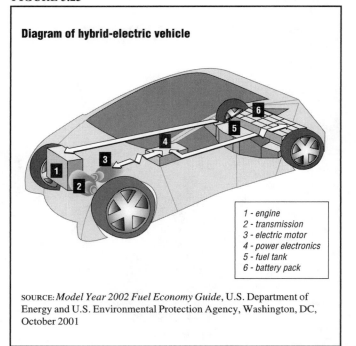

**Diagram of hybrid-electric vehicle**

1 - engine
2 - transmission
3 - electric motor
4 - power electronics
5 - fuel tank
6 - battery pack

SOURCE: *Model Year 2002 Fuel Economy Guide*, U.S. Department of Energy and U.S. Environmental Protection Agency, Washington, DC, October 2001

In fiscal year 2001 the federal government purchased 23,325 fleet vehicles. The vast majority (81 percent) were gasoline powered. Just over 2,500 of the vehicles were diesel powered. Another 1,870 of the vehicles used alternative fuels. Most of these vehicles were powered by ethanol-85.

## AIRPLANES

As air travel in affluent nations has increased, it has caused a number of environmental problems. The average American flew 1,739 miles a year by 2000. Europeans, though they flew fewer miles, had the world's most crowded skies, while the most rapid growth in flying was in Asia. Most air travel is done by a small portion of the world's population.

Flying carries an environmental price—it is the most energy-intensive form of transport. In much of the industrialized world, air travel is replacing more energy-efficient rail or bus travel. Although the fuel efficiency of jet engines has improved over the decades, overall jet fuel consumption has increased dramatically. According to the U.S. Bureau of Transportation's *National Transportation Statistics, 2003* (March 2004), jet fuel consumption has nearly doubled in just over 30 years, from 7,847 million gallons in 1970 to 14,017 gallons in 2001.

Another problem with air travel is its impact on global warming. Airplanes spew millions of tons of nitrous oxide into the air every year, much of it while cruising in the tropospheric zone five to seven miles above Earth where ozone is formed. According to the February 2000 U.S. General Accounting Office (GAO) report *Aviation's Effects on the*

TABLE 5.8

**Technical specifications and sales figures for hybrid car models, 1999–2002**

| | Honda Insight CVT[1] | Toyota Prius CVT[1] | Honda Civic Hybrid CVT SULEV[1] |
|---|---|---|---|
| Fuel economy (city/hwy) | 57/56 mpg | 52/45 mpg | 47/48 mpg |
| Fuel tank capacity | 10.6 gal. | 11.9 gal. | 11.9 gal. |
| Acceleration (0–60 mph) | 11.5 sec. | 12.3 sec. | 11.6 sec. |
| Emissions rating | SULEV | SULEV | SULEV |
| Aerodynamics | 0.25 Cd | 0.29 Cd | 0.28 Cd |
| Curb weight | 1,967 lbs. | 2,765 lbs. | 2,740 lbs. |
| Passenger capacity | 2 | 5 | 5 |
| Dimensions: | | | |
| Length | 155.1 in. | 169.6 in. | 174.8 in. |
| Width | 66.7 in. | 66.7 in. | 67.5 in. |
| Cargo capacity | 16.3 ft³ | 11.8 ft³ | 10.1 ft³ |
| Price | $21,280 | $20,480 | $20,550 |
| **Calendar year sales** | | | |
| 1999 | 17 | 0 | 0 |
| 2000 | 3,788 | 5,562 | 0 |
| 2001 | 4,726 | 15,556 | 0 |
| 2002 | 2,216 | 20,119 | ~12,000[2] |
| **Total** | **10,747** | **41,237** | **~12,000[2]** |

[1]Specifications are for the model containing a continuously variable transmission (CVT).
[2]Sales for the Civic Hybrid are not shown separately from other Civic models, but estimates of 2002 sales are approximately 12,000 vehicles since its March 2002 debut.
Note: SULEV = Super ultra low emission vehicle.

SOURCE: Stacy C. Davis and Susan W. Diegel, "Table 6.5: Sales and Specifications of Available Advanced Technology Vehicles," in *Transportation Energy Data Book: Edition 23*, U.S. Department of Energy, Office of Planning, Budget Formulation and Analysis, Energy Efficiency and Renewable Energy, Washington, DC, October 2003

*Global Atmosphere Are Potentially Significant and Expected to Grow,* the EPA estimates that air traffic accounts for about 3 percent of all U.S. greenhouse gases. The Intergovernmental Panel on Climate Change (IPCC) notes that emissions deposited directly into the atmosphere do greater harm than those released at the Earth's surface. (See Figure 5.26.)

## REDUCING EMISSIONS—BUT NOT ENOUGH

The most widespread technological inventions to reduce emissions have been electrostatic precipitators (electrical cleaning systems) and filters designed to control emissions from power plants. These reduce particulate emissions from smokestacks by 99.5 percent but do nothing about gaseous emissions. The primary way to reduce sulfur dioxide ($SO_2$) has been the use of scrubbers (an air pollution device that uses a spray of water or reactant to trap pollutants), which remove 95 percent of the $SO_2$ residue. The available technologies have limitations—they create scrubber ash, a hazardous waste, and do nothing to control carbon dioxide ($CO_2$) emissions.

With the aid of pollution-control equipment and improvements in energy efficiency, many industrial countries have reduced emissions. Since the passage of the CAA there has been a 90 percent reduction in emissions of hydrocarbons and CO from the average car in the United States and a 75 percent reduction in $NO_5$.

Other pollutants, however, have gone mostly unregulated, notably $CO_2$, the inevitable by-product of burning fossil fuels.

## THE COST OF AIR POLLUTION AND POLLUTION CONTROL

Air quality plays a major but complex role in public health. Among the factors that must be considered are the levels of pollutants in the air, the levels of individual exposure to these pollutants, individual susceptibility to toxic substances, and exposure time related to ill effects from certain substances. Blaming health effects on specific pollutants is also complicated by the health impact of nonenvironmental causes (such as heredity or poor diet).

Scientists do know that air pollution is related to a number of respiratory diseases, including bronchitis, pulmonary emphysema, lung cancer, bronchial asthma, eye irritation, weakened immune system, and premature lung tissue aging. In addition, lead contamination causes neurological and kidney disease and can be responsible for impaired fetal and mental development. The American Lung Association estimates the annual health costs of exposure to the most serious air pollutants at $40 to $50 billion.

Like most environmental issues, pollution involves limits. There is only so much air to receive automobile emissions and so much land on which freeways can be built. Many transportation analysts think that interest in public transportation is long overdue. Several major cities, including San Francisco, Chicago, and New York, have had positive environmental results with mass transit systems.

A major problem with the effort to reduce air pollution further—as well as some other types of pollution—is that most of the relatively cheap fixes have already been made, and many economists argue that the expensive ones may not be worth the price. The very premise of cleanup—that air pollution can be reduced to levels where it no longer poses any health risk at all—is questioned not just by industry but by observers as well.

Virtually all gains in the war on ozone have been achieved by reducing auto emissions. The costs for future air quality improvements, from some points of view, may exceed the value of any improvement, and the disparity may only get worse over time. However, some sources believe that there are other technologically easy—if politically unpopular—steps that could be taken to improve air quality. Such steps would include forcing light trucks, minivans, and SUVs to meet the same smog standards as standard passenger cars.

Other possible improvements could come from changes in "grandfather" clauses—that is, loopholes that exempt companies from compliance with laws because the companies existed prior to the law. Power plants rank first in grandfathered emissions. Other top industries

FIGURE 5.26

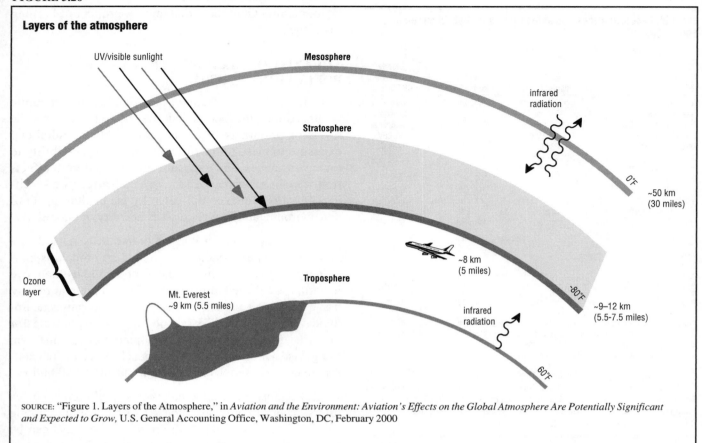

**Layers of the atmosphere**

SOURCE: "Figure 1. Layers of the Atmosphere," in *Aviation and the Environment: Aviation's Effects on the Global Atmosphere Are Potentially Significant and Expected to Grow,* U.S. General Accounting Office, Washington, DC, February 2000

affected by grandfather clauses include aluminum smelters, oil refineries, and carbon-black plants.

## THE CLEAN AIR ACT (CAA)—A HUGE SUCCESS

In 1970 the U.S. Congress passed the landmark CAA, proclaiming that it would restore urban air quality. It was no coincidence that the law was passed during a 14-day Washington, D.C., smog alert. Although the CAA has had mixed results, and many goals remain to be met, most experts credit it with making great strides toward cleaning up the air.

### The Clean Air Act Amendments (CAAA) of 1990

The overall goal of the CAAA is to reduce the pollutants in the air by 56 billion pounds a year—224 pounds for every man, woman, and child—when the law is fully phased in by 2005. Other aims were to cut acid rain in half by 2000, reduce smog and other pollutants, and protect the ozone layer by phasing out chlorofluorocarbons (CFCs) and related chemicals.

The CAAA also encouraged states to pursue market-based approaches to improve air quality. One such program, the Accelerated Vehicle Retirement program, commonly known as Cash for Clunkers, provides economic incentives for the owners of highly polluting vehi-

cles to retire their automobiles from use or repair them. The program gives pollution credits to private corporations for contributing funding to car dealers to entice car owners to trade in their old vehicles.

### Resistance to CAA Gets Federal Concessions

During the mid- to late 1990s, a number of states began balking at the strict auto emissions testing that seemed necessary to comply with the CAAA. Under the law, if a reduction does not come from auto emissions, it would have to be made up by other sources—for example, smokestack industries. The states are free to implement whatever methods they choose to cut pollution, but most states with serious air quality problems had previously chosen, with EPA encouragement, the stricter car inspection programs. This meant that many states were faced with testing that a great number of consumers considered overly restrictive and expensive.

The EPA had counted on enforcing the program through sanctions specified by the CAAA, which included cutting off highway money and other federal aid to the states. Some state legislatures, however, seemed willing to forego this aid in what some considered an act of civil disobedience. Rather than provoking a confrontation with the states, the EPA chose to allow greater flexibility in auto emissions testing.

In *The Benefits and Costs of the Clean Air Act, 1970 to 1990* (1997), the first report mandated by the CAA on the monetary costs and benefits of controlling pollution, the EPA concluded that the economic value of clean air programs was 42 times greater than the total costs of air pollution control during the 20-year period. The study found that numerous positive consequences occurred in the U.S. economy because of CAA programs and regulations. The CAA affected industrial production, investment, productivity, consumption, employment, and economic growth. In fact, the study estimated that total agricultural benefits from the CAA were almost $10 billion. The EPA compared benefits to direct costs or expenditures. The total costs of the CAA were $523 billion for the 20-year period; total benefits equaled $22.2 trillion—a net benefit of approximately $21.7 trillion.

In *Two Decades of Clean Air: EPA Assesses Costs and Benefits* (1998), the National Conference of State Legislatures used data from the EPA analysis and found that the act produced major reductions in pollution that causes illness and disease, smog, acid rain, haze, and damage to the environment.

*The Benefits and Costs of the Clean Air Act Amendments of 1990* (2000), the second mandated review of the CAA and the most comprehensive and thorough review ever conducted, showed similar results. Using a sophisticated array of computer models and the latest cost data, the EPA found that, by 2010, the act will have prevented 23,000 Americans from dying prematurely and averted more than 1.7 million asthma attacks. The CAA will prevent 67,000 episodes of acute bronchitis, 91,000 occurrences of shortness of breath, 4.1 million lost work days, and 31 million days in which Americans would have had to restrict activity because of illness. Another 22,000 respiratory-related hospital admissions will be averted, as well as 42,000 admissions for heart disease and 4,800 emergency room visits.

The EPA estimated that the benefits of CAA programs in reduction of illness and premature death alone will total about $110 billion. By contrast, the study found that the costs of achieving these benefits was only about $27 billion, a fraction of the value of the benefits. In addition, the study reported that there were other benefits that scientists and economists cannot quantify and express in monetary terms, such as controlling cancer-causing air toxins and bringing benefits to crops and ecosystems by reducing pollutants.

At the same time, many cities are still not in compliance with the law. One reason efforts to clean the air have been only partly successful is that they have focused on specific measures to combat individual pollutants rather than addressing the underlying social and economic structures that create the problem—for example, the distance between many Americans' residences and their places of work.

## Trading Pollution Credits

In another federal concession, the CAAA created pollution "credits," a free-market innovation that allowed companies that keep their emissions below standards to sell or trade their credits on the open market to other companies that do not keep their emissions below standards. This is often viewed essentially as permission to pollute. Companies can also choose to retire their credits permanently and thus reduce the potential of further pollution.

## THE ANTIREGULATORY REBELLION

The dissatisfaction with government regulation that developed in the 1980s grew even stronger in the 1990s. The Republican-led Congress took steps to put the brakes on what it considered growing environmental regulation. As a result, the funding was cut for environmental protection—including the CAA, the CAAA, the Clean Water Act (PL 92-500), the Safe Drinking Water Act (PL 93-523), and other environmental statutes. Subsequent fiscal budgets continued those cuts by many millions of dollars.

The CAA requires the EPA to review public health standards at least every five years to ensure they reflect the best current science. In 1997—in response to what many consider compelling scientific evidence of the harm caused to human health by ozone and fine particles —the EPA issued new, stricter air quality standards for ozone and PM.

This was the first revision in ozone standards in 20 years and the first-ever standard for fine particulates. The provisions tightened the standard for ground-level ozone from the level of 0.12 parts per million (ppm) at the highest daily measurement to 0.08 ppm average over an 8-hour period. The new PM standard included particles larger than 2.5 microns in diameter instead of the original standard of those larger than 10 microns. However, in May 1999 a three-judge federal appeals panel overturned the new standards. The EPA appealed, but in October 1999 the full U.S. Court of Appeals for the District of Columbia refused to overturn the decision. In January 2000 the American Lung Association petitioned the U.S. Supreme Court to review the decision. The Supreme Court ruled in February 2001 that the EPA had the authority under the CAA to set the new standards. Claims that the EPA's decision was arbitrary and capricious and not supported by evidence were struck down by the D.C. Circuit Court in late May 2002.

Despite improvements in ozone levels, few large urban areas in the United States comply with ozone standards. Under the proposed standard, many smaller population centers would also become noncompliant, and states would be forced to take further pollution-control measures to comply. Some critics contend that the standards are either unattainable or not worth the cost. The

Global Climate Coalition, a business and trade organization, claims that the proposal would damage the U.S. economy while doing little to help the environment.

In February 2002 President George W. Bush announced his plan to reduce air pollution from the power sector. The Clear Skies proposal would be a radical new way to regulate air pollution. Instead of adding new requirements, it would retain the major protections of the CAA and focus on a market-based approach to achieve significant emissions cuts in $NO_5$, $SO_2$, and mercury. Sources of emissions would be allowed to trade emission credits with each other. Thus, a facility well under the emissions cap could sell its credits to another facility that needed them.

## EFFECT OF UTILITY DEREGULATION ON AIR POLLUTION

Regulated for decades as "natural monopolies," by the year 2000 electric utilities faced a radical shift toward increased competition. As in the airline, trucking, natural gas, and telecommunications industries, more efficient technology and increasing demand for lower rates led regulators to consider some form of utility company deregulation. Thus, the Federal Energy Policy Act of 1992 (PL 102-486) gave other electricity generators access to the market.

Industrial customers eager to buy cheaper electricity from new sources or to build their own sources of power no longer viewed electricity as a nonnegotiable cost of doing business but saw it as a commodity they could either provide for themselves or for which they could shop. Under deregulation plans, utilities would be free to market electricity anywhere, and by the beginning of the twenty-first century more than half the state legislatures considered allowing utilities to compete for customers.

Environmental groups and some utilities have warned that by prompting increased competition deregulation may cause coal-fired plants to use more coal in order to produce more electricity, which would send more pollutants into the air. The Natural Resources Defense Council cautioned that deregulation could lead to as many as 500,000 tons of increased emissions of $NO_5$ per year. Responding to those concerns, former EPA administrator Carol Browner agreed that the open access rule could lead to future increases in $CO_2$ emissions and said that the EPA has to work closely with the DOE and the states to monitor the results of open access.

## FOSSIL FUEL USE IN THE DEVELOPING WORLD

Environmental pollution is worldwide, and environmental problems of the future are expected to become increasingly regional and global. Evidence mounts that the results of human activities—especially the gases produced by the combustion of coal, oil, and natural gas—may be causing atmospheric warming worldwide. Developing countries stand on the brink of economic growth that they hope will equal that of the developed world. That explosion in growth will undoubtedly be helped by fossil fuels, as was the case in the United States and Europe decades earlier. The filthiest smoke and water generally arise in the early stages of industrialization.

China especially faces a dilemma—coal harms the environment but it surely fuels economic growth. China's heavy reliance on coal, along with its inefficient and wasteful patterns of energy use, will make it the largest single producer of $CO_2$ by 2020, surpassing even the United States. Between 1970 and 1990 energy consumption in China rose 208 percent, compared with an average rise of 28 percent in developed countries during the same period. More than five million Chinese participate in coal extraction, feeding China's enormous and growing appetite for energy.

Five of China's largest cities are among the world's 10 most polluted cities. Polluted air reportedly kills 178,000 Chinese each year, primarily from emphysema and bronchitis. Children with sooty faces dodge traffic, and rain creates black rivers flowing down city streets.

China's situation is repeated, on a lesser scale, in India, Brazil, and the rest of the developing world, where meeting environmental goals is considered a rich country's luxury. Chinese officials believe, as do officials in many other developing nations, that developed countries cause most of the world's pollution. Chinese leadership believes the developed nations should be held responsible for the problems and, as a result, should help pay for cleaner coal-burning technologies in developing countries as well as financing hydroelectric plants, nuclear power stations, and alternative energy sources.

## FOREST FIRES—A CLEAR EXAMPLE OF TRANSBORDER POLLUTION

The burning of forests worldwide, either intentionally or accidentally, has consequences for nations many miles away. In 1997 fires in Indonesian forests purposefully started by humans practicing slash-and-burn agriculture (a method of cutting down and burning vegetation to clear the land) darkened the skies and blotted out the sun in seven Southeast Asian nations for many months. The haze caused the closure of airports, loss of work, shutdown of mines and factories, loss of investment and tourism, and illness among hundreds of thousands of people. Reduced sunlight slowed the growth of fruits and vegetables and caused reductions in corn, cocoa, and rice crops. Hundreds of people died from respiratory ailments, starvation, dysentery, and influenza.

In Mexico in 1998 scores of people died and at least 50 million were left to breathe in smoke from nearly 10,000

**TABLE 5.9**

### Public concern about air pollution, 2004

PLEASE TELL ME IF YOU PERSONALLY WORRY ABOUT THIS PROBLEM A GREAT DEAL, A FAIR AMOUNT, ONLY A LITTLE, OR NOT AT ALL. AIR POLLUTION?

| | Great deal % | Fair amount % | Only a little % | Not at all % | No opinion % |
|---|---|---|---|---|---|
| 2004 Mar 8–11 | 39 | 30 | 23 | 8 | * |
| 2003 Mar 3–5 | 42 | 32 | 20 | 6 | * |
| 2002 Mar 4–7 | 45 | 33 | 18 | 4 | * |
| 2001 Mar 5–7 | 48 | 34 | 14 | 4 | * |
| 2000 Apr 3–9 | 59 | 29 | 9 | 3 | * |
| 1999 Apr 13–14 | 52 | 35 | 10 | 3 | * |
| 1999 Mar 12–14 | 47 | 33 | 16 | 4 | * |
| 1997 Oct 27–28 | 42 | 34 | 18 | 5 | 1 |
| 1991 Apr 11–14 | 59 | 28 | 10 | 4 | * |
| 1990 Apr 5–8 | 58 | 29 | 9 | 4 | * |
| 1989 May 4–7 | 63 | 24 | 8 | 4 | * |

SOURCE: "Please tell me if you personally worry about this problem a great deal, a fair amount, only a little, or not at all. Air Pollution?," in *Poll Topics and Trends: Environment*, The Gallup Organization, Princeton, NJ, March 17, 2004 [Online] www.gallup.com [accessed March 30, 2004]

fires (most set for agricultural purposes). A cloud of haze and cinders hung over most of Mexico, Central America, and much of the United States for many weeks. Cities declared the air so unhealthy that they even closed airports.

## PUBLIC OPINION ABOUT AIR POLLUTION

Every year the Gallup Organization conducts a poll on the environment around the time of the nation's celebration of Earth Day. In the March 2004 poll, participants were asked about their level of concern related to particular environmental problems. Table 5.9 shows that 39 percent of those asked expressed a great deal of concern about air pollution, compared to 30 percent who expressed a fair amount of concern. Another 23 percent indicated a little concern, and 8 percent expressed no concern.

The percentage of poll respondents indicating a great deal of concern about air pollution has dropped dramatically in recent years from a high of 63 percent in 1989. In a poll conducted in 2003, Gallup asked people their opinion about some specific environmental proposals to help the nation's energy situation. The results are shown in Figure 5.27. The poll indicated that people strongly favor tougher enforcement of environmental regulations and higher emissions standards for automobiles and business and industry sources.

FIGURE 5.27

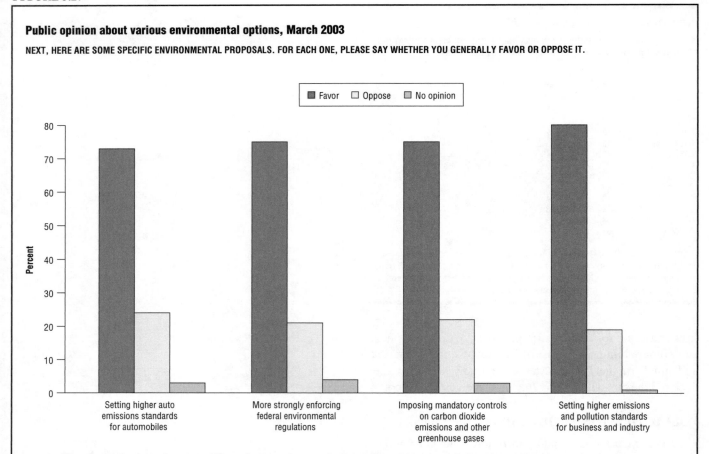

**Public opinion about various environmental options, March 2003**

NEXT, HERE ARE SOME SPECIFIC ENVIRONMENTAL PROPOSALS. FOR EACH ONE, PLEASE SAY WHETHER YOU GENERALLY FAVOR OR OPPOSE IT.

SOURCE: "Next I am going to read some specific environmental proposals. For each one, please say whether you generally favor or oppose it," in *Poll Topics and Trends: Environment*, The Gallup Organization, Princeton, NJ, March 17, 2004 [Online] www.gallup.com [accessed March 30, 2004]

## CHAPTER 6
# WATER ISSUES

Water is precious for many reasons. It is an essential resource for sustaining human, animal, and vegetative life. A living cell is mostly water. An adult human's body is about 65 percent water; blood is 90 percent water. Agriculture is absolutely dependent on water to produce food crops and livestock. Water is crucial to tourism, navigation, and industry. Enormous amounts are used to generate power, mine materials, and produce goods. Water is an ingredient, a medium, and a means of conveyance or cooling in most industrial processes. Water supplies a vital habitat for many of Earth's creatures, from the whale to the tadpole. There are entire ecosystems that are water-based.

All of these competing uses put an enormous strain on Earth's water supply. Overall, the amount of water on Earth remains constant, simply passing from one stage to another in a circular pattern known as the hydrologic cycle. Water in the atmosphere condenses and falls to Earth as precipitation, such as rain, sleet, or snow. Precipitation seeps into the ground, saturating the soil and refilling underground aquifers; it is drawn from the soil by vegetation for growth and returned into the air by plant leaves through the process of transpiration; and some precipitation flows into surface waters such as rivers, streams, lakes, wetlands, and oceans. Moisture evaporates from surface water back into the atmosphere to repeat the cycle. (See Figure 6.1.)

Humans have interrupted the cycle to accommodate the many water demands of modern life. Flowing rivers and streams are dammed up. Groundwater and surface water are pumped from their sources to other places. Water is either consumed or discharged back to the environment, usually not in the same condition. Water quality becomes increasingly important. There are two primary issues when it comes to water—availability and suitability.

## WATER AVAILABILITY

*Water must be considered as a finite resource that has limits and boundaries to its availability and suitability for use.*

— Wayne B. Solley, Robert R. Pierce, and Howard A. Perlman in *Estimated Use of Water in the United States in\ 1995,* U.S. Geological Survey, 1998

Although water covers nearly three-fourths of the planet, the vast majority of it is saline (water that contains at least 1000 milligrams of salt per liter of water). It is too salty to drink or nourish crops and too corrosive for many industrial processes. No cheap and effective method for desalinating large amounts of ocean water has yet to be discovered. This makes freshwater an extremely valuable commodity. While the overall water supply on Earth is enormous, freshwater is not often in the right place at the right time in the right amount to serve all of the competing needs.

Throughout history civilizations originated and declined based on the availability of water. Water supply in the United States is becoming a serious problem. The days of an unlimited bounty of water are over.

### Overall Water Use in 2000

Water use in the United States is monitored and reported by the U.S. Geological Survey (USGS) in its *Estimated Use of Water in the United States,* published at five-year intervals since 1950. The latest report available was published in 2004 and includes data through 2000.

For reporting purposes, water use in the United States is classified as in-stream or off-stream. In-stream use means the water is used at its source, usually a river or stream, for example, for the production of hydroelectric power at a dam. Off-stream use means the water is conveyed away from its source, although it may be returned later.

WATER USERS. The 2000 USGS report found that an estimated 408 billion gallons of water per day were withdrawn from surface and groundwater sources for off-stream use in 2000. (See Table 6.1.) Of this total, 47.8 percent was withdrawn for generation of thermoelectric

**FIGURE 6.1**

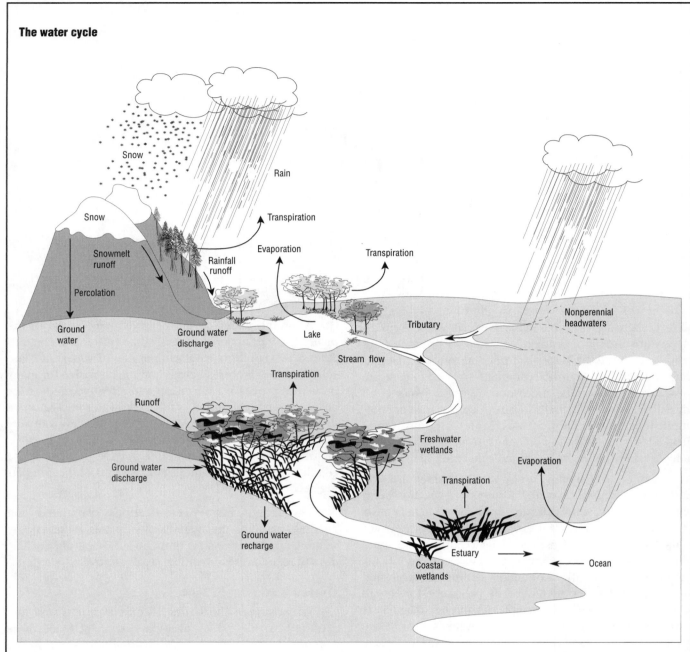

**The water cycle**

SOURCE: "The Water Cycle," in *National Water Quality Inventory: 1998 Report to Congress,* U.S. Environmental Protection Agency, Washington, DC, June 2000

power, approximately 33.6 percent was used for irrigation, and about 10.6 percent went to public water supply. Together these three uses accounted for about 92 percent of the total water used.

Minor uses included miscellaneous industrial (including commercial and mining), livestock and aquaculture, and self-supplied domestic (from private wells). Complete data were not available for all minor uses in 2000.

Together only three states—California, Texas, and Florida—accounted for 25 percent of all off-stream water

withdrawals in 2000. Irrigation and thermoelectric power generation were the primary users in these states.

In-stream water use for the generation of hydroelectric power at dams was not reported by USGS for 2000 but totaled 3.16 trillion gallons of water per day in the 1995 report. According to the U.S. Department of Energy, there are approximately 2,000 dams with hydroelectric generating capacity in the United States. Most are located in the Pacific Coast states of California, Oregon, and Washington. In-stream water usage is highest at dams along the

TABLE 6.1

**Trends in estimated water use, 1950–2000**

| | [1]1950 | [2]1955 | [3]1960 | [4]1965 | [4]1970 | [3]1975 | [3]1980 | [3]1985 | [3]1990 | [3]1995 | [3]2000 | Percentage change 1995–2000 |
|---|---|---|---|---|---|---|---|---|---|---|---|---|
| Population, in millions | 150.7 | 164.0 | 179.3 | 193.8 | 205.9 | 216.4 | 229.6 | 242.4 | 252.3 | 267.1 | 285.3 | +7 |
| Offstream use: | | | | | | | | | | | | |
|   Total withdrawals | 180 | 240 | 270 | 310 | 370 | 420 | 440 | 399 | 408 | 402 | 408 | +2 |
|     Public supply | 14 | 17 | 21 | 24 | 27 | 29 | 34 | 36.5 | 38.5 | 40.2 | 43.3 | +8 |
|   Rural domestic and livestock: | | | | | | | | | | | | |
|     Self-supplied domestic | 2.1 | 2.1 | 2.0 | 2.3 | 2.6 | 2.8 | 3.4 | 3.32 | 3.39 | 3.39 | 3.59 | +6 |
|     Livestock and aquaculture | 1.5 | 1.5 | 1.6 | 1.7 | 1.9 | 2.1 | 2.2 | [5]4.47 | 4.50 | 5.49 | [6] | — |
|   Irrigation | 89 | 110 | 110 | 120 | 130 | 140 | 150 | 137 | 137 | 134 | 137 | +2 |
|   Industrial: | | | | | | | | | | | | |
|     Thermoelectric power use | 40 | 72 | 100 | 130 | 170 | 200 | 210 | 187 | 195 | 190 | 195 | +3 |
|     Other industrial use | 37 | 39 | 38 | 46 | 47 | 45 | 45 | 30.5 | 29.9 | 29.1 | [7] | — |
| Source of water: | | | | | | | | | | | | |
|   Ground: | | | | | | | | | | | | |
|     Fresh | 34 | 47 | 50 | 60 | 68 | 82 | 83 | 73.2 | 79.4 | 76.4 | 83.3 | +9 |
|     Saline | [8] | 0.6 | 0.4 | 0.5 | 1.0 | 1.0 | 0.9 | 0.65 | 1.22 | 1.11 | 1.26 | +14 |
|   Surface: | | | | | | | | | | | | |
|     Fresh | 140 | 180 | 190 | 210 | 250 | 260 | 290 | 265 | 259 | 264 | 262 | −1 |
|     Saline | 10 | 18 | 31 | 43 | 53 | 69 | 71 | 59.6 | 68.2 | 59.7 | 61 | +2 |

[1]48 states and District of Columbia, and Hawaii
[2]48 states and District of Columbia
[3]50 states and District of Columbia, Puerto Rico, and U.S. Virgin Islands
[4]50 states and District of Columbia, and Puerto Rico
[5]From 1985 to present this category includes water use for fish farms
[6]Data not available for all states; partial total was 5.46
[7]Commercial use not available; industrial and mining use totaled 23.2
[8]Data not available

SOURCE: Susan S. Hutson, Nancy L. Barber, Joan F. Kenny, Kristin S. Linsey, Deborah S. Lumia, and Molly A. Maupin, "Table 14. Trends in Estimated Water Use in the United States, 1950–2000," in *Estimated Use of Water in the United States in 2000*, (Circular 1268), U.S. Department of the Interior, U.S. Geological Survey, Reston, VA, April 2004 [Online] http://water.usgs.gov/pubs/circ/2004/circ1268/htdocs/table14.html [accessed May 4, 2004]

Columbia River in the Pacific Northwest and along the Niagara and St. Lawrence River systems in New York.

**FRESHWATER AND SALINE.** In 2000 freshwater accounted for 345.3 billion gallons per day or 85 percent of total off-stream water withdrawals. Freshwater is used exclusively for public water supply, domestic self-supply (private wells), irrigation, livestock watering, and aquaculture. It is also an important source for thermoelectric power plants, industry, and mining. Most freshwater is obtained from surface water sources (rivers and lakes), as shown in Figure 6.2.

Irrigation and thermoelectric power plants are the largest users of off-stream freshwater, each accounting for approximately 40 percent of its use. However, the vast majority (around 97 percent) of the water withdrawn for thermoelectric power generation is used for cooling purposes and then discharged, meaning the actual amount of water consumed is much smaller. The United States Department of Agriculture (USDA) estimates that approximately 60 percent of the water withdrawn for irrigation purposes is consumed. This makes irrigation the largest consumer of freshwater.

Nearly all (98 percent) of the saline water used in 2000 came from surface water sources. Far less saline water than

freshwater was used in 2000. Only 15 percent of all water used was saline. Thermoelectric power plants are the largest user of saline water. They accounted for 96 percent of all saline water use in 2000. Again, most of this water was used and returned to the environment. Industry and mining each accounted for 2 percent of saline water use. Saline water is unsuitable for drinking and other domestic purposes, irrigation, aquaculture, or livestock watering.

In 2000 California and Texas accounted for 18 percent of all off-stream freshwater use. California and Florida accounted for 40 percent of all saline water use.

**SURFACE WATER AND GROUNDWATER.** The USGS estimates that 79 percent of all off-stream water used in 2000 was from surface water. The other 21 percent was from groundwater. Figure 6.3 shows the breakdown of surface water users. Thermoelectric power plants, irrigation, and public water supply were the primary users. Figure 6.4 shows the user breakdown for ground water. Irrigation and public supply were the primary users.

## Water Use Trends (1950–2000)

According to the USGS, total off-stream water withdrawals in the United States climbed steadily from 1950

FIGURE 6.2

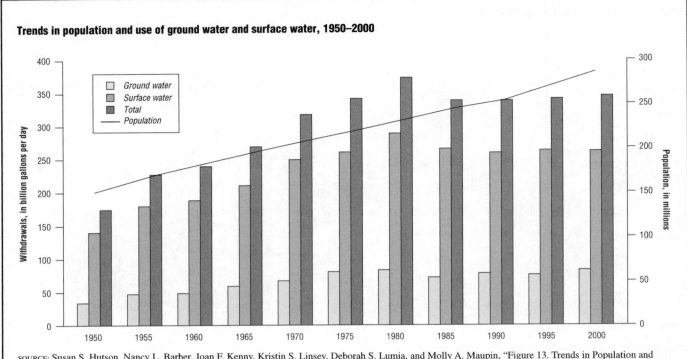

**Trends in population and use of ground water and surface water, 1950–2000**

SOURCE: Susan S. Hutson, Nancy L. Barber, Joan F. Kenny, Kristin S. Linsey, Deborah S. Lumia, and Molly A. Maupin, "Figure 13. Trends in Population and Freshwater Withdrawals by Source, 1950–2000," in *Estimated Use of Water in the United States in 2000*, (Circular 1268), U.S. Department of the Interior, U.S. Geological Survey, Reston, VA, March 2004 [Online] http://water.usgs.gov/pubs/circ/2004/circ1268/htdocs/figure13.html [accessed April 7, 2004]

to 1980, declined through 1985 and have remained relatively stable since then. (See Figure 6.5.)

Experts believe the general increase in water use from 1950 to 1980 and the decrease from 1980 to 2000 can be attributed to several factors:

- The expansion of irrigation systems and increases in energy development from 1950 to 1980 increased the demand for water.

- In some western areas the application of water directly to the roots of plants by center-pivot irrigation systems has replaced sprayer arms that project the water into the air, where much is lost to wind and evaporation.

- Higher energy prices in the 1970s and a decrease in groundwater levels in some areas increased the cost of irrigation water.

- A downturn in the farm economy reduced demands for irrigation water.

- New industrial technologies requiring less water, as well as improved efficiency, increased water recycling, higher energy prices, and changes in the law to reduce pollution decreased the demand for water.

- Increased awareness by the general public and active conservation programs reduced the demand for water.

Table 6.1 shows trends in U.S. population and off-stream water withdrawals for the period 1950–2000. The population increased by 89 percent over this time period, while water withdrawn increased by 127 percent. In 1950 the per capita (per person) off-stream water withdrawal was around 1,200 gallons per day. This value climbed steadily over the years, reaching a peak in 1975 of 1,940 gallons per day per person. Per capita use has since declined and was at 1,430 gallons per day per person in 2000.

Historically freshwater has accounted for 85–95 percent of all water used. The percentage was at the high end during the 1950s and has gradually decreased, leveling off around 85 percent from 1980 through 2000. The nation's saline water withdrawals have consistently been 98–99 percent from surface water sources.

Although in-stream water use for hydroelectric power is not covered in the 2000 USGS report, the 1995 report notes that in-stream withdrawals declined 4 percent between 1990 and 1995, from 3,290 to 3,160 billion gallons per day.

## The Freshwater Supply

Most great civilizations began and flourished on the banks of lakes and rivers. Throughout human history societies have depended on these surface water resources for food, drinking water, transportation, commerce, power, and recreation.

The withdrawal of surface water varies greatly depending on its location. In New England, for example,

FIGURE 6.3

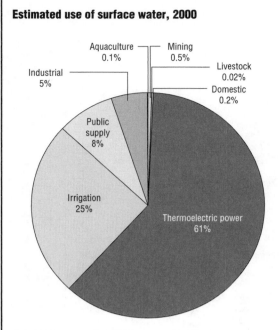

**Estimated use of surface water, 2000**

Aquaculture 0.1%
Mining 0.5%
Livestock 0.02%
Domestic 0.2%
Industrial 5%
Public supply 8%
Irrigation 25%
Thermoelectric power 61%

Note: Total may not sum to 100 because of rounding

SOURCE: Adapted from Susan S. Hutson, Nancy L. Barber, Joan F. Kenny, Kristin S. Linsey, Deborah S. Lumia, and Molly A. Maupin, "Table 3. Surface-Water Withdrawals by Water-Use Category, 2000," in *Estimated Use of Water in the United States in 2000*, (Circular 1268), U.S. Department of the Interior, U.S. Geological Survey, Reston, VA, March 2004 [Online] http://water.usgs.gov/pubs/circ/2004/circ1268/htdocs/table03.html [accessed April 7, 2004]

FIGURE 6.4

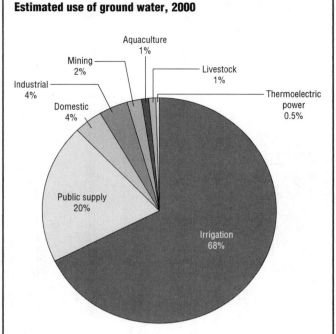

**Estimated use of ground water, 2000**

Aquaculture 1%
Mining 2%
Industrial 4%
Domestic 4%
Livestock 1%
Thermoelectric power 0.5%
Public supply 20%
Irrigation 68%

Note: Total may not sum to 100 because of rounding

SOURCE: Adapted from Susan S. Hutson, Nancy L. Barber, Joan F. Kenny, Kristin S. Linsey, Deborah S. Lumia, and Molly A. Maupin, "Table 4. Ground-Water Withdrawals by Water-Use Category, 2000," in *Estimated Use of Water in the United States in 2000*, (Circular 1268), U.S. Department of the Interior, U.S. Geological Survey, Reston, VA, March 2004 [Online] http://water.usgs.gov/pubs/circ/2004/circ1268/htdocs/table03.html [accessed April 7, 2004]

where rainfall is plentiful, less than 1 percent of the annual renewable water supply is used. In contrast, almost the entire annual supply is consumed in the area of the arid Colorado River Basin and the Rio Grande Valley.

**DAMS—UNEXPECTED CONSEQUENCES.** Dams have changed the natural water cycle. The huge dams built in the United States just before and after World War II substantially changed the natural flow of rivers. By reducing the amount of water available downstream and slowing stream flow, a dam not only affects a river but the river's entire ecological system.

Some 100,000 dams regulate America's rivers and creeks. Nationwide, reservoirs encompass an area equivalent to New Hampshire and Vermont combined. Of all the major rivers (more than 600 miles in length) in the 48 contiguous states, only the Yellowstone River still flows freely. America is second only to China in the use of dams. Worldwide, dams collectively store 15 percent of Earth's annual renewable water supply. Globally, water demand has more than tripled since the mid-twentieth century, and the rising demand has been met by building ever more and larger water supply projects.

Being a world leader in dams was a point of pride for the United States during the golden age of dam building, a

50-year flurry of architectural innovation that began with the construction of the massive Hoover Dam on the lower Colorado River in the 1930s and ended in approximately 1980. In the early years the Army Corps of Engineers built most dams for flood control; later projects served narrower interests, such as developers who wanted flood-plain land.

Dams epitomized progress, Yankee ingenuity, and humankind's mastery of nature. However, the very success of the dam-building endeavor accounted, in part, for its decline: by 1980 nearly all the nation's good sites—and many dubious ones—had been dammed. There were few appropriate places left in the United States to build a major dam.

Three other factors, however, accounted for most of the decline: public resistance to the enormous costs; a growing belief that politicians were foolishly spending taxpayers' money on "pork barrel" (local) projects, including dams; and a developing public awareness of the profound environmental degradation that dams can cause.

**WHERE HAVE ALL THE RIVERS GONE?** Dams provide a source of energy generation; flood control; irrigation;

FIGURE 6.5

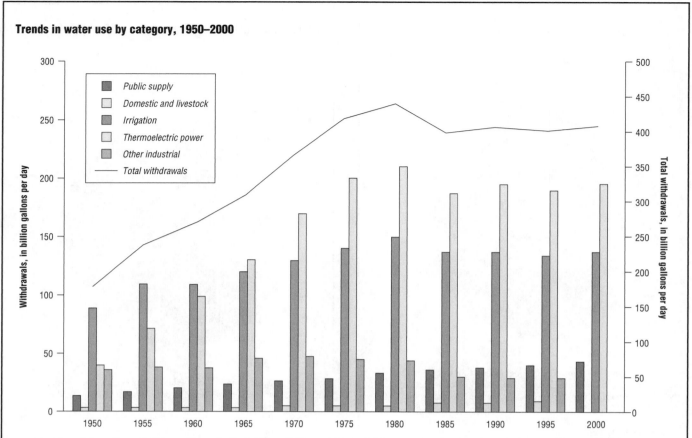

**Trends in water use by category, 1950–2000**

SOURCE: Adapted from Susan S. Hutson, Nancy L. Barber, Joan F. Kenny, Kristin S. Linsey, Deborah S. Lumia, and Molly A. Maupin, "Figure 14. Trends in Total Water Withdrawals by Water-Use Category, 1950–2000," in *Estimated Use of Water in the United States in 2000*, (Circular 1268), U.S. Department of the Interior, U.S. Geological Survey, Reston, VA, March 2004 [Online] http://water.usgs.gov/pubs/circ/2004/circ1268/htdocs/figure14.html

recreation for pleasure boaters, skiers, and anglers; and locks for the passage of barges and commercial shipping vessels. But dams alter rivers as well as the land abutting them, the water bodies they join, and the aquatic life they contain. All this results in profound changes in water systems and the ecosystems they support.

Many regions have fallen into a zero-sum game in which increasing the water supply to one user means taking it away from another. More water devoted to human activities means serious and potentially irreversible harm to natural systems. Many experts believe that the manipulation of river systems is wreaking havoc on the aquatic environment and its biological diversity. Hundreds of species or subspecies of fish are threatened or endangered because of habitat destruction. When rivers are dammed and water flow is stopped or reduced, wetlands dry up, species die, and nutrient loads carried by rivers into the sea are altered, with many negative consequences. Some rivers, including the large Colorado River, no longer reach the sea at all except in years of very high precipitation.

Concern for damage to the environment led Congress to pass the Grand Canyon Protection Act of 1992. The act directed the secretary of state to protect the Grand Canyon basin and its life forms and to monitor the effects of damming the Colorado River. Out of concern for any damage possibly being done to the canyon, for a two-week period in 1996 the Bureau of Reclamation conducted a controlled flood of the canyon by releasing water from the Glen Canyon Dam (up-canyon). The flooding created dozens of new beaches in the Grand Canyon, cleared out many harmful nonnative species, and invigorated fish habitats. The Environmental Protection Agency (EPA) reported that the release of water was significant in that "it was the first time in [U.S.] history that the economic agenda of a large water project was put aside purely for the good of the ecological resources downstream."

**EROSION IS DEVASTATING.** Deforestation and overgrazing have destroyed thousands of acres of vegetation that play a vital role in controlling erosion. Erosion leads to soil runoff into rivers and streams, causing disruption of stream flow. Destruction of vegetation reduces the amount of water released into the atmosphere by transpiration and less water in the atmosphere can mean less rainfall, which can, in turn, lead to desertification (transformation to desert) of once-fertile regions.

The most severe form of land degradation—desertification—is most acute in arid regions. Where land degradation has begun, the hydrologic cycle is disrupted, leaving water tables depleted and causing the sinking and drying of the land. Although desertification was long thought to be the result of droughts, there is much more involved in the process of degradation and desertification of grazing land, including:

• vegetation loss

• water erosion

• wind erosion

• salinization

• compaction of the land by machinery

• accumulation of toxic substances such as lead, chromium, pesticides, and industrial waste

A few hundred million years ago oceanic waters were still fresh enough to drink. It is the Earth that contains the mineral salts that one tastes in seawater. These salts are leached from soil and rock by runoff water. The runoff concentrations in rivers end up in the oceans or in salt lakes such as Mono Lake in California and the Great Salt Lake in Utah. These lakes are seven times saltier than the sea. Once in these bodies of water the salts have nowhere to go. Continuous runoff and evaporation of water leaves increasingly higher concentrations of salt, gradually causing the oceans or lakes to grow saltier. What is changing, however, is the drastic increase in concentrations of salt in the nation's rivers and on some of its prime agricultural land.

**GROUNDWATER.** Groundwater is water that fills pores or cracks in subsurface rocks. When rain falls or snow melts on the Earth's surface, water may run off into lower land areas or lakes and streams. Some is caught and diverted for human use. What is left is absorbed into the soil where it can be used by vegetation; seeps into deeper layers of soil and rock; or evaporates back into the atmosphere. (See Figure 6.6.)

Below the topsoil is an area called the unsaturated zone where, in times of adequate rainfall, the small spaces between rocks and grains of soil contain at least some water, while the larger spaces contain mostly air. After a major rain the zone may become saturated—that is, all the open spaces fill with water. During a drought, the area may become drained and almost completely dry.

With excessive rainfall, water will drain through the unsaturated zone (which has now absorbed as much water as it can hold) to the saturated zone. The saturated zone is always full of water—all the spaces between soil and rocks, and the rocks themselves, contain water. In the saturated zone water is under higher-than-atmospheric pressure. Thus, when a well is dug into the saturated zone, water flows from the area of higher pressure (in the

FIGURE 6.6

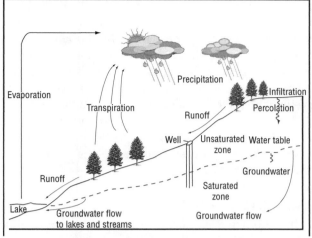

**Groundwater in the hydrologic cycle**

SOURCE: "Groundwater in the Hydrologic Cycle," in *Guide for Industrial Waste Management,* U.S. Environmental Protection Agency, Office of Solid Waste, Washington, DC, June 1999

ground) to the area of lower pressure (in the hollow well), and the well fills with water to the level of the existing water table (the level of groundwater). A well dug just into the unsaturated zone will not fill with water because the water in the unsaturated zone is at atmospheric pressure.

The water table is the level at which the unsaturated zone and the saturated zone meet. The water table is not fixed but may rise or fall, depending on water availability. In areas where the climate is fairly consistent, the level of the water table may vary little; in areas subject to extreme flooding and drought it may rise and fall substantially.

An aquifer is an underground formation that contains enough water to yield significant amounts when a well is sunk. The formation of an aquifer is actually a path of porous or permeable material through which substantial quantities of water flow relatively easily. The word "aquifer" comes from the Latin *aqua* (water) and *ferre* (to bear or carry). An aquifer can be a layer of gravel or sand, a layer of sandstone or cavernous limestone, a rubble zone between lava flows, or even a large body of massive rock, such as fractured granite.

Aquifers vary from a few feet thick to tens or hundreds of feet thick. They can be located just below the Earth's surface or thousands of feet beneath it, and one aquifer may be only a part of a large system of aquifers that feed into one another. They can cover a few acres of land or many thousands of square miles. Because runoff water can easily seep down to the water table, aquifers are susceptible to contamination.

Modern technological developments allow massive quantities of water to be pumped out of the ground. When large amounts of water are removed from the ground, underground aquifers can become depleted much more quickly than they can naturally be replenished. On almost every continent, many major aquifers are being drained faster than their natural rate of recharge. Depletion is most severe in India, China, the United States, North Africa, and the Middle East. In some areas this has led to the subsidence, or sinking, of the ground above major aquifers. Farmers in California's San Joaquin Valley began tapping the area's aquifer in the late nineteenth century. Since that time dehydration of the aquifer has caused the soil to subside by as much as 29 feet, cracking foundations, canals, and roadways. Removal of groundwater also disturbs the natural filtering process that occurs as water travels through rocks and sand.

## Focus on Irrigation

In 2000 irrigation accounted for 40 percent of all the freshwater withdrawn that year. It was by far the largest single user of groundwater and second-highest user of surface water (behind thermoelectric power plants). Because irrigation consumes more withdrawn water than do thermoelectric power plants, irrigation is actually the largest consumer of both surface water and groundwater.

Large-scale irrigation is concentrated in the Midwestern farm belt, southern Florida, the fertile valleys of California, and along the Mississippi River. According to the U.S. Department of Agriculture, more than 50 million acres were irrigated in 1997, primarily in western states. Figure 6.7 shows total irrigation withdrawals by state. California and Idaho withdrew 15–30 billion gallons of water per day for irrigation during 2000. Many states west of the Mississippi River withdrew at least 1 billion gallons per day for irrigation.

## A Water Crisis Looming in the West?

In much of the American West, millions of acres of profitable land overlie a shallow and impermeable clay layer, the residual bottom of an ancient sea, that is sometimes only a few feet below the Earth's surface. During the irrigation season, temperatures in much of the region fluctuate between 90 and 110 degrees Fahrenheit. The good water evaporates and polluted and saline water seep downward. Very little of this water seeps through the clay. As the water supplies are replenished with rainfall, the water table—now high in concentrations of salts and pollutants—rises back up through the root zone (the area containing plant roots), soaking the land and killing crops. (In general, high salt concentrations obstruct germination and impede the absorption of nutrients by plants.)

Several thousand acres in the West have already gone out of production and salt covers the ground like a dusting of snow; not even weeds can grow there. In the coming decades, as irrigation continues, that acreage is expected to

increase dramatically. It is this process rather than drought that is believed to have resulted in the decline of ancient civilizations such as Mesopotamia, Assyria, and Carthage.

**POPULATION PRESSURES.** Many of the fastest growing states are in the West. The U.S. Census Bureau expects population growth in California, Nevada, Arizona, and New Mexico to be in excess of 50 percent between 1995 and 2025. All of the states neighboring them are expected to grow by 31–50 percent.

This population growth in the West is expected to put enormous pressure on natural resources, including water, and to force huge changes in water consumption practices and prices.

**LINGERING DROUGHT.** The natural hydrologic cycle, already under pressure by such human uses as irrigation, is also under strain from years of drought. Although there is no set definition of what constitutes a drought, it is commonly used to describe a period of at least several months in which precipitation is significantly less than that normally expected based on historical records.

According to the Climate Prediction Center of the National Weather Service, drought affected approximately 30 percent of the country in the spring of 2004. Persistent drought was forecast for the entire states of Nevada, Utah, Arizona, Wyoming and parts of California, Oregon, Idaho, Montana, Colorado, and New Mexico. During the 1930s the so-called "Dust Bowl" drought engulfed up to 70 percent of the country.

**WATER 2025.** Although drought is a serious concern in the West, it is not the only worry related to water resources. In 2001 the Bush administration asked the Bureau of Reclamation to assess existing water supplies across the west and identify areas likely to experience severe water shortages within the coming decades. The result was a comprehensive report published in May 2003 titled *Water 2025: Preventing Crises and Conflict in the West.*

The report reviews the factors aggravating the water problems in the West, mainly booming population growth in the most arid regions, aging and poorly maintained water supply infrastructure, and continuing drought. However, the report notes that drought is not the chief cause of the region's water woes. It provides this stark assessment: "Today, in some areas of the West, existing water supplies are, or will be, inadequate to meet the water demands of people, cities, farms, and the environment even under normal water supply conditions."

The Water 2025 program proposes the following approaches to solving the West's looming water crises:

- Modernizing the existing water supply infrastructure

- Employing water conservation measures to more efficiently use existing water supplies

**FIGURE 6.7**

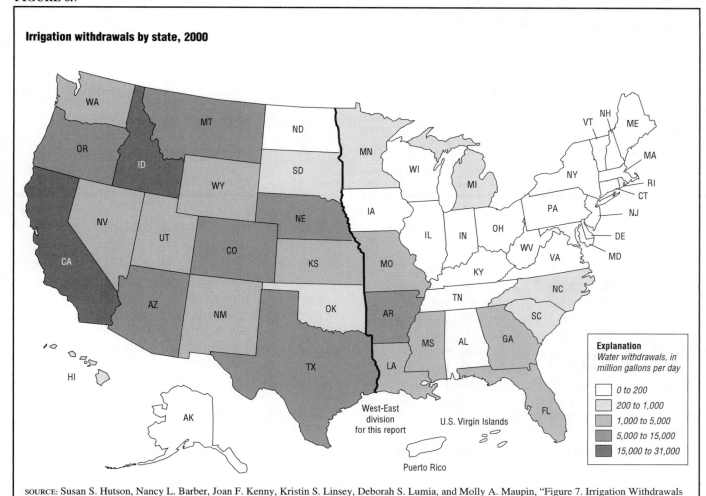

Irrigation withdrawals by state, 2000

SOURCE: Susan S. Hutson, Nancy L. Barber, Joan F. Kenny, Kristin S. Linsey, Deborah S. Lumia, and Molly A. Maupin, "Figure 7. Irrigation Withdrawals by Source and State, 2000," in *Estimated Use of Water in the United States in 2000,* (Circular 1268), U.S. Department of the Interior, U.S. Geological Survey, Reston, VA, March 2004 [Online] http://water.usgs.gov/pubs/circ/2004/circ1268/htdocs/figure07.html [accessed April 7, 2004]

- Establishing collaborative approaches and a market-based transfer system to minimize conflicts between water users

- Conducting research in promising water technology treatment options, such as desalination

The Bureau of Reclamation asked for $11 million in the fiscal year 2004 federal budget to fund the initiatives of the Water 2025 program.

## WATER SUITABILITY

The U.S. economy depends on water and studies have repeatedly shown a positive relationship between strong environmental standards and economic growth. Good water quality is important to local and national economic development.

Water is a powerful attraction for people. The EPA's *Liquid Assets 2000: America's Water Resources at a Turning Point* (May 2000) estimates that the travel, tourism, and recreation industries support jobs for more than 6.8 million people and generate annual sales in excess of $450 billion. The Center for Marine Conservation reports that federal, state, and local governments maintain more than 25,000 recreational facilities along U.S. coasts, while private organizations operate another 20,000. One-third of Americans visit a coast each year, spending about $44 billion. According to the U.S. Fish and Wildlife's 2001 *National Survey of Fishing, Hunting, and Wildlife-Associated Recreation,* these activities produced revenues of $108 billion that year.

More than 95 percent of U.S. foreign trade passes through U.S. harbors and ports. American farmers depend on water to produce and sell billions of dollars worth of food and fiber annually. Each year the Great Lakes, the Gulf of Mexico, and many other coastal areas produce billions of pounds of fish and shellfish. The National Marine Fisheries Service estimated the value of U.S. commercial fishing at about $3.2 billion in 2002. According to the

## TABLE 6.2

**Summary of quality of assessed surface waters, 2000**

| Waterbody type | Total size | Assessed for quality | | Rated good | | Rated good, but threatened | | Rated impaired | |
|---|---|---|---|---|---|---|---|---|---|
| | | Amount | % of total | Amount | % of assessed | Amount | % of assessed | Amount | % of assessed |
| Rivers and streams (miles) | 3,692,830 | 699,946 | 19% | 367,129 | 53% | 59,504 | 8% | 269,258 | 39% |
| Lake, reservoirs & ponds (acres) | 40,603,893 | 17,339,080 | 43% | 8,026,988 | 47% | 1,348,903 | 8% | 7,702,370 | 45% |
| Estuaries (square miles) | 87,369 | 31,072 | 36% | 13,850 | 45% | 1,023 | <4% | 15,676 | 51% |
| Great lakes shoreline (miles) | 5,521 | 5,066 | 92% | 0 | 0% | 1,115 | 22% | 3,951 | 78% |
| Ocean shoreline (miles) | 58,618 | 3,221 | 5% | 2,545 | 79% | 225 | 7% | 451 | 14% |

SOURCE: Adapted from "Figure 1. Summary of Quality of Assessed Rivers, Lakes, and Estuaries," in *Water Quality Conditions in the United States: A Profile from the 2000 National Water Quality Inventory*, U.S. Environmental Protection Agency, Office of Water, Washington, DC, August 2002

USGS, manufacturers use about 9 trillion gallons of fresh water every year.

Water is a fundamental need in every society. Families use water for drinking, cooking, and cleaning. Industry needs it to make chemicals, prepare paper, and clean factories and equipment. Cities use water to fight fires, clean streets, and fill public swimming pools. Farmers water their livestock, clean barns, and irrigate crops. Hydroelectric power stations use water to drive generators, while thermonuclear power stations need it for cooling. Water quality is important to all users, as differing levels of quality are required for different uses. While some industrial users can tolerate water containing high levels of contaminants, drinking water requirements are extremely strict.

### Clean Water Act

On June 22, 1969, the Cuyahoga River in Cleveland, Ohio, burst into flames, the result of oil and debris that had accumulated on the river's surface. This episode thrust the problem of water pollution into the public consciousness. Many people became aware—and wary—of the nation's polluted waters and, in 1972, Congress passed the Federal Water Pollution Control Act (PL 92-500) commonly known as the Clean Water Act.

The objective of the Clean Water Act was to "restore and maintain the chemical, physical, and biological integrity of the nation's waters." It called for ending the discharge of all pollutants into the navigable waters of the United States and to achieve "wherever possible, water quality which provides for the protection and propagation of fish, shellfish, and wildlife and provides for recreation in and on the water." The second provision was that waters be restored to "fishable/swimmable" condition.

Section 305(b) of the Clean Water Act requires states to assess the condition of their waters and report the extent to which the waters support the basic goals of the Clean Water Act and state water quality standards. Water quality standards are designed to protect designated uses (such as recreation, protection and propagation of aquatic

life, fish consumption, and drinking water supply) by setting criteria (for example, chemical-specific limits on discharges) and preventing any waters that do meet standards from deteriorating from their current condition.

Each state prepares and submits to the EPA a report documenting (1) the water quality of all navigable waters in the state, (2) the extent to which the waters provide for the protection and propagation of marine animals and allow recreation in and on the water, (3) the extent to which pollution has been eliminated or is under control, and (4) the sources and causes of the pollution. The act stipulates that these reports must be submitted to the EPA every two years.

### National Water Quality Inventory

Every two years the EPA releases the *National Water Quality Inventory*, which is prepared from the state assessments. The 2000 report was released in August 2002 and was the latest available in June 2004. The report summarizes information about the quality of the nation's rivers, streams, lakes, ponds, reservoirs, estuaries, wetlands, coastal waters, coral reefs, and groundwater. The EPA reports that wetlands, coastal waters, coral reefs, and groundwater are poorly represented in state monitoring programs because few states have adopted water quality standards for them.

SURFACE WATER QUALITY. In 2000 the states assessed surface water quality in rivers and streams, lakes (including the Great Lakes), ocean shoreline, and estuaries. Estuaries are areas where ocean and freshwater come together. Due to the tremendous resources required to assess all water bodies in the United States, only a small portion of each water body type is actually assessed for the report.

Table 6.2 summarizes the findings of the report for surface waters. It found that the states had assessed the quality of 19 percent of their river and stream miles, 43 percent of their lake acres (92 percent of Great Lakes shoreline), 36 percent of estuaries, and 5 percent of ocean shoreline.

Water bodies meeting applicable water quality standards for criteria and designated uses were rated "good."

TABLE 6.3

**Leading pollutants and sources causing impairment in assessed rivers, lakes, and estuaries, 2000**

| Rivers and streams | Lakes, ponds, and reservoirs | Estuaries |
|---|---|---|
| **Causes** | | |
| Pathogens (bacteria) | Nutrients | Metals (primarily mercury) |
| Siltation (sedimentation) | Metals (primarily mercury) | Pesticides |
| Habitat alterations | Siltation (sedimentation) | Oxygen-depleting substances |
| **Sources** | | |
| Agriculture | Agriculture | Municipal point sources |
| Hydrologic modifications | Hydrologic modifications | Urban runoff/ storm sewers |
| Habitat modifications | Urban runoff/ storm sewers | Industrial discharges |

*Excluding unknown, natural, and "other" sources.

SOURCE: "Figure 2. Leading Causes and Sources of Impairment in Assessed Rivers, Lakes, and Estuaries," in *Water Quality Conditions in the United States: A Profile from the 2000 National Water Quality Inventory*, U.S. Environmental Protection Agency, Office of Water, Washington, DC, August 2002

Those water bodies meeting water quality standards, but expected to degrade in the near future were rated "good, but threatened." Water bodies that did not meet water quality standards were rated "impaired."

Only slightly more than half of the assessed rivers and streams were rated good. Slightly less than half of assessed lakes, reservoirs, ponds, and estuaries were rated good. None of the Great Lakes shoreline assessed was found to be in good and unthreatened condition. In fact, 78 percent of the assessed Great Lakes shorelines were rated impaired and the remaining 22 percent were threatened. This is especially significant, because nearly all (92 percent) of the shorelines were assessed. By contrast, only 14 percent of ocean shoreline was rated impaired. However, only a tiny percentage (5 percent) of it was assessed.

Table 6.3 lists the leading pollutants and sources blamed for impairment of assessed rivers, lakes, and estuaries. The primary pollutants are pathogens, siltation, nutrients, metals, pesticides, oxygen-depleting substances, and habitat alterations.

Pathogens are bacteria and viruses that enter water bodies from animal waste, failing septic systems, urban runoff, storm sewers, and combined sewer overflows (CSOs). CSOs are sewer systems in which storm water flows into pipes already carrying raw sewage headed for a sewage treatment facility. During a large rainfall or snowmelt the system's capacity may be overwhelmed, causing the mixture of untreated sewage and storm water to bypass the sewage treatment facility and flow directly into receiving waters. (See Figure 6.8.) In 1994 the EPA ordered communities with CSOs to take "immediate and

FIGURE 6.8

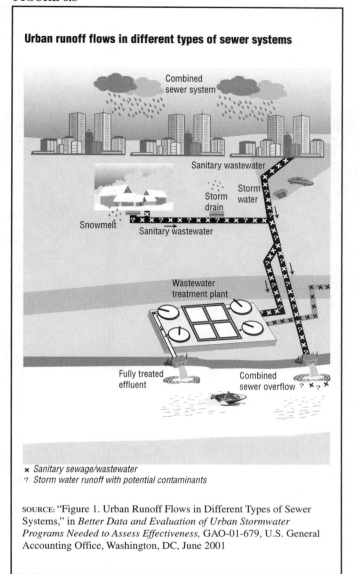

**Urban runoff flows in different types of sewer systems**

× Sanitary sewage/wastewater
? Storm water runoff with potential contaminants

SOURCE: "Figure 1. Urban Runoff Flows in Different Types of Sewer Systems," in *Better Data and Evaluation of Urban Stormwater Programs Needed to Assess Effectiveness,* GAO-01-679, U.S. General Accounting Office, Washington, DC, June 2001

long-term actions" to address CSO problems. However, according to a U.S. General Accounting Office (GAO) report presented to Congress in June 2001, CSOs are still used in approximately 900 cities in the United States.

Pathogens in water used for recreational purposes can pose human health risks due to incidental swallowing and skin contact. In addition, pathogens can accumulate in shellfish, prompting officials to issue warnings against eating shellfish taken from certain waters.

Siltation was another leading cause of surface water impairment in 2000. Siltation is the addition of sediment to water bodies. The sediment can be carried by wind or by runoff from construction sites and other nonvegetated lands, eroding banks, and road sanding operations. Siltation alters aquatic habitats by suffocating fish eggs, causing infection and disease among fish, scouring submerged aquatic vegetation, preventing sunlight from reaching aquatic plants, and burying habitat areas of

**FIGURE 6.9**

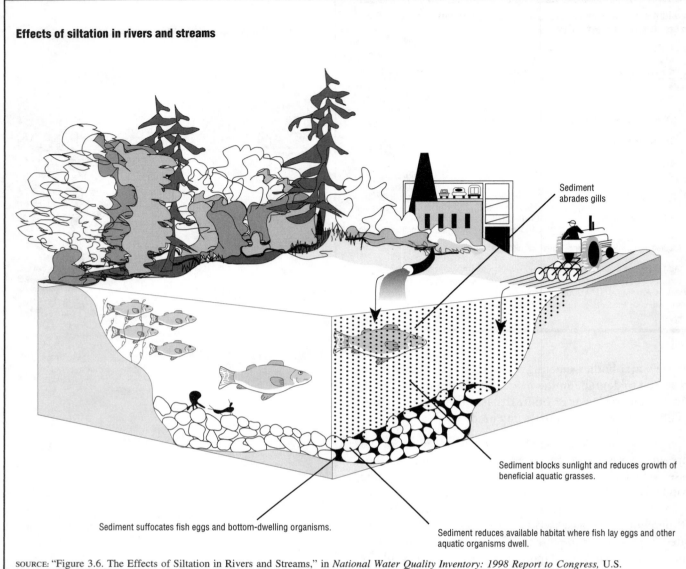

**Effects of siltation in rivers and streams**

Sediment abrades gills

Sediment blocks sunlight and reduces growth of beneficial aquatic grasses.

Sediment suffocates fish eggs and bottom-dwelling organisms.

Sediment reduces available habitat where fish lay eggs and other aquatic organisms dwell.

SOURCE: "Figure 3.6. The Effects of Siltation in Rivers and Streams," in *National Water Quality Inventory: 1998 Report to Congress,* U.S. Environmental Protection Agency, Washington, DC, June 2000

bottom-dwelling species. The loss of those species then impacts fish and other species that feed on them. Silt in the water can also interfere with drinking water treatment processes and recreational use of surface water bodies. (See Figure 6.9.)

Nutrients in water bodies are primarily nitrogen and phosphorus introduced via runoff of fertilizers and animal waste, failing septic systems, CSOs, and atmospheric deposition. Excessive nutrients pollute water bodies by spurring an overgrowth of plants, algae, and toxic and nontoxic blooms of planktonic marine organisms, such as "red tides" and *Pfiesteria*. This results in oxygen depletion and decomposition of plant matter. Fish suffocate, and unpleasant odor and taste result. (See Figure 6.10.)

Algal growth off the mouth of the Mississippi River has caused a massive area called an hypoxic zone, which is extremely low in oxygen and devoid of aquatic life. The hypoxic zone in the Gulf of Mexico averaged 8,500 square miles during the summer of 2002. It is blamed on excessive inflow of nutrients from fertilizer use in the agricultural areas bordering the Mississippi River. The EPA reports that a similar hypoxic zone in Long Island Sound may have killed millions of shellfish in the summer of 2000.

Metals are introduced to water bodies through industrial discharges, automobile fluid leaks, normal wear of automobile brake linings and tires, wear of metal roofs on buildings, and atmospheric deposits from power plants and waste incinerators. Many metals are toxic to aquatic organisms and pose a potential threat to humans via fish consumption. Mercury in particular can pose a serious health threat.

**FIGURE 6.10**

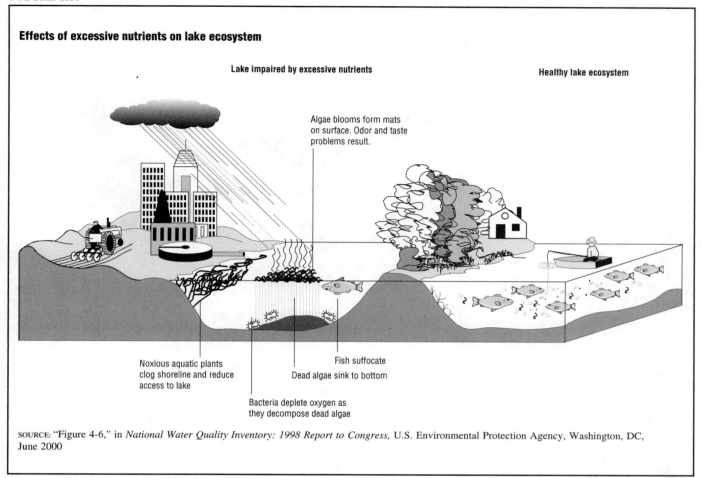

**Effects of excessive nutrients on lake ecosystem**

Lake impaired by excessive nutrients

Healthy lake ecosystem

Algae blooms form mats on surface. Odor and taste problems result.

Noxious aquatic plants clog shoreline and reduce access to lake

Fish suffocate

Dead algae sink to bottom

Bacteria deplete oxygen as they decompose dead algae

SOURCE: "Figure 4-6," in *National Water Quality Inventory: 1998 Report to Congress,* U.S. Environmental Protection Agency, Washington, DC, June 2000

Oxygen-depleting substances are a leading cause of estuary impairment. Estuaries, which are home to many shellfish species and are used as nursery areas for aquaculture enterprises, usually have large cities nearby. Urban sources, including municipal and industrial discharges, urban runoff, and storm sewers are the most prevalent sources of estuary pollution. Other causes of impairment include hydrological modifications (damming rivers and altering the flow of water) and habitat modifications.

The Great Lakes (Lakes Huron, Michigan, Superior, Ontario, and Erie) contain nearly 20 percent of the fresh surface water on Earth. However, they are impaired by a variety of contaminants introduced via storm water runoff, surface water and wastewater discharges, groundwater, and air deposition.

Only 15 states have coastal waters within their jurisdiction. The three leading pollutants blamed for ocean shoreline impairment were pathogens (bacteria), oxygen-depleting substances, and turbidity (high particle content in the water, causing muddy or cloudy appearance). The sources of those contaminants were urban runoff/storm sewers, nonpoint (having no specific point of release) sources, and land disposal.

**DO THE NATION'S SURFACE WATERS MEET THE FISH-ABLE/SWIMMABLE GOAL?** A goal of the Clean Water Act was to return U.S. waters to a fishable/swimmable condition. Meeting the fishable goal means providing a level of water quality that protects and promotes the population of fish, shellfish, and wildlife. As a result of polluted waters, fish often become contaminated. When humans eat these fish, they can suffer health effects from the toxins. In May 2003 the EPA published *Update: National Listing of Fish and Wildlife Advisories.* The report noted that 348 new fish advisories were issued in 2002, and 166 were rescinded, bringing the total number of advisories to 2,800. (See Figure 6.11.) This number is up from 2,618 total advisories in effect in 2001.

A total of 94,715 lakes were under advisory in 2002. The total lake area under advisory increased from 28 percent in 2001 to 33 percent in 2002, while the number of river miles under advisory increased from 13.7 percent in 2001 to 15.3 percent in 2002. In addition, 100 percent of the Great Lakes and their connecting waters and 71 percent of U.S. coastal waters of the 48 contiguous states were under advisory in 2002.

The USDA, EPA, and seven other federal agencies released the *National Coastal Condition Report* in Sep-

## FIGURE 6.11

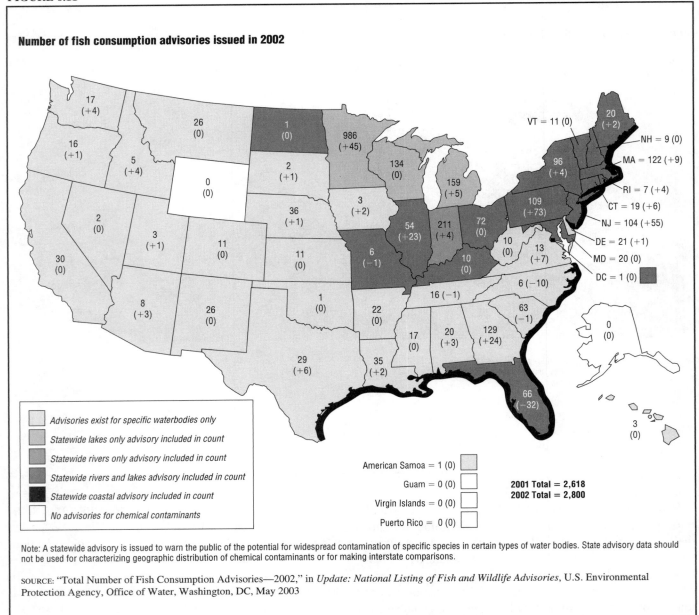

Number of fish consumption advisories issued in 2002

Advisories exist for specific waterbodies only

Statewide lakes only advisory included in count

Statewide rivers only advisory included in count

Statewide rivers and lakes advisory included in count

Statewide coastal advisory included in count

No advisories for chemical contaminants

American Samoa = 1 (0)

Guam = 0 (0)

Virgin Islands = 0 (0)

Puerto Rico = 0 (0)

**2001 Total = 2,618**
**2002 Total = 2,800**

Note: A statewide advisory is issued to warn the public of the potential for widespread contamination of specific species in certain types of water bodies. State advisory data should not be used for characterizing geographic distribution of chemical contaminants or for making interstate comparisons.

SOURCE: "Total Number of Fish Consumption Advisories—2002," in *Update: National Listing of Fish and Wildlife Advisories*, U.S. Environmental Protection Agency, Office of Water, Washington, DC, May 2003

tember 2001. The report examined the ecological health of the nation's coasts based on seven indicators:

- water clarity
- dissolved oxygen concentrations
- loss of coastal wetlands
- eutrophic condition
- sediment contamination
- benthic condition
- accumulation of contaminants in fish tissue

The report concluded that overall the nation's coasts were in fair to poor condition. Indicators receiving the best marks included water clarity and dissolved oxygen. The poorest rated indicators were coastal wetland loss, eutrophic (high level of nutrients) condition, and benthic (regarding the bottom of a body of water) condition.

In May 2003 the EPA reported in its BEACH Watch Program that 2,823 beaches provided information about beach advisories and closings for the 2002 swimming season. Of these beaches, 2,031 were coastal and 792 were on inland waterways. There were 709 beaches with at least 1 advisory or closing during 2002 (25 percent of those reporting). This percentage was down slightly from 2001 when 27 percent reported advisories or closings. The vast majority of the problems (75 percent) were attributed to elevated bacteria levels. Unfortunately, 43

**FIGURE 6.12**

**Examples of point and nonpoint sources of pollution**

Examples of point source pollution are indicated on the left side of the river.

Examples of nonpoint source pollution are indicated on the right side of the river.

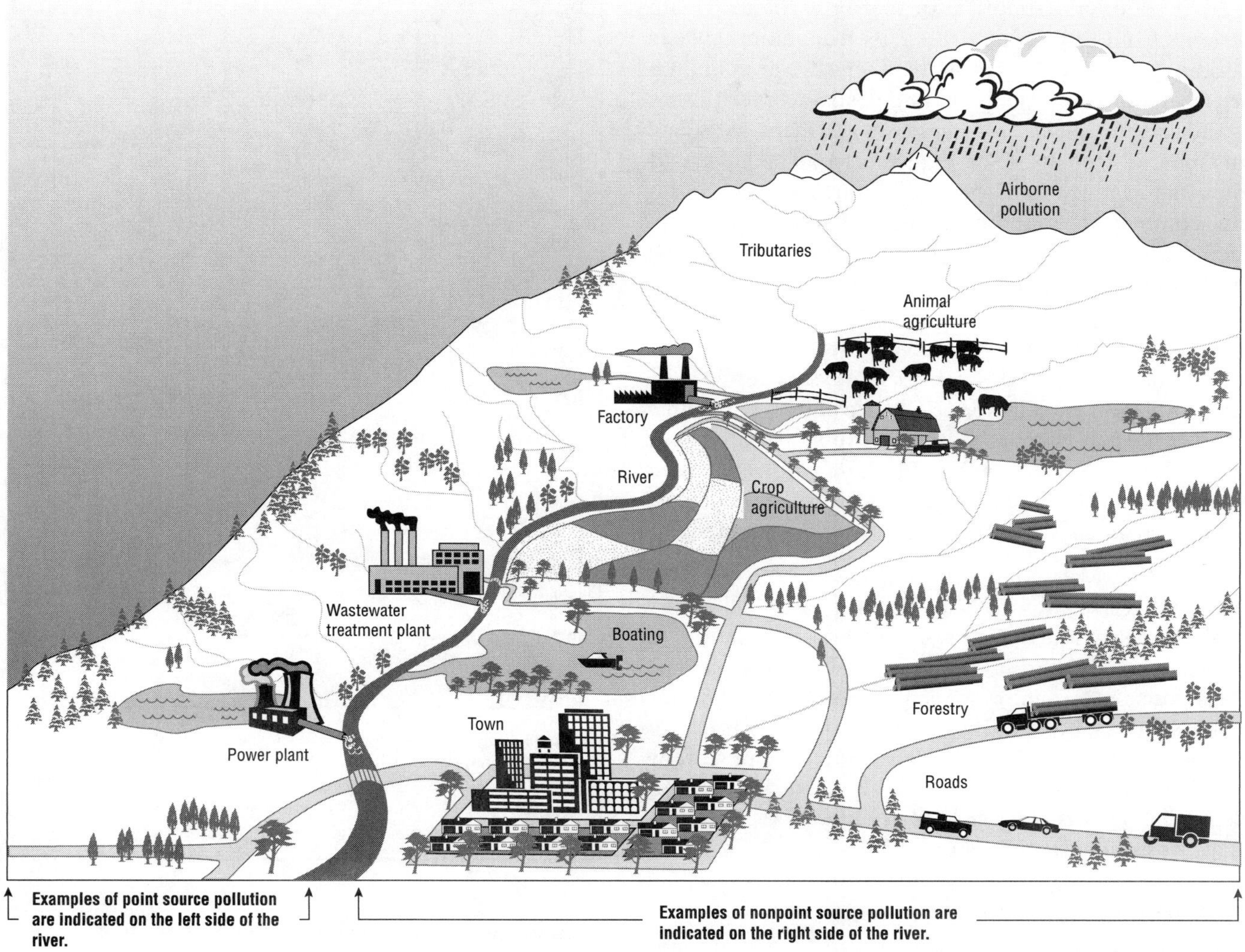

SOURCE: "Figure 3: Examples of Point and Nonpoint Sources of Pollution," in *Water Quality: Key EPA and State Decisions Limited by Inconsistent and Incomplete Data,* U.S. General Accounting Office, Washington, DC, March 2000

percent of the pollutant sources could not be identified. Sources that could be identified included storm water runoff (21 percent), wildlife (11 percent), boat discharges (3 percent), and a number of sewage-related causes.

**FOCUS ON WATER POLLUTION SOURCES.** The main reason that a body of water cannot support its designated uses is that it has become polluted. There are a vast number of pollutants that can make water "impaired," but in order to control a specific pollutant, it is necessary to find out where it is coming from. Although there are many ways in which contaminants can enter waterways, sources

of pollution are generally categorized as point sources and nonpoint sources.

Figure 6.12 shows examples of point and nonpoint sources of pollution. Point sources are those that disperse pollutants from a specific source or area, such as a sewage drain or an industrial discharge pipe. Pollutants commonly discharged from point sources include bacteria (from wastewater treatment plants and sewer overflow), toxic chemicals, and heavy metals from industrial plants. Point sources are regulated under the National Pollutant Discharge Elimination System (NPDES). Any facility using point sources to discharge to receiving waters must obtain an NPDES permit for them.

Nonpoint sources are those that are spread out over a large area and have no specific outlet or discharge point. These include agricultural and urban runoff, runoff from mining and construction sites, and accidental or deliberate spills. Agricultural runoff is primarily associated with nutrients from fertilizers, pathogens from animal waste operations, and pesticides. Urban runoff can contain a variety of contaminants, including pesticides, fertilizers, chemicals and metals, oil and grease, sediment, salts, and atmospheric deposits. (See Figure 6.13.) The EPA estimates that as much as 65 percent of surface water pollutants come from nonpoint sources. These sources are much more difficult to regulate than point sources and may require a new approach to water protection.

## Groundwater Quality

In "The Hidden Freshwater Crisis" (December 9, 2000, http://www.worldwatch.org/) Worldwatch Institute researcher Payal Sampat warned, "We're polluting our cheapest and most easily accessible supply of water. Most groundwater is still pristine, but unless we take immediate action, clean groundwater will not be there when we need it."

The EPA's *National Water Quality Inventory* for 2000 reported that 39 states assessed water quality in aquifers in their states and identified sources of contamination. In general, groundwater quality in the nation is good. The EPA reports that groundwater can support its many different uses but is potentially threatened by a variety of sources. The leading sources were underground storage tanks containing toxic chemicals, septic systems, landfills, spills, fertilizers, large industrial facilities, and hazardous waste sites. (See Figure 6.14.)

## The Federal Role in Protecting Groundwater

Parts of several federal laws help to protect groundwater. The 1972 Clean Water Act provides guidance and money to the states to help develop groundwater programs. The Safe Drinking Water Act of 1974 (SDWA; PL 93-523) and the Safe Drinking Water Act Amendments of 1996 (PL 104-182) require communities to test their water to make sure it is safe and help communities finance projects needed to comply with SDWA regulations. The 1976 Resource Conservation and Recovery Act includes many programs designed to clean up hazardous waste, landfills, and underground storage tanks. New storage tanks must be made of strong plastics that will not rust or leak contaminants into the water table.

The Comprehensive Environmental Response, Compensation, and Liability Act of 1980 (PL 96-510) and the Superfund Amendments and Reauthorization Act of 1986 (PL 99-499; PL 99-563; and PL 100-202) require the cleanup of hazardous wastes that can seep into the

FIGURE 6.13

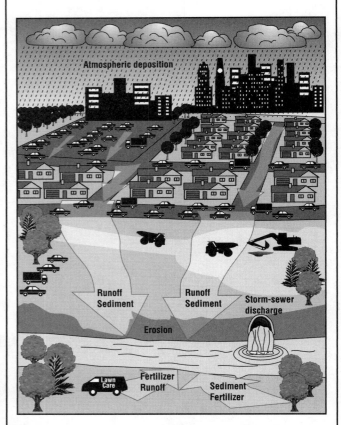

**Contributions of air and water pollution from certain patterns of development**

SOURCE: "Figure 2. Contributions of Air and Water Pollution from Certain Patterns of Development," in *Federal Incentives Could Help Promote Land Use That Protects Air and Water Quality*, GAO-02-12, U.S. General Accounting Office, Washington, DC, October 2001

groundwater. These two laws also require that cities and industry build better-managed and better-constructed garbage dumps and landfills for hazardous materials so that groundwater will not be polluted in the future.

The Federal Insecticide, Fungicide, and Rodenticide Act (61 Stat 163; amended 1988, PL 100-532) regulates dangerous chemicals used on farms. The act requires the EPA to register the pesticides farmers use against insects, rats, mice, and so on. If the EPA thinks the pesticides might be dangerous to the groundwater, it can refuse to register them.

THE STATUS OF THE CLEAN WATER ACT. Both political conservatives and environmentalists credit the Clean Water Act with reversing, in a single generation, what had been a decline in the health of the nation's water since the mid-nineteenth century. In the 1990s, however, some politicians proposed legislation to change the Clean Water Act, giving more authority to the states and more weight to economic

FIGURE 6.14

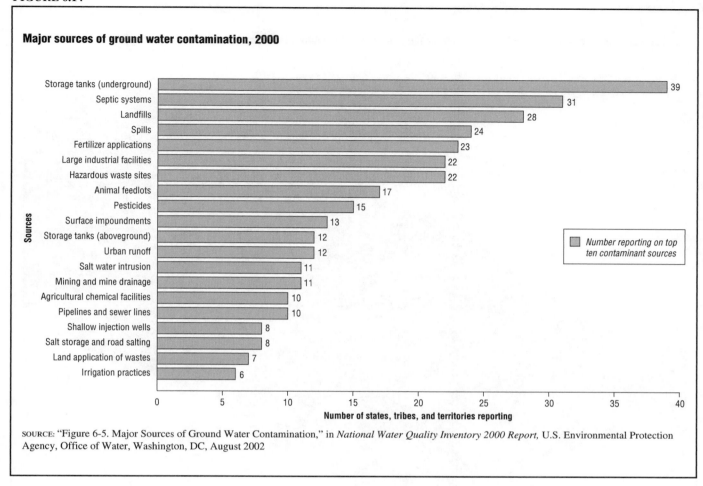

**Major sources of ground water contamination, 2000**

SOURCE: "Figure 6-5. Major Sources of Ground Water Contamination," in *National Water Quality Inventory 2000 Report,* U.S. Environmental Protection Agency, Office of Water, Washington, DC, August 2002

considerations. These politicians and their supporters (coalitions of industry, agriculture, and state and local governments) argue that enough has been accomplished and that now it is time to make the law more flexible. They claim that the huge cost of maintaining clean water risks making the United States noncompetitive in the international market. Government regulations, they think, demand more than is necessary to maintain drinkable water.

## The Future of Water Management

In June 2001 the EPA issued a report titled *Protecting and Restoring America's Watersheds*. A watershed is defined as a "land area that drains to a body of water such as a stream, lake, wetland, or estuary." In other words, a watershed is determined geologically and hydrologically, rather than politically. Figure 6.15 shows a watershed example and the many issues and processes that affect it.

The EPA believes that the nation's water quality problems cannot be solved by further regulating point-source discharges. Instead, the agency advocates a comprehensive approach that crosses jurisdictional boundaries and addresses all of the air, water, land, and social and economic issues that affect a particular watershed. The water-

shed approach would balance competing needs for drinking water, recreation, navigation and flood control, agriculture and forestry, aquatic ecosystems, hydropower, and other uses. Currently, these uses are managed by a variety of agencies at the federal, state, and local level.

Actually, the idea is not new. It was originally suggested in 1890 by John Wesley Powell, then director of the USGS. He suggested that the West be divided into watershed units that would be governing bodies and coordinate management of the natural resources within their jurisdiction. More than a century later, the idea is receiving serious consideration.

## WHERE WATER IS POWER—INTERNATIONAL WATER WARS?

As it becomes rarer, because of the growing population and the pollution of water supplies, usable water is expected to become a commodity, like iron or oil, leading some experts to predict that, at some point, water will be more expensive than oil. Of the 200 largest river systems in the world, 120 flow through 2 or more countries. All are potential objects of world political power struggles over this critical resource.

FIGURE 6.15

**Land drawing demonstrating watershed approach for the management of water resources**

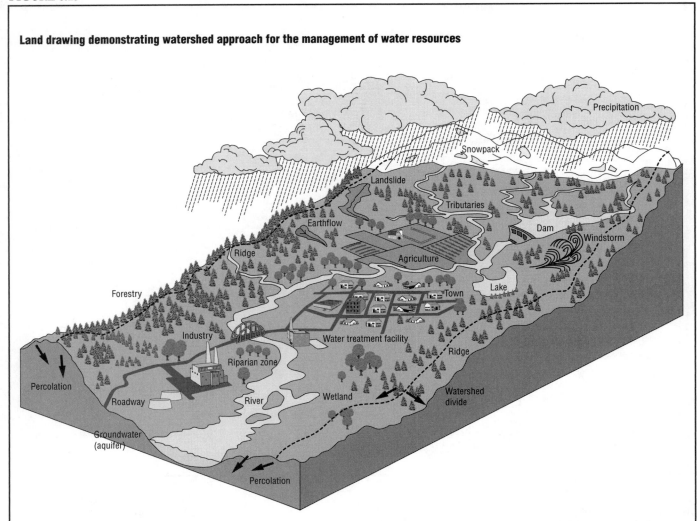

SOURCE: "Figure 1. The area hydrologically defined by a watershed is affected by many processes and issues. A 'watershed approach' coordinates their management," in *Protecting and Restoring America's Watersheds,* U.S. Environmental Protection Agency, Office of Water, Washington, DC, June 2001

Three areas of the world are particularly short of water—Africa, the Middle East, and South Asia. Other dry areas include the southwestern United States, parts of South America, and large areas of Australia.

Middle Eastern countries are especially threatened as growing nations compete for a shrinking water supply. Freshwater has never come easily to this area. Rainfall occurs only in winter and drains quickly through the parched land. Most Middle Eastern countries are joined by common aquifers. The United Nations (UN) has cautioned that future wars in the Middle East could be fought over water.

The oil-rich Middle Eastern nation of Kuwait has little water but has the money to secure it. In order to use seawater, Kuwait has constructed large-scale, oil-powered water desalination plants. Saudi Arabia, farther down the Arabian Peninsula, leads the world in water desalination. As of

March 2004 it has 30 desalination plants that produce nearly 600 million gallons of water per day. This output accounts for 30 percent of global desalinated water production. Desalinated water meets 70 percent of the country's drinking water needs. The remainder comes from groundwater. Saudi Arabia is a leader in the pumping of fossil water—water accumulated in an earlier geologic age lying deep in aquifers beneath Africa and the Middle East.

The United States and Mexico have been bickering over a 1944 treaty that granted Mexico water from the Colorado River in exchange for a lesser amount from the Rio Grande to be given to Texas farmers. U.S. officials claim that Mexico owes them more than 1.3 million acre-feet (or 423 trillion gallons) of water. In March 2002 the issue was raised by President George W. Bush with Mexican president Vicente Fox. Mexican officials blame nine years of drought for their inability to turn over the water. The mighty Rio Grande was once a navigable waterway that marked the 2,000-mile border between the 2 countries. By 2002 it was merely a trickle. Dams, inefficient irrigation methods, drought, and invasion of nonnative weeds are blamed for diminishing the river.

## OCEAN PROTECTION

Throughout history humans have used the oceans virtually as they pleased. Ocean waters have long served as highways and harvest grounds. Now, however, humankind is at a threshold. Marine debris (garbage created by humans) is a problem of global proportions and is extremely evident in countries like the United States where there is extensive recreational and commercial use of coastal waterways. The oceans have become overfished and badly polluted. An estimated half of fish species in the United States are overexploited—more are being caught than can be replenished by natural reproduction. Coastal wetlands are decreasing. Louisiana, for example, loses about 50 square miles of estuaries each year. Sewage, wastewater, and runoff foul the remaining wetlands and ocean waters and trash is piling up on many shorelines. Toxic chemicals contaminate fish caught for food.

### International Convention for the Prevention of Pollution from Ships

Established in 1973, the International Convention for the Prevention of Pollution from Ships regulates numerous materials that are dumped at sea. The international treaty has been in effect in the United States only since its ratification in 1998. Although 83 countries have ratified the treaty, they have not necessarily complied, as evidenced by the current level of marine debris.

### Ocean Dumping Act

Congress enacted the Marine Protection, Research, and Sanctuaries Act in 1972 (PL 92-532) to regulate intentional ocean disposal of materials and to authorize research. Title 1 of the act, known as the Ocean Dumping Act, contains permit and enforcement provisions for ocean dumping. Four federal agencies have authority under the act—the EPA, the U.S. Army Corps of Engineers, the National Oceanic and Atmospheric Administration, and the U.S. Coast Guard. Title 1 prohibits all ocean dumping, except that allowed by permits, in any ocean waters under U.S. jurisdiction by any U.S. vessel or by any vessel sailing from a U.S. port. The act bans dumping of radiological, chemical, and biological warfare agents, high-level radioactive waste, and medical wastes. In 1997 Congress amended the act to ban dumping of municipal sewage sludge and industrial waste.

The act authorizes the EPA to assess civil penalties of up to $50,000 for each violation, as well as criminal penalties (seizure and forfeiture of vessels). For dumping of medical wastes the act authorizes civil penalties of up to $125,000, criminal penalties of up to $250,000 and five years in prison, or both.

In July 1999 the world's second largest cruise line pleaded guilty in federal court to criminal charges of dumping oil and hazardous chemicals in U.S. waters and lying about it to the Coast Guard. Royal Caribbean agreed to pay a record $18 million fine, the largest ever paid by a cruise line for polluting waters, in addition to the $9 million in criminal fines the company agreed to pay in a previous plea agreement. Six other cruise lines have pleaded guilty to illegal waste dumping since 1993 and have paid fines ranging up to $1 million. The cases have focused attention on the difficulties of regulating the fast growing cruise line industry in which most major ships sailing out of U.S. ports are registered in foreign countries.

### Oil Pollution Act

In 1989 the oil freighter *Exxon Valdez* ran into a reef in Prince William Sound, Alaska, spilling more than 11 million gallons of oil into one of the richest and most ecologically pristine areas in North America. An oil slick the size of Rhode Island killed wildlife and marine species. A $5 billion damage penalty was levied against Exxon, whose ship captain was found to be at fault in the wreck.

In response to the *Valdez* oil spill, Congress passed the Oil Pollution Act of 1990 (PL 101-380) that went into effect in 1993. The law requires companies involved in storing and transporting petroleum to have standby plans for cleaning up oil spills on land or in water. Under the act a company that does not adequately take care of a spill is vulnerable to almost unlimited litigation and expense. The law makes the Coast Guard responsible for approving cleanup plans and procedures for coastal and seaport oil spills, while the EPA oversees cleanups on land and in inland waterways. The law also requires that oil tankers be built with double hulls to better secure the oil in event of a hull breach.

## DRINKING WATER

### Drinking Water Legislation

Almost any legislation concerning water affects drinking water, either directly or indirectly. The following pieces of legislation are aimed specifically at providing safe drinking water for the nation's residents.

SAFE DRINKING WATER ACT OF 1974. The SDWA mandated that the EPA establish and enforce minimum national drinking water standards for all public water systems—community and noncommunity—in the United States. The law also required the EPA to develop guidelines for water treatment and to set testing, monitoring, and reporting requirements.

To address pollution of surface water supplies to public systems, the EPA established a permit system requiring any facility that discharges contaminants directly into surface waters (lakes and rivers) to apply for a permit to discharge a set amount of materials—and that amount only. It also created groundwater regulations to govern underground injection of wastes.

Congress intended that, after the EPA had set regulatory standards, each state would run its own drinking water program. Since 1974, 54 states and territories have been granted "primacy"; that is, they have been given the primary responsibility for enforcing the requirements of the SDWA. In order to be granted primacy a state must adopt drinking water standards at least as stringent as the national standards (those established by the EPA), and it must be able to conduct monitoring and enforcement programs that meet federal standards.

The EPA established the Primary Drinking Water Standards by setting maximum containment levels (MCLs) for contaminants known to be detrimental to human health. All public water systems in the United States are required to meet primary standards. Secondary standards cover non-health-threatening aspects of drinking water such as odor, taste, staining properties, and color. Secondary standards are recommended but not required.

SOLE-SOURCE AQUIFERS. Under the SDWA, the EPA has the authority to designate certain groundwater supplies as the sole source of drinking water for a community (referred to as "sole-source aquifers") and to determine if federal financially assisted projects may contaminate these aquifers. If the EPA determines that contamination could occur, no commitment of federal financial assistance—such as grants, contracts, loan guarantees, and so on—can be made for that project.

As of April 2004 the EPA had designated 73 sole-source aquifers nationwide. To be designated as a sole-source aquifer for an area, at least 50 percent of the population in a given area must depend on the aquifer for drinking water; a significant public health hazard would result if the aquifer were contaminated, and no reasonable alternative drinking water supplies exist.

1986 AMENDMENTS TO THE SAFE DRINKING WATER ACT. The 1986 amendments to the SDWA required that the EPA set MCLs for an additional 53 contaminants by June 1989, 25 more by 1991, and 25 every 3 years thereafter. The amendments also required the EPA to issue a maximum contaminant level goal (MCLG) along with each MCL. An MCLG is a health goal equal to the maximum level of a pollutant not expected to cause any health problems over a lifetime of exposure. The EPA is mandated by law to set MCLs as close to MCLGs as technology and economics will permit.

The 1986 amendments banned the use of lead pipe and lead solder in new public drinking water systems and

in the repair of existing systems. In addition, the EPA had to specify criteria for filtration of surface water supplies and to set standards for disinfection of all surface and groundwater supplies. The EPA was required to take enforcement action, including filing civil suits against violators of drinking water standards, even in states granted primacy if those states did not adequately enforce regulations. Violators became subject to fines up to $25,000 daily until violations were corrected.

WATER QUALITY CONTROL ACT OF 1987. Section 304 (1) of the revised Clean Water Act of 1987 (PL 100-4) determines the state of the nation's water quality and reviews the effectiveness of the EPA's regulatory programs designed to protect and improve that water quality. Section 308—known as the Water Quality Control Act—requires that the administrator of the EPA report annually to Congress on the effectiveness of the water quality improvement program.

The main purpose of the Water Quality Control Act is to identify water sources that need to be brought up to minimum standards and to establish more stringent controls where needed. States are now required to develop lists of contaminated waters as well as lists of the sources and amounts of pollutants causing toxic problems. In addition, each state is required to develop "individual control strategies" for dealing with these pollutants.

LEAD CONTAMINATION CONTROL ACT OF 1988. The Lead Contamination Control Act of 1988 (PL 100-572) strengthened the controls on lead contamination set out in the 1986 amendments to the SDWA. It requires the EPA to provide guidance to states and localities in testing for and remedying lead contamination in drinking water in schools and day care centers. The act also contains requirements for the testing, recall, repair, and/or replacement of water coolers with lead-lined storage tanks or parts containing lead. It attaches civil and criminal penalties to the manufacture and sale of water coolers containing lead.

The ban on lead states that plumbing must be lead-free. In addition, each public water system must identify and notify anyone whose drinking water may be contaminated with lead, and the states must enforce the lead ban through plumbing codes and the public-notice requirement. The federal government gave the EPA the power to enforce the lead ban law by authorizing the agency to withhold up to 5 percent of federal grant funds to any state that does not comply with the new rulings.

REINVENTING DRINKING WATER LAW—1996 AMENDMENTS TO THE SAFE DRINKING WATER ACT. In 1996 Congress passed a number of significant amendments (PL 104-182) to the SDWA. The law changed the relationship between the federal government and the states in administering drinking water programs, giving states greater flexibility and more responsibility.

The centerpiece of the law is the State Revolving Fund (SRF), a mechanism for providing low-cost financial aid to local water systems to build the treatment plants necessary to meet state and federal drinking water standards. The law also requires states to train and certify operators of drinking water systems. If they do not, states risk losing up to 20 percent of their federal grants. The law requires states to approve the operation of any new water supply system, making sure it complies with the technical, managerial, and financial requirements. The 1996 SDWA gives the EPA discretion in regulating only those contaminants that may be harmful to health, and requires the EPA to select at least five contaminants every five years for consideration for new standards. A further change is that the EPA, when proposing a regulation, now must determine—and publish—whether or not the benefits of a new standard justify the costs.

Furthermore, the law affirms Americans' "right to know" the quality of their drinking water and mandates notification. Water suppliers must promptly (within 24 hours) alert consumers if water becomes contaminated by something that can cause illness and must advise as to what precautions can be taken. In 1998 states began to compile information about individual systems, which the EPA now summarizes in an annual compliance report. As of October 1999 water systems have been required to make that data available to the public. Large suppliers have to mail their annual safety reports to customers, while smaller systems can post the reports in a central location or publish it in a local newspaper. (Information on individual water systems is available on the EPA website at http://www.epa.gov.)

In 1996 Congress directed the EPA to issue a new standard for arsenic in drinking water by January 1, 2001. The existing standard at that time was 50 parts per billion (ppb). The EPA proposed a standard of 5 ppb in June 2000. However, this was too late to resolve scientific and public debate about the new standard in time to meet the January 1 deadline, so Congress extended the deadline. A new standard of 10 ppb became effective in February 2002, but public water systems were given until January 2006 to meet it.

### Sources of Drinking Water—Public and Private Supply

According to the EPA there were 161,201 public water supply systems in operation in 2003, serving 302.9 million people. (See Table 6.4.) These included systems that served homes, businesses, schools, hospitals, and recreational parks. About 273.3 million Americans got their water from a community water system. Those who did not get their water from a public system were for the most part in rural areas and got their water from private

**TABLE 6.4**

**Inventory of public water systems, 2003**

| | Community water systems | | |
| --- | --- | --- | --- |
| | Ground water | Surface water | Total |
| No. of systems | 41,499 | 11,864 | 53,363 |
| % of systems | 78% | 22% | |
| Population served | 86.3 million | 187 million | 273.3 million |
| % of population served that depends on each water type | 32% | 68% | |
| | Non-transient non-community water systems | | |
| | Ground water | Surface water | Total |
| No. of systems | 18,908 | 778 | 19,686 |
| % of systems | 96% | 4% | |
| Population served | 5.6 million | 730 thousand | 6.3 million |
| % of population served that depends on each water type | 88% | 12% | |
| | Transient non-community water systems | | |
| | Ground water | Surface water | Total |
| No. of systems | 86,061 | 2,091 | 88,152 |
| % of systems | 98% | 2% | |
| Population served | 10.5 million | 12.8 million | 23.3 million |
| % of population served that depends on each water type | 45% | 55% | |

SOURCE: Adapted from "Public Water System Inventory Data," in *Factoids: Drinking Water and Ground Water Statistics for 2003*, U.S. Environmental Protection Agency, Office of Ground Water and Drinking Water, Washington, DC, January 2004

wells. Although most systems obtain their water from groundwater, most people receive drinking water from surface water sources. This is because a relatively small number of public systems served large metropolitan areas.

The EPA and state health or environmental departments regulate public water supplies. Public supplies are required to ensure that the water meets certain government-defined health standards. The SDWA governs this regulation. The law mandates that all public suppliers test their water on a regular basis to check for the existence of contaminants and treat their water supplies constantly to take out or reduce certain pollutants to levels that will not harm human health.

Private water supplies, usually wells, are not regulated under the SDWA. System owners are solely responsible for the quality of the water provided from private sources. However, many states have programs designed to help well owners protect their water supplies. Usually, these state-run programs are not regulatory but provide safety information. This type of information is vital because private wells are often shallower than those used by public suppliers. The shallower the well the greater is the potential for contamination.

### Chemicals and Contaminants in Drinking Water

All drinking water contains minerals dissolved from the Earth. In small amounts some of these are acceptable because they enhance the quality of the water (for example, by giving it a pleasant taste). A few, such as zinc and selenium, in very small amounts, contribute to good health. Other naturally occurring minerals are not desirable because they may cause a bad taste or odor (as excessive amounts of iron, manganese, or sulfur often do) or because they may be harmful to health.

The health effects from drinking contaminated water can occur either over a short or long period of time, depending on the type of pollutant. Short-term, or acute, reactions are those that occur within a few hours or days after drinking tainted water. Long-term, or chronic, effects occur after water with relatively low doses of a pollutant has been consumed for several years or even over a lifetime. Fortunately, the ability to detect contaminants has improved over the past few decades. Scientists can now identify specific pollutants in terms of 1 part contaminant in 1 billion parts of water. In some cases contaminants can be measured in the trillionths.

Water supplies may contain a wide variety of contaminants that can cause serious health risks. While bacterial infections generally make their presence known quickly by causing illness with fairly obvious symptoms, the effects of noxious chemicals may not be apparent for months, or even years, after exposure. Some pollutants are known carcinogens (cancer-causing agents), while others are suspected of causing birth defects, miscarriages, and heart disease. In many cases the effects occur only after prolonged exposure, but no one can say for sure what is a safe level of exposure.

**LEAD.** When water leaves the treatment plant it is relatively free of lead, but it can pick up the metal from lead pipes (and pipes with lead solder) in distribution systems or homes, resulting in tap water containing significant amounts of this highly toxic substance. Unlike many water contaminants, lead has been extensively studied for its prevalence and effects on human health and for ways to eliminate it from the water supply.

Most people understand that drinking water contaminated with lead is very dangerous. Ingested lead causes extremely serious health problems, especially in children, whose developing bodies absorb and retain more lead than adults' bodies do. According to EPA reports, even very low level exposures can result in lowered intelligence, impaired learning and language skills, loss of hearing, reduced attention spans, and poor school performance. High levels damage the brain and the central nervous system, thereby interfering with both learning and physical development. While drinking water is not the only means of lead contamination (lead exists in some paints, soils, and older food containers), the EPA estimates that it accounts for between 10 and 20 percent of total lead poisoning in young children.

Pregnant women are another high-risk group. Lead is believed to cause miscarriages, premature births, and impaired fetal development. It has also been linked to high blood pressure, fatigue, and hearing loss.

Lead is rarely found in either the surface water or groundwater that are the sources of drinking water for most Americans. Lead usually enters the water supply after it leaves the treatment plant. The industries that release the most lead into the environment include lead smelting and refining, copper smelting, steelworks, manufacturers of storage batteries and china plumbing fixtures, iron foundries, and copper mining.

Normally, areas served by lead service lines or residences containing lead interior piping or copper piping with lead solder installed after 1982 are considered high risk. Since 1986 it has been illegal to use lead solder that contains more than 0.2 percent lead. But ripping out and replacing lead piping is extremely expensive. As an alternative some cities are using chemicals to make the water less acidic so that it picks up less lead.

In 1991 the EPA's Lead and Copper Rule set a maximum contaminant level goal for lead at 0 milligrams per liter (mg/L), with water systems required to take action if lead levels reach 0.015 mg/L. The EPA believes this is the lowest level to which water systems can reasonably be required to control the contaminant. If a water supply is found to exceed that amount in more than 10 percent of the homes served, it must be tested twice a year. If levels remain above the standard, the supplier must take steps to reduce those levels. The water supplier must also notify the public about the elevated levels via newspapers, radio, television, and other means and must inform its consumers of additional measures they may take to reduce the levels in their homes.

**NITRATES AND NITRITES.** Nitrates and nitrites are nitrogen-oxygen chemicals that combine with organic and inorganic compounds. Once taken into the body, nitrates are converted into nitrites. Nitrates are most frequently used as fertilizer. Primary sources include human sewage and livestock manure, especially from feedlots. Since they are soluble, nitrates can easily migrate into groundwater.

Nitrates in drinking water are an immediate threat to small children. In some babies high levels of nitrates react with red blood cells to cause an anemic condition commonly known as "blue baby." The MCL for nitrates has been set at 10 parts per million (ppm) and for nitrites at 1 ppm. If contaminant levels exceed these standards, a water provider must take steps to reduce the levels—either through ion exchange, reverse osmosis, or electrodialysis—and must notify the public of their presence.

**MERCURY.** Mercury is unique among metals in that it can evaporate when released into water or soil. Large amounts of mercury are released naturally from the Earth's crust. Metal smelters, cement manufacture, landfills, sewage, and combustion of fossil fuels are also important sources of mercury release. Mercury is especially dangerous when released into water because it

TABLE 6.5

## Waterborne pathogens found in human waste and associated diseases

| Type | Organism | Disease | Effects |
|------|----------|---------|---------|
| Bacteria | Escherichia coli (enteropathogenic) | Gastroenteritis | Vomiting, diarrhea, death in susceptible populations |
| | Legionella pneumophila | Legionellosis | Acute respiratory illness |
| | Leptospira | Leptospirosis | Jaundice, fever (Well's disease) |
| | Salmonella typhi | Typhoid fever | High fever, diarrhea, ulceration of the small intestine |
| | Salmonella | Salmonellosis | Diarrhea, dehydration |
| | Shigella | Shigellosis | Bacillary dysentery |
| | Vibrio cholerae | Cholera | Extremely heavy diarrhea, dehydration |
| | Yersinia enterolitica | Yersinosis | Diarrhea |
| Protozoans | Balantidium coli | Balantidiasis | Diarrhea, dysentery |
| | Cryptosporidium | Cryptosporidiosis | Diarrhea |
| | Entamoeba histolytica | Amoebiasis (amoebic dysentery) | Prolonged diarrhea with bleeding, abscesses of the liver and small intestine |
| | Giardia lamblia | Giardiasis | Mild to severe diarrhea, nausea, indigestion |
| | Naegleria fowleri | Amebic Meningoencephalitis | Fatal disease; inflammation of the brain |
| Viruses | Adenovirus (31 types) | Conjunctivitis | Eye, other infections |
| | Enterovirus (67 types, e.g., polio-, echo-, and Coxsackie viruses) | Gastroenteritis | Heart anomalies, meningitis |
| | Hepatitis A | Infectious hepatitis | Jaundice, fever |
| | Norwalk agent | Gastroenteritis | Vomiting, diarrhea |
| | Reovirus | Gastroenteritis | Vomiting, diarrhea |
| | Rotavirus | Gastroenteritis | Vomiting, diarrhea |

SOURCE: "Table 3-20. Waterborne Pathogens Found in Human Waste and Associated Diseases," in *Onsite Wastewater Treatment Systems Manual*, U.S. Environmental Protection Agency, Office of Research and Development, Office of Water, Washington, DC, February 2002

tends to accumulate in the tissues of fish. When tainted fish are eaten by humans, mercury poisoning is often the result. The MCL for mercury has been set at 2 ppb.

**MICROBIOLOGICAL ORGANISMS.** Many kinds of biological organisms exist in drinking water. These include certain types of bacteria, viruses, or parasites. (See Table 6.5.) These tiny organisms get into the water supplies when the water is contaminated with human or animal wastes. The bacteria, viruses, and parasites that contaminate drinking water can cause flu-like symptoms, including headaches, vomiting, diarrhea, abdominal pain, and dehydration. Although usually not life threatening, they can be debilitating and uncomfortable for their victims.

One type of microscopic parasite that is a common cause of illness is *Giardia lamblia*. Once thought to be harmless, it is now believed to be the most frequent cause of waterborne epidemics in the United States. Its symptoms include mild-to-severe gastrointestinal pain, vomiting, and diarrhea. *Giardia* is particularly threatening because it can enter the water supply in any number of ways, including as sewage overflow. It is also easily transmitted from person to person making places with high concentrations of people (for example, day care centers and schools) particularly susceptible to outbreaks.

*Cryptosporidium* is a one-celled, infectious parasite that frequently contaminates the water supply. Although there are tests for *Cryptosporidium,* current testing methods cannot determine with certainty whether *Cryptosporidium* detected in drinking water is alive or whether

it can affect humans. In addition, the technology often requires several days to get results, by which time the tested water has already been used by the public and is no longer in the community's water pipes. Consequently, water utilities do not routinely test to detect its presence. (A water utility may voluntarily test for the microorganism, and it is also possible that a state may require water systems to test for it. Otherwise, it is unlikely that a given water system tests for *Cryptosporidium*.)

Because *Cryptosporidium* is highly resistant to chlorination, disinfection of water is not a reliable method of preventing exposure to it. The Centers for Disease Control and Prevention (CDC) and the EPA report that the organism can be killed by boiling water for one minute. Outbreaks of *Cryptosporidium* have generally occurred when turbidity (cloudiness of water due to high particle content) reached 0.9 to 2.0 nephelometric turbidity units.

Coliform bacteria from human or animal wastes can also pose serious health problems. Waterborne diseases such as typhoid, cholera, infectious hepatitis, and dysentery have all been traced to untreated drinking water.

### Modern Drinking Water Treatment

The water treatment process begins with choosing the highest quality source available. Raw water must be transported from the source to the treatment plant while groundwater is usually pumped directly into the plant. In many cases the only treatment needed before the water is distributed to consumers is disinfection. Groundwater is naturally filtered as it seeps through layers of rock and soil. However, sometimes it must be treated to remove contaminants that may have percolated down from the surface to the aquifer. In addition, some groundwater must have certain minerals or gases removed to make the water less "hard" (high in natural minerals). Hard water can clog pipes, stain fixtures, and make soap hard to lather.

Surface water is sent to the water treatment plant through aqueducts or pipes. An initial screen at the intake pipe removes large objects. The water is then aerated to eliminate gases and add oxygen. Once the water is inside the plant, chemicals may be added both to clean the water and also to make it more palatable. If the water is hard, lime or soda is added to remove the calcium and magnesium. Chlorine or other such disinfectants may be used as well.

The water is mixed well with the various chemicals, then sent to sedimentation basins where the heavy particles (floc) settle to the bottom and are removed. The water is then sent to filtration beds for polishing—the removal of any remaining small particles and disease-causing protozoa, bacteria, and viruses. (Although filtration removes some viruses, most pass through the filtration process.)

Additional treatment may be required if the raw water contains high levels of toxic chemicals. The pollutants that are the most difficult to detect and remove are often the ones with the greatest potential for severe health effects for large portions of the population.

At various points in the treatment process, the water is monitored by computers and other technological procedures. As the water leaves the treatment plant, chlorine is added as a disinfectant to keep it free of organisms as it travels to customers.

The water then goes to reservoirs where it is stored until needed. These reservoirs may be elevated towers, where gravity brings the water to the consumer without unnecessary energy expense, or ground-level containers that require pumps to move the water. (See Figure 6.16.) The water that flows from the tap should be clear, tasteless, and safe to drink.

## Chemicals Deliberately Added to Drinking Water

Water purification facilities deliberately add certain chemicals to drinking water to destroy contaminants that may cause illness and to improve the taste, smell, and look of the water.

CHLORINE. The most extensively used disinfectant in the United States is chlorine, which is used to kill infectious microorganisms and parasites. Disinfection with chlorine or similar chemicals can prevent outbreaks of salmonellosis, dysentery, and *Giardia*. Chlorination first began in the early 1900s as an attempt to eliminate cholera and typhoid.

Chlorination is not risk free, however. Chlorine reacts with organic chemicals to form trihalomethanes (THMs) that have been shown to cause cancer in laboratory animals. In 1979 the EPA established regulations limiting the amount of THMs to 0.1 mg/L for water supplies serving 10,000 people or more. The EPA and the medical community continue to study the effects of chlorine. Most concerns about chlorine involve possible damage to the Earth's ozone layer. Some water systems now use ozone gas instead of chlorine. Ozone, bubbled through water, can kill more microorganisms than chlorine and may present less risk.

FLUORIDE. Fluoride was first added to drinking water in 1945 to prevent tooth decay. Since that time most community water systems in the United States have introduced water fluoridation. Because the fluoridation of drinking water proved effective in reducing dental cavities, researchers also developed other methods to deliver fluoride to the public (toothpastes, rinses, and dietary supplements). The widespread use of these products has assured that virtually all Americans have been exposed to fluoride. The American Dental Association estimated in 1992 that each $1 expenditure for water fluoridation results in a savings of approximately $80 in dental treatment costs.

There have been concerns about the effects of fluoridation since it was first introduced. The Public Health Service has recommended further assessment of potential problems, although it notes that fluoridation is believed to be greatly beneficial for a number of bone-related conditions.

## How Clean Is Our Drinking Water?

Safe drinking water is a cornerstone of public health. Fortunately, the nation's drinking water is generally safe. The vast majority of U.S. residents receive water from systems that have no reported violations of MCLs and no flaws in treatment techniques, monitoring, or reporting. Nevertheless, numerous studies have found some deficiencies in those systems, and recent measurements suggest areas of potential danger.

In accordance with the 1996 SDWA amendments, public water systems are mandated to submit compliance reports on the quality of their drinking water. Figure 6.17 shows the states with systems that violated water quality levels or treatment standards in fiscal year 2002.

## Incidence of Disease Caused by Tainted Water—CDC Surveillance Report

It is difficult to know how many illnesses are caused by contaminated water. People may not know the source of many illnesses and may attribute them to food (which may also have been in contact with polluted water), chronic illness, or other infectious agents. The EPA notes that some researchers think that the actual number of drinking water–related diseases may be 25 times the reported number. They believe most are not reported

**FIGURE 6.16**

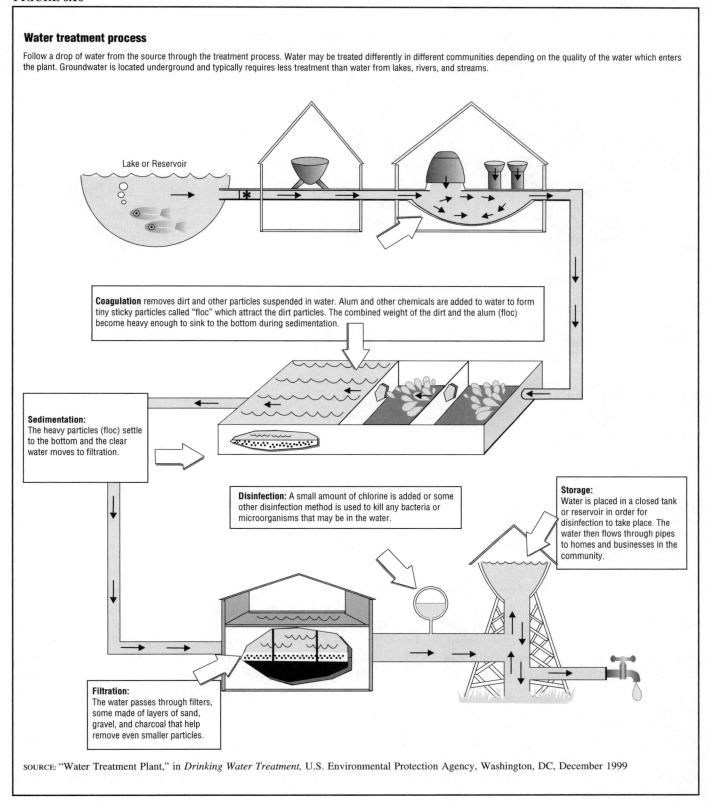

**Water treatment process**

Follow a drop of water from the source through the treatment process. Water may be treated differently in different communities depending on the quality of the water which enters the plant. Groundwater is located underground and typically requires less treatment than water from lakes, rivers, and streams.

Lake or Reservoir

**Coagulation** removes dirt and other particles suspended in water. Alum and other chemicals are added to water to form tiny sticky particles called "floc" which attract the dirt particles. The combined weight of the dirt and the alum (floc) become heavy enough to sink to the bottom during sedimentation.

**Sedimentation:**
The heavy particles (floc) settle to the bottom and the clear water moves to filtration.

**Disinfection:** A small amount of chlorine is added or some other disinfection method is used to kill any bacteria or microorganisms that may be in the water.

**Storage:**
Water is placed in a closed tank or reservoir in order for disinfection to take place. The water then flows through pipes to homes and businesses in the community.

**Filtration:**
The water passes through filters, some made of layers of sand, gravel, and charcoal that help remove even smaller particles.

SOURCE: "Water Treatment Plant," in *Drinking Water Treatment,* U.S. Environmental Protection Agency, Washington, DC, December 1999

because victims believe them to be "stomach upsets" and simply treat themselves.

Since 1971 the CDC and the EPA have collected and reported data that relate to waterborne-disease outbreaks.

The latest report available from the CDC, *Surveillance for Waterborne-Disease Outbreaks—United States, 1999–2000* (November 2002), includes data about outbreaks associated with drinking water, recreational water, and occupational exposure. The CDC defines an outbreak as

FIGURE 6.17

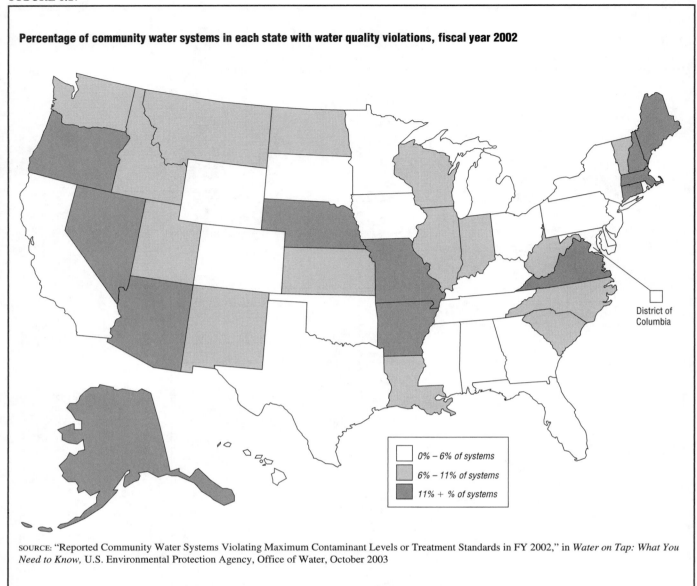

**Percentage of community water systems in each state with water quality violations, fiscal year 2002**

District of Columbia

| | 0% – 6% of systems |
| | 6% – 11% of systems |
| | 11% + % of systems |

SOURCE: "Reported Community Water Systems Violating Maximum Contaminant Levels or Treatment Standards in FY 2002," in *Water on Tap: What You Need to Know,* U.S. Environmental Protection Agency, Office of Water, October 2003

an incident in which at least two people develop a similar illness that evidence indicates was probably caused by ingestion of drinking water or exposure to water in recreational or occupational settings.

According to the CDC, from January 1999 through December 2000 there were 39 outbreaks that sickened 2,068 people and caused two deaths associated with drinking water reported in 25 states. The specific microbe or chemical cause of the outbreaks was identified in 22 of the cases. Pathogens were blamed in 20 of the cases, while chemicals were blamed in the other two cases. Infectious organisms were suspected of being the cause in the remaining 17 cases. Most (72 percent) of the 39 outbreaks were linked to ingestion of groundwater, primarily from private wells not regulated by the EPA.

The CDC report notes that the proportion of drinking water outbreaks associated with surface water increased from 12 percent during 1997–1998 to 18 percent during 1999–2000. The proportion linked with groundwater sources increased by 87 percent from the previous reporting period.

The CDC found evidence that during 1999–2000 recreational water exposure in 23 states caused 59 outbreaks that sickened 2,093 people and caused four deaths. The illnesses involved included gastroenteritis, dermatitis, primary amebic meningoencephalitis, leptospirosis, Pontiac fever, and chemical keratitis. Most of the victims were exposed at swimming pools, interactive fountains, or hot tubs. Two outbreaks associated with occupational exposure caused leptospirosis and Pontiac fever.

## Milwaukee—"The Nation's Worst Drinking Water Disaster"

In April 1993, 403,000 residents of Milwaukee became victims of what is considered the worst drinking water disaster the nation has experienced. *Cryptosporidium* flourished in the city water supply, which had been turbid for several days. For a week more than 800,000 residents were without potable (drinkable) tap water. By the end of the disaster more than 40 people lost their lives because of the outbreak. In addition to the human suffering, the disease cost an estimated $37 million in lost wages and productivity.

Among the possible causes for the outbreak were the advanced age and flawed design of the Milwaukee water plant, which returned dirty water back to the reservoir. Other explanations included failure of plant personnel to react quickly when turbidity levels rose; critical monitoring equipment that was broken at the time turbidity levels peaked; a water intake point that was vulnerable to contamination; and a slaughterhouse, feedlot, and sewage treatment plant that were located upriver from the plant. Water treatment experts blamed the complacency of officials and false assumptions based on a history of quality water dispersal.

As a result of the disaster, Milwaukee launched one of the most aggressive drinking water programs in the country. Each week the city monitors for *Cryptosporidium* and has set a zero standard for the parasite. It has also adopted a turbidity standard five times tougher than federal regulations. Turbidity, although harmless in itself, is often a precursor to the presence of organisms such as *Cryptosporidium*.

## PUBLIC OPINION ABOUT WATER ISSUES

In March 2004 the Gallup Organization conducted its annual poll on environmental topics. As shown in Figure 1.8 in Chapter 1 the pollsters found that water issues top the list of Americans' environmental concerns. The percentage of people expressing a great deal of worry about a particular environmental problem was highest for pollution of drinking water. More than half of those asked (53 percent) indicated they feel a great deal of concern about it. Pollution of surface waters ranked second with 48 per-

**TABLE 6.6**

**Public concern about pollution of drinking water, 2004**

PLEASE TELL ME IF YOU PERSONALLY WORRY ABOUT THIS PROBLEM A GREAT DEAL, A FAIR AMOUNT, ONLY A LITTLE, OR NOT AT ALL. POLLUTION OF DRINKING WATER?

| | Great deal % | Fair amount % | Only a little % | Not at all % | No opinion % |
|---|---|---|---|---|---|
| 2004 Mar 8–11 | 53 | 24 | 17 | 6 | * |
| 2003 Mar 3–5 | 54 | 25 | 15 | 6 | – |
| 2002 Mar 4–7 | 57 | 25 | 13 | 5 | * |
| 2001 Mar 5–7 | 64 | 24 | 9 | 3 | * |
| 2000 Apr 3–9 | 72 | 20 | 6 | 2 | * |
| 1999 Apr 13–14 | 68 | 22 | 7 | 3 | * |
| 1991 Apr 11–14 | 67 | 19 | 10 | 3 | 1 |
| 1990 Apr 5–8 | 65 | 22 | 9 | 4 | * |

SOURCE: "Please tell me if you personally worry about this problem a great deal, a fair amount, only a little, or not at all. Pollution of drinking water?," in *Poll Topics and Trends: Environment*, The Gallup Organization, Princeton, NJ, March 17, 2004 [Online] www.gallup.com [accessed March 30, 2004]

cent. Maintenance of the nation's fresh water supply for household needs ranked fourth with 47 percent. Gallup concluded that Americans are significantly more worried about water issues than other environmental issues, such as air pollution or plant and animal resources.

As shown in Table 6.6, concern about drinking water was down in 2004 compared to all previous years. In 1990 nearly two-thirds of the people asked expressed a great deal of worry about drinking water pollution. The percentage increased to 72 percent in 2000 and then decreased in each subsequent year. Concern about surface water pollution also shows a downward trend over the years. (See Table 6.7.) The percentage of people expressing a great deal of concern about this issue in 1989 was at a high of 72 percent. In 2004 only 48 percent of those asked expressed this opinion.

When asked about fresh water supplies for household needs, Gallup poll respondents have indicated varying levels of concern over the years. Table 6.8 shows that the percentage of people feeling a great deal of concern about this issue increased from 42 percent in 2000 to 50 percent in 2002 and then decreased to 47 percent in 2004.

TABLE 6.7

**Public concern about pollution of rivers, lakes, and reservoirs, 2004**

I'M GOING TO READ YOU A LIST OF ENVIRONMENTAL PROBLEMS. AS I READ EACH ONE, PLEASE TELL ME IF YOU PERSONALLY WORRY ABOUT THIS PROBLEM A GREAT DEAL, A FAIR AMOUNT, ONLY A LITTLE, OR NOT AT ALL. FIRST, HOW MUCH DO YOU PERSONALLY WORRY ABOUT—*[RANDOM ORDER]*?

|  | Great deal % | Fair amount % | Only a little % | Not at all % | No opinion % |
|---|---|---|---|---|---|
| 2004 Mar 8–11 | 48 | 31 | 16 | 5 | * |
| 2003 Mar 3–5 | 51 | 31 | 13 | 5 | -- |
| 2002 Mar 4–7 | 53 | 32 | 12 | 3 | * |
| 2001 Mar 5–7 | 58 | 29 | 10 | 3 | * |
| 2000 Apr 3–9 | 66 | 24 | 8 | 2 | * |
| 1999 Apr 13–14 | 61 | 30 | 7 | 2 | * |
| 1999 Mar 12–14 | 55 | 30 | 12 | 3 | * |
| 1991 Apr 11–14 | 67 | 21 | 8 | 3 | 1 |
| 1990 Apr 5–8 | 64 | 23 | 9 | 4 | -- |
| 1989 May 4–7 | 72 | 19 | 5 | 3 | 1 |

SOURCE: "Please tell me if you personally worry about this problem a great deal, a fair amount, only a little, or not at all. Pollution of rivers, lakes, and reservoirs?," in *Poll Topics and Trends: Environment,* The Gallup Organization, Princeton, NJ, March 17, 2004 [Online] www.gallup.com [accessed March 30, 2004]

TABLE 6.8

**Public concern about maintenance of the nation's supply of fresh water for household needs, 2004**

PLEASE TELL ME IF YOU PERSONALLY WORRY ABOUT THIS PROBLEM A GREAT DEAL, DEAL, A FAIR AMOUNT, ONLY A LITTLE, OR NOT AT ALL. MAINTENANCE OF THE NATION'S SUPPLY OF FRESH WATER FOR HOUSEHOLD NEEDS?

|  | Great deal % | Fair amount % | Only a little % | Not at all % | No opinion % |
|---|---|---|---|---|---|
| 2004 Mar 8–11 | 47 | 25 | 20 | 8 | * |
| 2003 Mar 3–5 | 49 | 28 | 15 | 8 | * |
| 2002 Mar 4–7 | 50 | 28 | 17 | 5 | * |
| 2001 Mar 5–7 | 35 | 34 | 19 | 10 | 2 |
| 2000 Apr 3–9 | 42 | 31 | 14 | 12 | 1 |

SOURCE: "Please tell me if you personally worry about this problem a great deal, a fair amount, only a little, or not at all. Maintenance of the nation's supply of fresh water for household needs?," in *Poll Topics and Trends: Environment*, The Gallup Organization, Princeton, NJ, March 17, 2004 [Online] www.gallup.com [accessed March 30, 2004]

# CHAPTER 7
# ACID RAIN

## WHAT IS ACID RAIN?

Acid rain is the common name for acidic deposits that fall to Earth from the atmosphere. The term was coined in 1872 by English chemist Robert Angus Smith to describe the acidic precipitation in Manchester, England. Today scientists study both wet and dry acidic deposits. Although there are natural sources of acid in the atmosphere, acid rain is primarily caused by emissions of sulfur dioxide ($SO_2$) and nitrous oxide ($N_2O$) from electric utilities burning fossil fuels, especially coal. These chemicals are converted to sulfuric acid and nitric acid in the atmosphere and can be carried by the winds for many miles from where the original emissions took place. (See Figure 7.1.)

Wet deposition occurs when the acid falls in rain, snow, or ice. Dry deposition is caused by very tiny particles (or particulates) in combustion emissions. They may stay dry as they fall or pollute cloud water and precipitation. Moist deposition occurs when the acid is trapped in cloud or fog droplets. This is most common at high altitudes and in coastal areas. Whatever its form, acid rain can create dangerously high levels of acidic impurities in water, soil, and plants.

### Measuring Acid Rain

The acidity of any solution is measured on a potential hydrogen (pH) scale numbered from zero to 14, with a pH value of seven considered neutral. Values higher than seven are considered more alkaline or basic (the pH of baking soda is eight); values lower than seven are considered acidic (the pH of lemon juice is two). The pH scale is a logarithmic measure. This means that every pH change of one is a ten-fold change in acid content. Therefore, a decrease from pH seven to pH six is a ten-fold increase in acidity; a drop from pH seven to pH five is a 100-fold increase in acidity; and a drop from pH seven to pH four is a 1,000-fold increase. (See Figure 7.2.)

Pure, distilled water has a neutral pH of seven. Normal rainfall has a pH value of about 5.6. It is slightly acidic because it accumulates naturally occurring sulfur oxides ($SO_5$) and nitrogen oxides ($NO_5$) as it passes through the atmosphere. Acid rain has a pH of less than 5.6.

Figure 7.3 shows the average rainfall pH measured during 2002 at various field laboratories around the country by the National Atmospheric Deposition Program, a cooperative project between many state and federal government agencies and private entities. Rainfall was most acidic in the Mid-Atlantic states, particularly New York, Pennsylvania, Maryland, Ohio, West Virginia, and in portions of western Virginia, North Carolina, eastern Tennessee, and Kentucky. Unfortunately, the areas with lowest rainfall pH contain some of the country's most sensitive natural resources—the Appalachian Mountains, Adirondack Mountains, Chesapeake Bay, and Great Smoky Mountains National Park.

## SOURCES OF SULFATE AND NITRATE IN THE ATMOSPHERE

### Natural Sources

Natural sources of sulfate in the atmosphere include ocean spray, volcanic emissions, and readily oxidized hydrogen sulfide released from the decomposition of organic matter found in the Earth. Natural sources of nitrogen or nitrates include $NO_5$ produced by microorganisms in soils, by lightning during thunderstorms, and by forest fires. Scientists generally speculate that one-third of the sulfur and nitrogen emissions in the United States comes from these natural sources (this is a rough estimate as there is no way to measure natural emissions as opposed to those that are manmade.)

### Sources Caused by Human Activity

The primary anthropogenic (human-caused) contributors to acid rain are $SO_2$ and $NO_5$, resulting from the burning of fossil fuels, such as coal, oil, and natural gas.

FIGURE 7.1

## Origins of acid rain

A combination of natural and manmade activities result in the deposition of acidic compounds.

SOURCE: "Figure 1. Origins of Acid Rain," in *Progress Report on the EPA Acid Rain Program,* U.S. Environmental Protection Agency, Washington, DC, November 1999

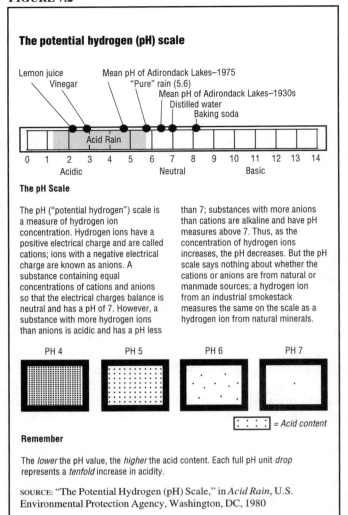

FIGURE 7.2

## The potential hydrogen (pH) scale

### The pH Scale

The pH ("potential hydrogen") scale is a measure of hydrogen ion concentration. Hydrogen ions have a positive electrical charge and are called cations; ions with a negative electrical charge are known as anions. A substance containing equal concentrations of cations and anions so that the electrical charges balance is neutral and has a pH of 7. However, a substance with more hydrogen ions than anions is acidic and has a pH less than 7; substances with more anions than cations are alkaline and have pH measures above 7. Thus, as the concentration of hydrogen ions increases, the pH decreases. But the pH scale says nothing about whether the cations or anions are from natural or manmade sources; a hydrogen ion from an industrial smokestack measures the same on the scale as a hydrogen ion from natural minerals.

### Remember

The *lower* the pH value, the *higher* the acid content. Each full pH unit *drop* represents a *tenfold* increase in acidity.

SOURCE: "The Potential Hydrogen (pH) Scale," in *Acid Rain*, U.S. Environmental Protection Agency, Washington, DC, 1980

Figure 5.15 in Chapter 5 shows the breakdown of U.S. $SO_2$ emissions by source from 1983 to 2002. Fuel combustion by fossil-fueled electric utilities historically has been by far the greatest source of these emissions, accounting for 85 percent of them in 2002. Lesser sources included transportation vehicles and industrial processes.

$NO_5$ emission sources are shown in Figure 5.6 in Chapter 5. Transportation vehicles are the primary source, accounting for 56 percent of the total in 2002. Fuel combustion in power plants is another major source, accounting for 37 percent of the total. Emissions by industry and miscellaneous sources (for example, agriculture) accounted for only 7 percent of the total. Agricultural emissions of nitrogen compounds are due to windblown fertilizers.

Nitrogen pollution of waters has historically been blamed on surface runoff from fertilizer, animal waste, sewage, and industrial waste. Although these are still significant causes, scientists have come to believe that airborne nitrates account for one-fourth of all nitrogen, the second most prevalent cause after fertilizer. Scientists also blame ammonia emissions, which come largely from agricultural activities such as manure handling and fertilizing, for contributing to acid rain. According to the U.S. Geological Survey (USGS), ammonium levels in precipitation increased throughout the 1990s across most of the country. The average increase was 24 percent.

## NATURAL FACTORS THAT AFFECT ACID RAIN DEPOSITION

### Air Movement

Several factors contribute to the impact of acid rain on an area. Transport systems—primarily the movement of air—distribute acid emissions in definite patterns around the planet. The movement of air masses transports emitted pollutants many miles, during which the pollutants are transformed into sulfuric and nitric acid by mixing with clouds of water.

In the United States a typical transport pattern occurs from the Ohio River Valley to the northeastern United States and southeastern Canada, as prevailing winds tend to move from west to east and from south to north. About one-third of the total sulfur compounds deposited over the eastern United States originates from sources in the Midwest more than 300 miles away.

### Climate

In drier climates, such as those of the western United States, windblown alkaline dust moves more freely through

FIGURE 7.3

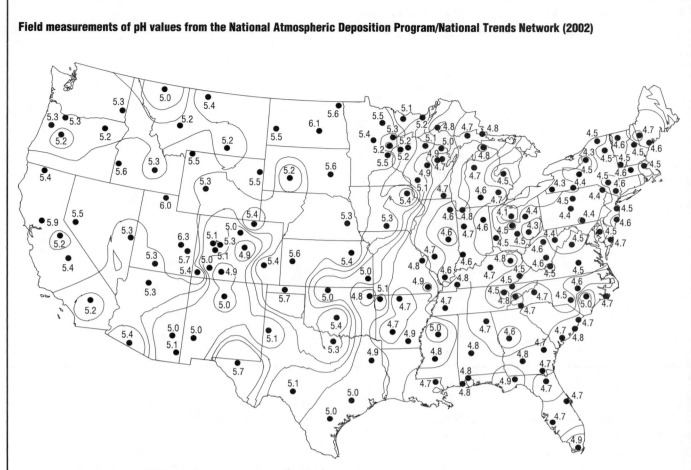

**Field measurements of pH values from the National Atmospheric Deposition Program/National Trends Network (2002)**

SOURCE: "Hydrogen Ion Concentration as pH from Measurements Made at the Field Laboratories, 2002," in *2003 Isopleth Maps,* National Atmospheric Deposition Program, Champaign, IL, 2004 [Online] http://nadp.sws.uiuc.edu/isopleths/maps2002/phfield.pdf [accessed April 13, 2004]

the air and tends to neutralize atmospheric acidity. The effects of acid rain can be greatly reduced by the presence of basic (also called alkali) substances. Sodium, potassium, and calcium are examples of basic chemicals. When a basic and an acid chemical come into contact, they react chemically and neutralize each other. On the other hand, in more humid climates where there is less dust, such as along the eastern seaboard, precipitation is more acidic.

### Topography/Geology

Areas most sensitive to acid rain contain hard, crystalline bedrock and very thin surface soils. When no alkaline-buffering particles are in the soil, runoff from rainfall directly affects surface waters, such as mountain streams. In contrast, a thick soil covering or soil with a high buffering capacity, such as flat land, neutralizes acid rain better. Lakes tend to be most susceptible to acid rain because of low alkaline content in lake beds. A lake's depth, its watershed (the area draining into the lake), and the amount of time the water has been in the lake are also factors.

The lakes and forests in and near the Adirondack Mountains in upstate New York are an example of what occurs in areas that do not have carbonate rock to quickly neutralize acid. Approximately half the lakes above the altitude of 2,000 feet have a pH of less than 5.0. Ninety percent of these lakes contain no aquatic life.

The states bordering and east of the Mississippi River contain approximately 17,000 lakes and 112,000 miles of streams. An estimated 25 percent of the land contains soil and bedrock that allow acidity to travel through underground water to these lakes and streams. Approximately half of these bodies of water have such a limited ability to neutralize acid that acid-laden pollutants will eventually cause acidification.

### EFFECTS OF ACID RAIN ON OUR ENVIRONMENT

In nature, the combination of rain and oxides is part of a natural balance that nourishes plants and aquatic life. However, when the balance is upset, the results to the environment can be harmful and destructive. (See Table 7.1.)

TABLE 7.1

**Effect of acid rain on human health and selected ecosystems and anticipated recovery benefits**

| Human health and ecosystem | Effects | Recovery benefits |
|---|---|---|
| Human health | In the atmosphere, sulfur dioxide and nitrogen oxides become sulfate and nitrate aerosols, which increase morbidity and mortality from lung disorders, such as asthma and bronchitis, and impacts to the cardiovascular system. | Decrease emergency room visits, hospital admissions, and deaths. |
| Surface waters | Acidic surface waters decrease the survivability of animal life in lakes and streams and in the more severe instances eliminate some or all types of fish and other organisms. | Reduce the acidic levels of surface waters and restore animal life to the more severely damaged lakes and streams. |
| Forests | Acid deposition contributes to forest degradation by impairing trees' growth and increasing their susceptibility to winter injury, insect infestation, and drought. It also causes leaching and depletion of natural nutrients in forest soil. | Reduce stress on trees, thereby reducing the effects of winter injury, insect infestation, and drought, and reduce the leaching of soil nutrients, thereby improving overall forest health. |
| Materials | Acid deposition contributes to the corrosion and deterioration of buildings, cultural objects, and cars, which decreases their value and increases costs of correcting and repairing damage. | Reduce the damage to buildings, cultural objects, and cars, and reduce the costs of correcting and repairing future damage. |
| Visibility | In the atmosphere, sulfur dioxide and nitrogen oxides form sulfate and nitrate particles, which impair visibility and affect the enjoyment of national parks and other scenic views. | Extend the distance and increase the clarity at which scenery can be viewed, thus reducing limited and hazy scenes and increasing the enjoyment of national parks and other vistas. |

SOURCE: "Appendix I: Effect of Acid Rain on Human Health and Selected Ecosystems and Anticipated Recovery Benefits," in *Acid Rain: Emissions Trends and Effects in the Eastern United States*, U.S. General Accounting Office, Washington, DC, March 2000

## Aquatic Systems

Although pH levels vary considerably from one body of water to another, a typical pH range for the lakes and rivers in the United States is six to eight.

Low pH levels kill fish eggs, frog eggs, and fish food organisms. The degree of damage depends on several factors, one of which is the buffering capacity of the watershed soil—the higher the alkalinity, the more slowly the lakes and streams acidify. The exposure of fish to acidified freshwater lakes and streams has been intensely studied since the 1970s. Scientists distinguish between sudden shocks and chronic (long-term) exposure to low pH levels.

Sudden, short-term shifts in pH levels result from snowmelts, which release acidic materials accumulated during the winter, or sudden rainstorms that can wash residual acid into streams and lakes. The resulting acid shock can be devastating to fish and their ecosystems. At pH levels below 4.9, damage occurs to fish eggs. At acid levels below 4.5, some species of fish die. Below pH 3.5, most fish die within hours. (See Table 7.2.)

Mountainous streams in New York, North Carolina, Pennsylvania, Tennessee, and Arkansas have shown an acidity during rainstorms and snowmelts of three to 20 times that experienced during the rest of the year. Because many species of fish hatch in the spring, even mild increases in acidity can harm or kill the new life. Temporary increases in acidity also affect insects and other invertebrates, such as snails and crayfish, on which the fish feed.

Gradual decreases of pH levels over time affect fish reproduction and spawning. Moderate levels of acidity in water can confuse a salmon's sense of smell, which it uses to find the stream from which it came. Atlantic salmon are unable to find their home streams and rivers because of acid rain. In addition, excessive acid levels in female fish cause low amounts of calcium, thereby preventing the production of eggs. Even if eggs are produced, their development is often abnormal. Over time the fish population decreases while the remaining fish population becomes older and larger.

Increased acidity can also cause the release of aluminum and manganese particles stored in a lake or river bottom. High concentrations of these metals are toxic to fish.

In 1988 the Environmental Defense Fund (EDF), an environmental watch group, sounded one of the first alarms that the coastal waters of the eastern United States were receiving large inputs of nitrogen. The nitrogen led to an excessive growth of algae on the surface of the water. This in turn resulted in the loss of oxygen and light to the water and the long-term decline of marine life. The EDF concluded that the major sources of the nitrogen were human activities—the runoff of fertilizer, animal waste from farms, and discharge from sewage treatment plants and industrial facilities. Researchers noted that the significant decline of the Chesapeake Bay and other estuaries could also be attributed to the increase in $NO_5$ from automobiles and electric power plants, along with toxic chemicals, pesticides, and wetland destruction.

TABLE 7.2

**Generalized short-term effects of acidity on fish**

| pH range | Effect |
|---|---|
| 6.5–9 | No effect |
| 6.0–6.4 | Unlikely to be harmful except when carbon dioxide levels are very high (>1000 mg l$^{-1}$) |
| 5.0–5.9 | Not especially harmful except when carbon dioxide levels are high (>20 mg l$^{-1}$) or ferric ions are present |
| 4.5–4.9 | Harmful to the eggs of salmon and trout species (salmonids) and to adult fish when levels of $Ca^{2+}$, $Na^+$ and $Cl^-$ are low |
| 4.0–4.4 | Harmful to adult fish of many types which have not been progressively acclimated to low pH |
| 3.5–3.9 | Lethal to salmonids, although acclimated roach can survive for longer |
| 3.0–3.4 | Most fish are killed within hours at these levels |

SOURCE: "Generalized Short-Term Effects of Acidity on Fish," in *National Water Quality Inventory: 1998 Report to Congress,* U.S. Environmental Protection Agency, Washington, DC, June 2000

FIGURE 7.4

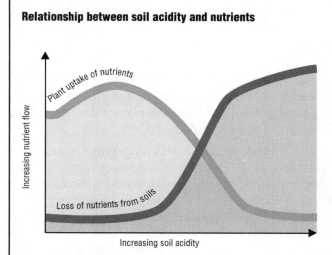

**Relationship between soil acidity and nutrients**

Note: As soil and deposition acidity increases, nutrients, such as base cations, are leached from the soil and are not available for plant growth.

SOURCE: "Figure 18. Relationship between Soil Acidity and Nutrients," in *Progress Report on the EPA Acid Rain Program,* U.S. Environmental Protection Agency, Air and Radiation, Washington, DC, November 1999

During the 1990s acidic and polluted waters caused the disappearance of many aquatic species, leaving gaping holes in the food chain and diminishing the biological balance and diversity that keeps Earth genetically healthy. According to the American Fisheries Society and the Environmental Protection Agency (EPA), many species of freshwater fish have become extinct since the late 1970s, and additional species have become endangered, threatened, or listed as "of special concern" for their ultimate survival.

**Soil and Vegetation**

Acid rain is believed to harm vegetation by changing soil chemistry. Soils exposed to acid rain can gradually lose valuable nutrients, such as calcium, magnesium, and potassium, and become too concentrated with dissolved inorganic aluminum, which is toxic to vegetation. Long-term changes in soil chemistry may have already affected sensitive soils, particularly in forests. Forest soils saturated in nitrogen cannot retain other nutrients required for healthy vegetation. Subsequently, these nutrients are washed away. The EPA reports that nitrogen saturation has already been found in a number of regions, including northeastern forests, the Colorado Front Range, and mountain ranges near Los Angeles, California. The same effects have been reported in Canada and Europe. Nutrient-poor trees are more vulnerable to climatic extremes, pest invasion, and the effects of other air pollutants, such as ozone.

Some researchers believe that acid rain disrupts soil regeneration, which is the recycling of chemical and mineral nutrients through plants and animals back to the Earth. They also believe acids suppress decay of organic matter, a natural process needed to enrich the soils. Valuable nutrients like calcium and magnesium are normally bound to soil particles and are, therefore, protected from being rapidly washed into groundwater. Acid rain, however, may accelerate the process of breaking these bonds to rob the soil of these nutrients. This, in turn, decreases plant uptake of vital nutrients. (See Figure 7.4.)

Acid deposition can cause leafy plants such as lettuce to hold increased amounts of potentially toxic substances like the mineral cadmium. Research has also found a decrease in carbohydrate production in the photosynthesis process of some plants exposed to acid conditions. Research is underway to determine whether acid rain could ultimately lead to a permanent reduction in tree growth, food crop production, and soil quality. Effects on soils, forests, and crops are difficult to measure because of the numerous species of plants and animals, the slow rate at which ecological changes occur, and the complex interrelationships between plants and their environment.

TREES. The effect of acid rain on trees is influenced by many factors. Some trees adapt to environmental stress better than others; the type of tree, its height, and its leaf structure (deciduous or evergreen) influence how well it will adapt to acid rain. Acid rain may affect trees in at least two ways: in areas with high evaporation rates, acids will concentrate on leaf surfaces; in regions where a dense leaf canopy does not exist, more acid may seep into the Earth to affect the soil around the tree's roots.

Scientists believe that acid rain directly harms trees by leaching calcium from their foliage and indirectly harms them by lowering their tolerance to other stresses. Trees are exposed to many natural threats, including drought, ice storms, invasive species, and forest fires. These stresses, combined with increased air and water pollution, can prove too much for sensitive tree species.

A 1994 joint report of the European Commission and the UNECE surveyed 102,300 trees at 26,000 sampling plots in 35 European countries and found that almost one-quarter of the trees in Europe were defoliated by more than 25 percent. The report showed that forest damage is a problem in virtually all European countries. The most severely affected country was the Czech Republic, where 53 percent of all trees had suffered moderate or severe defoliation or died. The least affected was Portugal, where 7.3 percent of trees were damaged.

In 1998 the National Acid Precipitation Assessment Program (NAPAP) identified forest ecosystems in the United States that are most at risk to acid rain damage due to natural sensitivity and high acid deposition rates. (See Figure 7.5.) The EPA blames acid deposition, along with other pollutants and natural stress factors, for increased death and decline of northeastern red spruce at high elevations (for example, in the Adirondacks) and decreased growth of red spruce in the southern Appalachians. Acid rain is also closely linked to the decline of sugar maple trees in Pennsylvania.

In *Soil Calcium Depletion Linked to Acid Rain and Forest Growth in the Eastern United States*, the USGS reported in March 1999 that calcium levels in forest soils had declined at locations in ten states in the eastern United States. Calcium is necessary to neutralize acid rain and is an essential nutrient for tree growth. Sugar maple and red spruce trees, in particular, showed reduced resistance to stresses such as insect defoliation and low winter temperatures. Although the specific relationships among calcium availability, acid rain, and forest growth are uncertain, Gregory Lawrence, a scientist and coauthor of the report, speculated: "Acid rain releases aluminum from the underlying mineral soil layer.... The result is that aluminum replaces calcium, and the trees have a harder time trying to get the needed calcium from the soil layer."

In 2001 the Hubbard Brook Research Foundation reported that more than half of large-canopy red spruce trees in the Adirondack Mountains and the Green Mountains had died since the 1960s. Acid rain was considered the primary cause.

According to the EPA, acid rain has also been implicated in impairing the winter hardening process of some trees, making them more susceptible to cold-weather damage. In some trees, the roots are prone to damage because the movement of acidic rain through the soil releases aluminum ions, which are toxic to plants.

One area in which acid rain has been linked to direct effects on trees is from moist deposition via acidic fogs and clouds. The concentrations of acid and $SO_5$ in fog droplets are much greater than in rainfall. In areas of frequent fog, such as London, significant damage has occurred to trees and other vegetation because the fog condenses directly on the leaves.

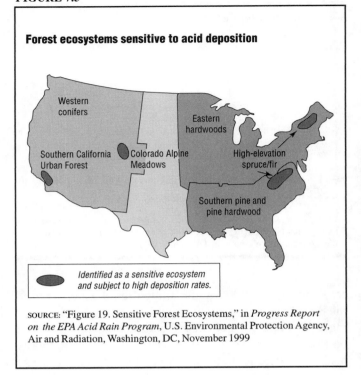

**FIGURE 7.5**

Forest ecosystems sensitive to acid deposition

Western conifers

Eastern hardwoods

Southern California Urban Forest

Colorado Alpine Meadows

High-elevation spruce/fir

Southern pine and pine hardwood

Identified as a sensitive ecosystem and subject to high deposition rates.

SOURCE: "Figure 19. Sensitive Forest Ecosystems," in *Progress Report on the EPA Acid Rain Program*, U.S. Environmental Protection Agency, Air and Radiation, Washington, DC, November 1999

The Forest Health Monitoring Program is a joint effort supported by the U.S. Department of Agriculture (USDA) and private and academic entities. The program monitored precipitation pH and the deposition of $SO_5$, nitrate, ammonium, and total nitrogen in U.S. forests from 1979 to 1995. It estimates that nearly half of forest area in the North and just over 20 percent of forest area in the South are covered by relatively high $SO_5$ deposition. Northern forests were much more exposed to nitrate deposition (40 percent) than were southern forests (less than 1 percent). High ammonium deposition was a problem for more than 62 percent of forests in the North, but less than 20 percent of forests in the South.

**Birds**

Increased freshwater acidity harms some species of migratory birds. Experts believe the dramatic decline of the North American black duck population since the 1950s is due to decreased food supplies in the acidified wetlands. The U.S. Fish and Wildlife Service reports that ducklings in wetlands created by humans in Maryland are three times more likely to die before adulthood if raised in acidic waters.

Acid rain leaches calcium out of the soil and robs snails of the calcium they need to form shells. Because titmice and other species of songbirds get most of their calcium from the shells of snails, the birds are also perishing. The eggs they lay are defective—thin and fragile. The chicks either do not hatch or have bone malformations and die.

In 2002 researchers at Cornell University released the results of a large-scale study showing a clear link between acid rain and widespread population declines in a songbird called the wood thrush. The scientists believe that calcium depletion has had a negative impact on the birds' food source, mainly snails, earthworms, and centipedes. The birds may also be ingesting high levels of metals that are more likely to leach out of overly acidic soils. Declining wood thrush populations were most pronounced in the higher elevations of the Adirondack, Great Smoky, and Appalachian mountains. The researchers warned that acid rain may also be contributing to population declines in other songbird species.

## Materials

Investigations into the effects of acid rain on objects such as stone buildings, marble statues, metals, and paints only began in the 1990s. A joint study conducted by the EPA, the Brookhaven National Laboratory, and the Army Corps of Engineers in 1993 found that acid rain was causing $5 billion worth of damage annually in a 17-state region. Two-thirds of the damage was created by pollution whose source was less than 30 miles away. Many of the country's historical monuments and buildings are located in eastern states that have been most hard-hit by acid rain.

Acid rain is suspected, in part, of damaging the Statue of Liberty and the Egyptian pyramids. Examination of the 700-year-old, 37-foot-tall bronze Great Buddha of Kamakura, an important symbol of Japanese culture, shows pock marks and rust stains, the result of acid rain.

New kinds of protective chemicals that adhere to limestone and marble are helping to save some of the world's decomposing monuments from acid rain and other pollutants. These chemicals, called consolidants, were developed in the 1960s in response to widespread water damage to stone buildings in Venice. Among the monuments getting close attention are the Taj Mahal in India; the Parthenon in Athens, Greece; the Lincoln Memorial in Washington, D.C.; and the Alamo in Texas. Experts report, however, that these chemicals have many limitations. They are toxic and difficult to apply, and their effects are only temporary, yet they permanently alter the nature of the stone. Most important is that their long-term effects are uncertain. For those reasons their use was banned on the Acropolis in Athens, Greece.

AUTOMOTIVE COATINGS. Reports of damage to automotive coverings have been increasing. The general consensus within the automobile industry is that the damage is caused by some form of "environmental fallout"—the term used in the automobile industry. Automakers suspect acid rain damage to automobile paint, especially to many newer models that have clear protective overcoats. Chemical analyses of the damaged areas of some car finishes have showed elevated levels of $SO_5$, implicating acid rain.

The auto industry began using clear-coat finishes in the mid-1980s. Although the new high-gloss paints look better, complaints are mounting over marred surfaces, especially on dark- or metallic-colored cars in the northeastern and southeastern United States. Automakers believe that when acid rain falls on autos the moisture evaporates, leaving a permanent blemish caused by sulfuric acid and nitric acid—the composition of acid rain. Some car dealers now offer optional protective sealants at added expense to buyers. Higher-priced cars often include protective sealants in the purchase price.

## Human Health

Acid rain has several direct and indirect effects on human health. Particulates are extremely small pollutant particles that can threaten human health. Particulates related to acid rain include fine particles of $SO_5$ and nitrates. These particles can travel long distances and, when inhaled, penetrate deep into the lungs. Studies of death rates across the United States, such as that reported in "Lung Cancer, Cardiopulmonary Mortality, and Long-Term Exposure to Fine Particulate Air Pollution" (*Journal of the American Medical Association,* March 6, 2002), have found some correlation between elevated mortality levels and high $SO_5$ levels. Acid rain and the pollutants that cause it can lead to the development of bronchitis and asthma in children. Acid rain is also believed to be responsible for increasing health risks to those over the age of 65; those with asthma, chronic bronchitis, and emphysema; pregnant women; and those with histories of heart disease.

## THE POLITICS OF ACID RAIN

Scientific research on acid rain was sporadic and largely focused on local problems until the late 1960s, when Scandinavian scientists began more systematic studies. Acid precipitation in North America was not identified until 1972, when scientists found that precipitation was acidic in eastern North America, especially in northeastern and eastern Canada. In 1975 the First International Symposium on Acid Precipitation and the Forest Ecosystem convened in Columbus, Ohio, to define the acid rain problem. Scientists used the meeting to propose a precipitation-monitoring network in the United States that would cooperate with the European and Scandinavian networks and to set up protocols for collecting and testing precipitation.

In 1977 the Council on Environmental Quality was asked to develop a national acid rain research program. Several scientists drafted a report that eventually became the basis for NAPAP. This initiative eventually translated into legislative action with the Energy Security Act (PL 96-264) in June 1980. Title VII of the Energy Security Act (the Acid Precipitation Act of 1980) produced a formal proposal that created NAPAP and authorized federally financed support.

The first international treaty aimed at limiting air pollution was the UNECE Convention on Long-Range Transboundary Air Pollution, which went into effect in 1983. It was ratified by 38 of the 54 UNECE members, which included not only European countries but also Canada and the United States. The treaty targeted sulfur emissions, requiring that countries reduce emissions 30 percent from 1980 levels—the so-called "30 percent club."

The early acid rain debate centered almost exclusively on the eastern United States and Canada. The controversy was often defined as a problem of property rights. The highly valued production of electricity in coal-fired utilities in the Ohio River Valley caused acid rain to fall on land in the Northeast and Canada. An important part of the acid rain controversy in the 1980s was the adversarial relationship between U.S. and Canadian government officials over emission controls of $SO_2$ and $NO_2$. More of these pollutants crossed the border into Canada than the reverse. Canadian officials very quickly came to a consensus over the need for more stringent controls, while this consensus was lacking in the United States.

Throughout the 1980s the major lawsuits involving acid rain all came from eastern states, and the states that passed their own acid rain legislation were those in the eastern part of the United States.

Legislative attempts to restrict emissions of pollutants were often defeated after strong lobbying by the coal industry and utility companies. Those industries advocated further research for pollution-control technology rather than placing restrictions on utility company emissions.

## THE ACID RAIN PROGRAM—CLEAN AIR ACT AMENDMENTS, TITLE IV

In 1980 Congress established NAPAP to study the causes and effects of acid deposition. About 2,000 scientists worked with an elaborate multimillion-dollar computer model in an eight-year, $570 million undertaking. In 1988 NAPAP produced an overwhelming 6,000-page report on its findings, including:

- Acid rain had adversely affected aquatic life in about 10 percent of eastern lakes and streams.

- Acid rain had contributed to the decline of red spruce at high elevations by reducing that species' cold tolerance.

- Acid rain had contributed to erosion and corrosion of buildings and materials.

- Acid rain and related pollutants had reduced visibility throughout the Northeast and in parts of the West.

The report concluded, however, that the incidence of serious acidification was more limited than originally feared. The Adirondacks area of New York was the only region showing widespread, significant damage from acid at that time.

Results indicated that electricity-generating power plants were responsible for two-thirds of $SO_2$ emissions and one-third of $NO_5$ emissions. In response, Congress created the Acid Rain Program under Title IV (Acid Deposition Control) of the 1990 Clean Air Act Amendments (PL 101-549).

The goal of the Acid Rain Program is to reduce annual emissions of $SO_2$ and $NO_5$ from electric power plants nationwide. The program set a permanent cap on the total amount of $SO_2$ that could be emitted by these power plants. That cap was set at 8.95 million tons (approximately half the number of tons of $SO_2$ emitted by these plants during 1980). The program also established $NO_5$ emissions limitations for certain coal-fired electric utility plants. The objective of the $NO_5$ program was to achieve and maintain a two-million-ton reduction in $NO_5$ emission levels by the year 2000 compared to the emissions that would have occurred in 2000 if the program had not been implemented.

The reduction was implemented in two phases. Phase 1 began in 1995 and covered 263 units at 110 utility plants in 21 states with the highest levels of emissions. Most of these units were at coal-burning plants in eastern and midwestern states. They were mandated to reduce their annual $SO_2$ emissions by 3.5 million tons. An additional 182 units joined Phase 1 voluntarily, bringing the total of Phase 1 units to 445.

Phase 2 began in 2000. It tightened annual emission limits on the Phase 1 group and set new limits for more than 2,000 cleaner and smaller units in all 48 contiguous states and the District of Columbia.

### A New Flexibility in Meeting Regulations

Traditionally, environmental regulation has been achieved by the "command and control" approach, in which the regulator specifies how to reduce pollution, by what amount, and what technology to use. Title IV, however, gave utilities flexibility in choosing how to achieve these reductions. For example, utilities could reduce emissions by switching to low-sulfur coal, installing pollution-control devices called scrubbers, or shutting down plants.

Utilities took advantage of their flexibility under Title IV to choose less costly ways to reduce emissions, such as switching from high- to low-sulfur coal, and they have been achieving sizable reductions in their $SO_2$ emissions. Fifty-five percent of Phase 1 plants opted to switch to low-sulfur coal, 16 percent chose to install scrubbers, and only 3 percent initially planned to purchase allowances (which allow plants to emit extra $SO_2$). Not surprisingly, the market for low-sulfur coal is growing as a result of Title IV, and the market for high-sulfur coal is decreasing.

FIGURE 7.6

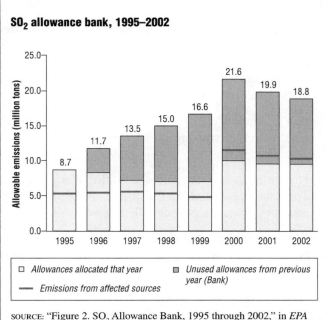

SO₂ allowance bank, 1995–2002

SOURCE: "Figure 2. SO₂ Allowance Bank, 1995 through 2002," in *EPA Acid Rain Program 2002 Progress Report*, U.S. Environmental Protection Agency, Office of Air and Radiation, Clean Air Markets Division, Washington, DC, November 2003

FIGURE 7.7

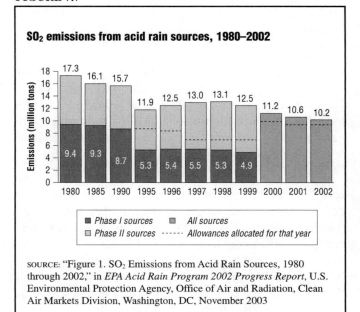

SO₂ emissions from acid rain sources, 1980–2002

SOURCE: "Figure 1. SO₂ Emissions from Acid Rain Sources, 1980 through 2002," in *EPA Acid Rain Program 2002 Progress Report*, U.S. Environmental Protection Agency, Office of Air and Radiation, Clean Air Markets Division, Washington, DC, November 2003

## Allowance Trading

Title IV also allows electric utilities to trade allowances to emit SO₂. Utilities that reduce their emissions below the required levels can sell their extra allowances to other utilities to help them meet their requirements.

Title IV allows companies to buy, sell, trade, and bank pollution rights. Utility units are allocated allowances based on their historic fuel consumption and a specific emissions rate. Each allowance permits a unit to emit one ton of SO₂ during or after a specific year. For each ton of SO₂ discharged in a given year, one allowance is retired and can no longer be used. Companies that pollute less than the set standards will have allowances left over. They can then sell the difference to companies that pollute more than they are allowed, bringing them into compliance with overall standards. Companies that clean up their pollution would recover some of their costs by selling their pollution rights to other companies.

The EPA holds an allowance auction each year. The sale offers allowances at a fixed price. This use of market-based incentives under Title IV is regarded by many as a major new method for controlling pollution.

From 1995 to 1998 there was considerable buying and selling of allowances among utilities. Because the utilities that participated in Phase 1 reduced their sulfur emissions more than the minimum required, they did not use as many allowances as they were allocated for the first four years of the program. Those unused allowances could be used to offset SO₂ emissions in future years. From 1995 to 1998 a total of 30.2 million allowances were allocated to utilities nationwide; almost 8.7 million, or 29 percent, of the allowances were not used, but were carried over (banked) for subsequent years.

Figure 7.6 shows the status of the allowance bank from 1995 through 2002. In 2002 a total of 9.54 million allowances were allocated. Another 9.30 million banked allowances were carried over from previous years. The allowance bank reached a maximum during 2000 and began to decline after that. The EPA expects that the allowance bank will gradually be depleted.

## EMISSIONS AND DEPOSITION

Each year the EPA publishes a report detailing the progress achieved by the Acid Rain Program. The latest report is titled *Acid Rain Program, 2002 Progress Report* and was published in November 2003.

The report notes that in 2002 there were 3,208 electric generating units subject to the SO₂ provisions of the Acid Rain Program. They emitted 10.2 million tons of SO₂ into the air as shown in Figure 7.7. The EPA expects that the 8.95-million-ton annual cap will be achieved by the year 2010. SO₂ emissions from sources covered by the program decreased by 41 percent between 1980 and 2002.

The downward trend in SO₂ emissions was accompanied by a decrease in SO₂ concentrations measured in the air and in sulfate deposition recorded at monitoring sites operated by the National Atmospheric Deposition Program. Between 1990 and 2002 average SO₂ concentrations in the atmosphere decreased by 54 percent. Wet sulfate deposition across the Northeast and Midwest declined by approximately 50 percent.

Between 1990 and 2002 $NO_5$ emissions from power plants subject to the Acid Rain Program decreased from 6.7 million tons per year to 4.5 million tons per year. In 2000 the program achieved its goal of reducing emissions by at least two million tons; 8.1 million tons were originally predicted in 1990 to be emitted in the year 2000 without the program in place.

Decreased $NO_5$ emissions have not resulted in uniformly lower levels of $NO_5$ in the atmosphere or in deposits measured at recording stations. The EPA reports that concentrations of wet nitrates in the atmosphere generally remained constant between 1989 and 2002 across much of the country. In a few areas, concentrations actually increased. Progress for wet nitrate deposition was a little more promising. Large decreases in deposition were reported across the Northeast and the state of Michigan. Unfortunately, most of the Midwest and the mid-Atlantic regions showed little to no significant improvement.

## ARE ECOSYSTEMS RECOVERING?

### Recovery Times

The EPA reports that ecosystems harmed by acid rain deposition can take a long time to fully recover even after harmful emissions cease. The most chronic aquatic problems can take years to be resolved. Forest health is even slower to improve following decreases in emissions, taking decades to recover from damage by acid deposition. Finally, soil nutrient reserves (such as calcium) can take centuries to replenish.

### Recent Studies Show Mixed Results

According to the USGS in *Trends in Precipitation Chemistry in the United States, 1983–1994: An Analysis of the Effects in 1995 of Phase 1 of the Clean Air Act Amendments of 1990, Title IV,* rainwater tested at 109 test sites across the United States was less acidic in 1995 than in 1983, particularly along the Ohio River Valley and in the Mid-Atlantic region. $SO_5$ had declined at 92 percent of the sites. The USGS attributed the improvement to the standards put in place by the Clean Air Act Amendments Title IV program. For nitrates, approximately as many sites showed decreased levels as reported increased levels. Overall, nitrate levels rose slightly, with the largest increases occurring in the western states.

In April 1999 NAPAP released findings from the study *National Acid Precipitation Assessment Program Biennial Report to Congress: An Integrated Assessment.* The study warned that, despite important strides in reducing air pollution, acid rain remained a serious problem in sensitive areas. The report provided additional evidence that acid

rain is more "complex and intractable than was believed 10 years ago." Among the findings were the following:

- New York's Adirondack Mountain waterways suffer from serious levels of acid. Even though sulfur levels are declining, nitrogen levels there are climbing. The agency predicted that by 2040 about half the region's 2,800 lakes and ponds will be too acidic to sustain life.

- The Chesapeake Bay is suffering from excess nitrogen, which is causing algae blooms that suffocate other life forms.

- High elevation forests in Colorado, West Virginia, Tennessee, and southern California are nearly saturated with nitrogen, a key ingredient in acid rain.

- High elevation lakes and streams in the Sierra Nevada, the Cascades, and the Rocky Mountains may be on the verge of "chronically high acidity."

The report concluded that further reductions in sulfur and nitrogen would be needed. The report also found, however, that the 1990 Clean Air Act Amendments have reduced sulfur emissions and acid rain in much of the United States. Some scientists believe that the problems associated with acid rain are theoretically reversible. That is, recovery is possible if a threshold of damage is not passed.

In March 2000 the EPA and the U.S. General Accounting Office (GAO) concluded in *Acid Rain: Emissions Trends and Effects in the Eastern United States* that some surface waters in New England harmed by acid rain were showing signs of recovery. However, ecosystems considered most severely affected—such as the Adirondacks—were not yet showing improvement. The GAO reported that acidified lakes in the Adirondack Mountains were taking longer to recover than lakes elsewhere and might not recover fully or at all without further reductions in acid deposition. Recovery was considered dependent on improving the nearby soil condition.

In early 2001 the EPA released a report on progress made by the United States and Canada on cross-border air pollution. The study, *U.S.-Canada 2000 Air Quality Agreement Progress Report,* is the fifth biennial report related to the 1991 agreement between the two countries. The report says that $SO_5$ deposition was reduced by up to 25 percent between 1995 and 1998 over a large area of the eastern United States. Most of the reduction was in the Northeast, where many sensitive ecosystems are located. $SO_5$ concentrations in lakes and streams decreased all over North America. Declines in nitrate concentrations were much smaller and rarer. Only one region, Vermont/Quebec, showed recovery as evidenced by decreasing acidity or increasing alkalinity.

In 2001 the Hubbard Brook Research Foundation released *Acid Rain Revisited,* which examined progress since 1990. The report concluded that acid rain was still a significant problem in the Northeast, despite declines in sulfur emissions. Researchers urged for tighter controls on emissions of $NO_5$ and ammonia, two problems that have not been well addressed by the Acid Rain Program.

The report noted that many ecosystems in the northeast have reached or passed their tolerance for acid input, making recovery unlikely under the existing emissions reductions scheme. The researchers called for an additional 40 to 80 percent reduction in sulfur emission from electric utilities in addition to what is mandated now. Reductions at the 80 percent level are predicted to allow recovery of acidic streams to nonacidic status in approximately 20 to 25 years.

## A GLOBAL PROBLEM

Because of the transport properties of acid rain, it is not a localized problem. Emissions can originate hundreds of miles from where acid deposition occurs. Canadian authorities estimate that more than 30 percent of the acid rain that falls in Canada is due to U.S. emission sources.

In Europe pollutants are carried from the smokestacks of the United Kingdom over Sweden. In southwestern Germany many trees of the famed Black Forest are dying from the effects of acid rain transported to the region by wind. Germans have coined a word for the phenomenon, *waldsterben* (forest death).

Acid rain is a growing problem in Asia. According to the International Institute for Applied Systems Analysis, $SO_2$ emissions in Asia are surpassing those in Europe and North America. In China acid rain is implicated in large die-offs in southwestern forests. A study by China's National Environment Protection Agency found that farmland is also affected by acid rain, so crops are at risk as well. According to Todd Johnson et al., in *Clear Water, Blue Skies: China's Environment in the New Century,* (Washington, DC: World Bank, 1997), the World Bank estimated annual forest and crop losses at $5 billion. Researchers believe it will be difficult to control pollution from China, the world's biggest consumer of coal, as that nation goes through an accelerated economic expansion that involves increased coal consumption.

Scientists estimate that about one-third of Japan's sulfur deposition comes from China. Atmospheric acidity levels are highest in the winter and early spring. During this time, huge air masses from continental Asia move to Japan, propelled by the prevailing monsoons. As in other coun-

tries, forests in Japan have experienced abnormally high death rates, particularly in stands of red pine and Japanese cedar. Japanese laws governing the emission of gases that acidify rain are among the strictest in the world. Nonetheless, Japan's rain is increasingly acidic. In Kawasaki, where $NO_5$ levels are posted every day outside City Hall, the rain is sometimes as acidic as grapefruit juice.

In 2001 atmospheric scientists at Princeton University said that acid rain in Asia could triple over the next 30 years due to large expected increases in industrial emissions of $NO_5$. Already, nearly 25 percent of China's $NO_5$ emissions return to Earth in acid rain. Chinese emissions are blamed for more than 27 percent of $NO_5$ acid rain in Japan and more than half in North Korea.

## PUBLIC OPINION ABOUT ACID RAIN

Every year the Gallup Organization polls Americans about their attitudes regarding environmental issues. The most recent poll was conducted in March 2004. As shown in Figure 1.8 in Chapter 1, acid rain ranked last among the environmental problems considered during the poll. Only 20 percent of respondents expressed a great deal of worry about acid rain. Table 7.3 shows a dramatic decline in concern about acid rain since the late 1980s. In 1989 Gallup found that 41 percent of respondents felt a great deal of concern about acid rain and 11 percent felt none at all. By 2004 only 20 percent of people polled were concerned a great deal about acid rain and more than a quarter of those asked expressed no concern about the acid rain issue.

**TABLE 7.3**

**Public concern about acid rain, 2004**

PLEASE TELL ME IF YOU PERSONALLY WORRY ABOUT THIS PROBLEM A GREAT DEAL, A FAIR AMOUNT, ONLY A LITTLE, OR NOT AT ALL. ACID RAIN?

| | Great deal % | Fair amount % | Only a little % | Not at all % | No opinion % |
|---|---|---|---|---|---|
| 2004 Mar 8–11 | 20 | 26 | 27 | 26 | 1 |
| 2003 Mar 3–5 | 24 | 26 | 27 | 21 | 2 |
| 2002 Mar 4–7 | 25 | 23 | 31 | 19 | 2 |
| 2001 Mar 5–7 | 28 | 28 | 26 | 16 | 2 |
| 2000 Apr 3–9 | 34 | 31 | 19 | 15 | 1 |
| 1999 Apr 13–14 | 29 | 35 | 23 | 11 | 2 |
| 1991 Apr 11–14 | 34 | 30 | 20 | 14 | 3 |
| 1990 Apr 5–8 | 34 | 30 | 18 | 14 | 4 |
| 1989 May 4–7 | 41 | 27 | 19 | 11 | 3 |

SOURCE: "Please tell me if you personally worry about this problem a great deal, a fair amount, only a little, or not at all. Acid rain?," in *Poll Topics and Trends: Environment*, The Gallup Organization, Princeton, NJ, March 17, 2004 [Online] www.gallup.com [accessed March 30, 2004]

## CHAPTER 8
# TOXINS IN EVERYDAY LIFE

Many of the substances naturally found in the environment or released by modern, industrialized society are harmful to humans and other living creatures. Common toxins include heavy metals (lead, cadmium, aluminum, mercury, and manganese), chlorine, organic chemicals (such as pesticides and herbicides), and radiation.

These substances may be found in the home, workplace, or backyard, in the food and water people eat and drink, and even in medications. Most of the chemicals attacked in Rachel Carson's 1962 book *Silent Spring* (Boston: Houghton Mifflin) are still used in the United States, and some that have been banned have been replaced by substances even more toxic.

## WHY ARE TOXINS TOXIC?

A toxin is a substance—bacterial, viral, chemical, metal, fibrous, or radioactive—that poisons or harms a living organism. A toxin may cause immediate, acute symptoms such as gastroenteritis, or cause harm after long-term exposure such as living in a lead- or radon-contaminated home for many years. Some toxins can have both immediate and long-term effects: Living in an environment with cigarette smoke may trigger an acute asthma attack or, after many years' exposure, it may contribute to lung cancer. Although the effects of a toxin may not show up for years, these effects may, nevertheless, be serious. Toxins are often grouped according to their effect on living creatures. They may be called carcinogens, mutagens, and teratogens:

- A carcinogen is any substance that causes cancerous growth.

- A mutagen is an agent capable of producing genetic change.

- A teratogen is a substance that produces malformations or defective development.

The risks posed by environmental contamination may not be blatantly obvious. For example, people or animals with impaired immune systems and who are exposed to these contaminants may take longer (or even be unable) to recover from infectious diseases. Tracking this problem to environmental pollutants, however, can be difficult.

## CHEMICAL TOXINS

Hundreds of chemicals released into the air, water, and the food chain are harmful to the environment—to humans as well as other living things. Under the Federal Insecticide, Fungicide, and Rodenticide Act (61 Stat 163; amended 1988, PL 100-532), the Environmental Protection Agency (EPA) is charged with reviewing chemical studies and taking appropriate action, including banning dangerous substances. The EPA may regulate the manufacture, importation, processing, distribution, use, and disposal of any chemical that poses a risk to human health or the environment. Regulatory tools used by the EPA range from requiring a substance to bear a warning label, to placing a total ban on production and importation. The EPA reviews more than 2,000 new chemicals each year, and research and testing continually find new substances to add to this list.

Among the most common contaminants identified as harmful are DDT, kelthane, lindane, some synthetic pyrethroids, dioxins, polychlorinated biphenyls (PCBs), furans, some heavy metals such as lead and cadmium, and some plastics. Solvents are common in industrial applications but are also found in such products as furniture polish, bathroom tile cleaners, disinfectants, and shoe polish.

Of the thousands of chemicals currently produced in the United States, many have yet to be tested to see whether they might cause cancer, birth defects, infertility, or abnormal growth in children. Some observers charge that the EPA follow-up on warnings of adverse effects has been slow and, sometimes, has enabled dangerous pesticides to remain on the market.

## Toxic Release Inventory

In 1984 a deadly cloud of chemicals was released from the Union Carbide pesticide plant in Bhopal, India, following an explosion in the plant. The methyl isocyanate gas killed approximately 3,000 people and injured 200,000 others. Shortly after, a similar chemical release occurred in West Virginia, where a cloud of gas sent 135 people to the hospital with eye, throat, and lung irritation complaints. There were no fatalities. Such incidents fueled the demand by workers and the general public for information about hazardous materials in their areas. As a result Congress passed the Emergency Planning and Community Right-to-Know Act of 1986 (EPCRA; PL 99-499).

The EPCRA established, among other things, the Toxics Release Inventory (TRI), a publicly available database (http://www.epa.gov/tri) that contains information on toxic chemical releases by various facilities. More than 650 toxic chemicals are on the TRI list.

Certain manufacturing facilities (which are called "original" industries under the program) have had to file TRI reports since 1987. In 1998 the program was expanded to include certain facilities within a group of industries called the "new" industries. These "new" industries include metal and coal mining, electric utilities burning coal or oil, chemical wholesale distributors, petroleum terminals, bulk storage facilities, Resource Conservation and Recovery Act subtitle C hazardous water treatment and disposal facilities, solvent recovery services, and federal facilities. The requirements only apply to facilities that use certain thresholds of toxic chemicals and that employ ten or more full-time workers.

The *2001 Toxics Release Inventory (TRI) Public Data Release Report* was published in July 2003. Facilities reported a total of 6.16 billion pounds of chemical releases in 2001. On-site releases to air, land, and water accounted for 91 percent of all releases, while off-site releases (when a facility sends toxic chemicals to another facility where they are then released) accounted for the remainder. Most (56.2 percent) releases were to landfills or surface impoundments. Air emissions accounted for just over 27 percent of the total. Releases via underground injection to deep wells and releases to surface water accounted for nearly 4 percent each. (See Figure 5.18 in Chapter 5.)

In 2001 the metal mining industry accounted for the largest amount of total emissions (45 percent), followed by electric utilities (17 percent), and the chemical industry (9 percent). (See Table 8.1.) The 20 chemicals with the largest releases in 2001 are listed in Table 8.2. These 20 chemicals accounted for the vast majority (88 percent) of the total releases. The metal compounds were primarily released via landfills or some other type of land release.

The organic chemicals, acids, and chlorine were mainly released in air emissions, while nitrate compounds were primarily discharged to surface waters.

The TRI report for 2001 focused on a group of chemicals called persistent bioaccumulative toxic (PBT) chemicals. These are toxic chemicals that persist in the environment a relatively long time and accumulate in the tissues of plants, animals, and humans. The TRI describes releases of 454 million pounds of PBT chemicals in 2001, including dioxins, lead, mercury, polycyclic aromatic compounds, PCBs, pesticides, and other complex organic compounds. (See Table 8.3.)

## Pesticides and Herbicides

Pesticides are chemicals used to kill or control insects. Herbicides are chemicals used to kill or control vegetation, particularly weeds. These are a unique group of chemicals because they are specifically formulated to be toxic (to some living things) and are deliberately introduced into the environment. Due to these two facts they are closely regulated. The EPA reviews every pesticide for every particular use.

According to the August 2002 EPA report *Pesticides Industry Sales and Usage: 1998 and 1999 Market Estimates* by David Donalson et al., approximately 70 percent of all pesticides sold in the United States are used for agricultural purposes. Agricultural pesticide use costs $11 billion annually. Every dollar spent on agricultural pesticides is estimated to return about $4 in crops saved. Agricultural dependence on chemicals developed since the 1940s. Prior to that time natural methods, such as crop rotation, mechanical weed control, and other practices, were used to control pests and weeds.

Agricultural use of pesticides grew steadily through the 1960s and 1970s and leveled off at about one million pounds of active ingredient per year. This level has remained relatively constant since the 1980s. Pesticide use in the 1960s centered around organochlorines, including DDT, aldrin, and toxaphene. The environmental dangers and health hazards of these chemicals gradually led to their replacement with other types of pesticides, mainly pyrethroids.

Pesticides have been detected in surface waters, groundwater, and even rainfall. In 2001 the U.S. Geological Survey (USGS) announced that testing performed for the National Water Quality Assessment Program had found at least one pesticide in more than 95 percent of stream samples and in more than 60 percent of shallow agricultural wells. Although the concentrations were generally low and below drinking water standards, their effects on the environment and human health are not known for certain.

The study also found sediments contaminated with persistent organochlorine pesticides, such as DDT, dield-

TABLE 8.1

**Releases of toxic chemicals by industry, 2001**

| SIC code industry | Total facilities Number | Total forms Number | Total air emissions Pounds | Surface water discharges Pounds | Underground injection — Class I wells Pounds | Underground injection — Class II-V wells Pounds | On-site land releases — RCRA Subtitle C landfills Pounds | On-site land releases — Other on-site land releases Pounds | Total on-site releases Pounds | Off-site releases — Transfers off-site to disposal Pounds | Total on- and off-site releases Pounds |
|---|---|---|---|---|---|---|---|---|---|---|---|
| 10 Metal mining | 89 | 658 | 2,854,512 | 427,312 | 0 | 21,629,444 | 0 | 2,757,118,053 | 2,782,029,322 | 524,935 | 2,782,554,257 |
| 12 Coal mining | 88 | 348 | 768,762 | 760,047 | 14,387 | 149,109 | 0 | 14,414,576 | 16,106,880 | 10,830 | 16,117,710 |
| 20 Food | 1,688 | 3,611 | 56,136,054 | 55,154,841 | 56,231 | 677 | 254 | 7,553,308 | 118,901,365 | 6,161,820 | 125,063,185 |
| 21 Tobacco | 31 | 67 | 2,491,746 | 532,888 | 0 | 0 | 0 | 215,914 | 3,240,548 | 322,970 | 3,563,517 |
| 22 Textiles | 289 | 671 | 5,740,423 | 175,447 | 0 | 0 | 0 | 295,953 | 6,211,823 | 749,261 | 6,961,084 |
| 23 Apparel | 16 | 41 | 343,274 | 5 | 0 | 0 | 0 | 85 | 343,364 | 57,862 | 401,226 |
| 24 Lumber | 1,006 | 2,499 | 30,478,706 | 19,931 | 0 | 0 | 4,092 | 394,721 | 30,897,450 | 518,539 | 31,415,989 |
| 25 Furniture | 282 | 632 | 7,823,922 | 557 | 0 | 0 | 0 | 32 | 7,824,511 | 184,300 | 8,008,810 |
| 26 Paper | 50 | 3,128 | 157,150,488 | 16,536,395 | 0 | 0 | 31,330 | 16,168,805 | 189,887,018 | 5,767,146 | 195,654,165 |
| 27 Printing | 231 | 478 | 19,290,560 | 305 | 0 | 0 | 0 | 4,531 | 19,295,396 | 424,007 | 19,719,402 |
| 28 Chemicals | 3,618 | 20,355 | 227,840,792 | 57,577,475 | 167,868,820 | 95,407 | 1,174,731 | 46,748,593 | 501,305,818 | 81,278,999 | 582,584,818 |
| 29 Petroleum | 542 | 4,299 | 48,169,295 | 17,091,882 | 1,973,515 | 56,363 | 41 | 780,800 | 68,071,896 | 3,312,421 | 71,384,317 |
| 30 Plastics | 1,822 | 3,909 | 77,101,109 | 71,092 | 0 | 0 | 72,149 | 841,059 | 78,085,410 | 10,450,023 | 88,535,433 |
| 31 Leather | 60 | 149 | 1,207,544 | 124,727 | 0 | 0 | 0 | 8,596 | 1,340,867 | 1,273,813 | 2,614,681 |
| 32 Stone/clay/glass | 1,027 | 2,800 | 31,265,083 | 162,077 | 0 | 54,431 | 37,180 | 3,863,435 | 35,382,207 | 5,078,467 | 40,460,674 |
| 33 Primary metals | 1,941 | 7,445 | 57,612,307 | 44,670,397 | 842,853 | 5 | 5,736,662 | 177,961,804 | 286,824,028 | 271,764,651 | 558,588,679 |
| 34 Fabricated metals | 2,959 | 8,089 | 40,447,264 | 1,743,317 | 1 | 2,172 | 110,340 | 511,183 | 42,814,277 | 21,204,117 | 64,018,394 |
| 35 Machinery | 1,143 | 2,881 | 8,279,835 | 18,063 | 0 | 0 | 56,873 | 2,394,310 | 10,749,081 | 4,610,570 | 15,359,651 |
| 36 Electrical equip. | 1,831 | 3,883 | 12,721,139 | 2,936,994 | 0 | 0 | 19,017 | 681,803 | 16,358,954 | 7,575,105 | 23,934,059 |
| 37 Transportation equip. | 1,348 | 4,872 | 66,691,884 | 198,256 | 750 | 0 | 45,805 | 727,692 | 67,664,387 | 12,965,556 | 80,629,943 |
| 38 Measure./photo. | 375 | 749 | 7,166,923 | 1,424,454 | 0 | 0 | 301 | 13,009 | 8,604,688 | 770,558 | 9,375,245 |
| 39 Miscellaneous | 312 | 683 | 6,764,766 | 36,566 | 0 | 0 | 14,102 | 1,865 | 6,817,299 | 1,616,458 | 8,433,757 |
| — Multiple codes 20–39 | 1,317 | 4,869 | 66,687,839 | 16,511,697 | 1,723 | 10 | 263,125 | 5,360,924 | 88,825,318 | 15,100,990 | 103,926,308 |
| — No codes 20–39 | 348 | 887 | 3,369,277 | 1,065,165 | 0 | 0 | 3,884 | 8,204,785 | 12,643,111 | 1,120,451 | 13,763,561 |
| 491/493 Electric utilities | 732 | 6,634 | 717,575,860 | 3,519,693 | 0 | 4 | 1,383,707 | 266,699,366 | 989,178,631 | 73,068,649 | 1,062,247,281 |
| 5169 Chemical wholesale distributors | 475 | 3,335 | 1,254,310 | 1,856 | 5 | 0 | 5 | 1,074 | 1,257,250 | 211,020 | 1,468,270 |
| 5171 Petroleum bulk terminals/ bulk storage | 596 | 4,779 | 21,164,969 | 11,177 | 0 | 100 | 26 | 11,215 | 21,187,488 | 153,163 | 21,340,651 |
| 7389/4953 Hazardous waste/solvent recovery | 223 | 2,762 | 974,414 | 23,498 | 22,678,278 | 0 | 129,266,508 | 15,482,910 | 168,425,606 | 51,446,405 | 219,872,011 |
| **Total** | **24,896** | **95,513** | **1,679,373,058** | **220,796,115** | **193,436,563** | **21,987,723** | **138,220,131** | **3,326,460,403** | **5,580,273,993** | **577,723,085** | **6,157,997,078** |

Notes: Off-Site releases include metals and metal category compounds transferred off-site for solidification/stabilization and for wastewater treatment, including to publicly owned treatment works (POTWs). Off-site releases do not include transfers to disposal sent to other Toxics Release Inventory (TRI) facilities that reported the amount as an on-site release. RCRA=Resources Conservation and Recovery Act.

SOURCE: "Table ES-3: TRI On-site and Off-site Releases by Industry, 2001," in *2001 Toxics Release Inventory Public Data Release,* U.S. Environmental Protection Agency, Office of Waste, Washington, DC, 2003

**TABLE 8.2**

**Top 20 chemicals with the largest toxics release inventory (TRI) releases, 2001**

| CAS number | Chemical | Total air emissions Pounds | Surface water discharges Pounds | On-site releases — Underground injection Class I wells Pounds | Class II-V wells Pounds | On-site land releases — RCRA Subtitle C landfills Pounds | Other on-site land releases Pounds | Total on-site releases Pounds | Off-site releases — Transfers off-site to disposal Pounds | Total on- and off-site releases Pounds |
|---|---|---|---|---|---|---|---|---|---|---|
| — | Copper compounds | 1,416,114 | 418,663 | 176,425 | 717,758 | 4,123,867 | 994,136,994 | 1,000,989,821 | 17,096,211 | 1,018,086,032 |
| — | Zinc compounds | 6,159,692 | 2,172,966 | 537,865 | 9,119,066 | 19,080,399 | 719,530,209 | 756,600,197 | 205,468,291 | 962,068,488 |
| 7647-01-0 | Hydrochloric acid | 587,134,079 | 2,445 | 46,398 | 2,172 | 0 | 188,608 | 587,373,702 | 746,442 | 588,120,144 |
| — | Lead compounds | 1,254,381 | 362,122 | 206,138 | 6,026,677 | 5,986,663 | 358,128,545 | 371,964,527 | 50,213,158 | 422,177,685 |
| — | Manganese compounds | 1,877,226 | 6,236,782 | 11,316,486 | 1,165,027 | 3,562,861 | 328,138,941 | 352,297,323 | 47,995,623 | 400,292,945 |
| — | Arsenic compounds | 177,832 | 141,172 | 65,713 | 1,466,028 | 3,130,653 | 373,104,273 | 378,085,671 | 3,000,912 | 381,086,583 |
| — | Nitrate compounds | 515,899 | 192,795,905 | 36,885,869 | 1,643 | 87,500 | 13,419,054 | 243,685,870 | 14,174,527 | 257,860,397 |
| — | Barium compounds | 2,163,853 | 1,475,918 | 15,514 | 1,965,433 | 2,351,397 | 198,668,435 | 206,640,550 | 45,504,419 | 252,144,968 |
| 67-56-1 | Methanol | 175,844,606 | 3,380,299 | 13,377,821 | 80,001 | 97,719 | 1,354,297 | 194,134,743 | 2,829,290 | 196,964,033 |
| 7664-41-7 | Ammonia | 122,057,546 | 6,621,166 | 22,930,936 | 43,301 | 15,065 | 3,605,984 | 155,273,998 | 3,247,048 | 158,521,046 |
| 7664-93-9 | Sulfuric acid | 146,397,844 | 694 | 679,045 | 0 | 66,810 | 481,965 | 147,626,358 | 129,606 | 147,755,964 |
| — | Chromium compounds | 671,106 | 178,352 | 2,188,916 | 51,250 | 2,676,859 | 94,041,777 | 99,808,260 | 16,725,832 | 116,534,092 |
| — | Vanadium compounds | 1,399,953 | 639,856 | 882,575 | 0 | 51,005 | 78,884,807 | 81,858,196 | 6,641,947 | 88,500,143 |
| 108-88-3 | Toluene | 71,539,704 | 75,909 | 264,765 | 365 | 82,903 | 44,754 | 72,008,400 | 1,780,473 | 73,788,873 |
| 7664-39-3 | Hydrogen fluoride | 67,248,474 | 21,549 | 4,400,000 | 0 | 30,407 | 249,168 | 71,949,598 | 136,470 | 72,086,068 |
| — | Nickel compounds | 1,004,693 | 244,846 | 710,080 | 270,609 | 5,600,859 | 45,383,246 | 53,214,334 | 13,753,100 | 66,967,433 |
| 7439-96-5 | Manganese | 904,434 | 165,392 | 0 | 0 | 3,796,840 | 12,535,661 | 17,402,327 | 31,582,035 | 48,984,362 |
| 100-42-5 | Styrene | 46,466,141 | 2,993 | 394,001 | 0 | 106,595 | 58,625 | 47,028,355 | 1,953,117 | 48,981,473 |
| 1330-20-7 | Xylene (mixed isomers) | 47,081,406 | 21,972 | 80,521 | 550 | 133,072 | 44,222 | 47,361,743 | 1,158,921 | 48,520,664 |
| 110-54-3 | n-Hexane | 47,644,345 | 10,531 | 69,663 | 0 | 343 | 9,600 | 47,734,482 | 427,958 | 48,162,440 |
| | **Subtotal (top 20 chemicals)** | **1,328,959,329** | **214,969,533** | **95,208,731** | **20,909,880** | **50,981,817** | **3,222,009,165** | **4,933,038,455** | **464,565,380** | **5,397,603,835** |
| | **Total (all chemicals)** | **1,679,373,058** | **220,796,115** | **193,436,563** | **21,987,723** | **138,220,131** | **3,326,460,403** | **5,580,273,993** | **577,723,085** | **6,157,997,078** |

Notes: Off-site releases include metals and metal category compounds transferred off-site for solidification/stabilization and for wastewater treatment, including to publicly owned treatment works (POTWs). Off-site releases do not include transfers to disposal sent to other Toxics Release Inventory (TRI) facilities that reported the amount as an on-site release. RCRA=Resources Conservation and Recovery Act.

SOURCE: "Table ES-4: Top 20 Chemicals with the Largest Total Releases, 2001," in *2001 Toxics Release Inventory Public Data Release*, U.S. Environmental Protection Agency, Office of Waste, Washington, DC,

# TABLE 8.3

## Toxics release inventory (TRI) releases of persistant bioaccumulative toxic (PBT) chemicals

| CAS number | Chemical | Total forms Number | On-site releases | | | | | | | Off-site releases Transfers off-site to disposal Pounds | Total on- and off-site releases Pounds |
| | | | Total air emissions Pounds | Surface water discharges Pounds | Underground injection | | On-site land releases | | Total on-site releases Pounds | | |
| | | | | | Class I wells Pounds | Class II-V wells Pounds | RCRA subtitle C landfills Pounds | Other on-site land releases Pounds | | | |
|---|---|---|---|---|---|---|---|---|---|---|---|
| | **Dioxin and dioxin-like compounds*** | 1,320 | 6.37 | 4.08 | 0.14 | 0.19 | 21.97 | 94.39 | 127.14 | 200.88 | 328.01 |
| — | Dioxin and dioxin-like compounds (in grams)* | 1,320 | 2,887.566 | 1,850.869 | 63.881 | 84.270 | 9,963.843 | 42,807.558 | 57,657.988 | 91,100.805 | 148,758.793 |
| | **Lead and lead compounds** | 8,561 | 1,633,121.66 | 413,419.80 | 206,138.00 | 6,026,683.34 | 18,610,199.14 | 360,809,675.39 | 387,699,237.33 | 55,292,470.94 | 442,991,708.27 |
| 7439-92-1 | Lead | 4,201 | 378,740.71 | 51,297.37 | 0.00 | 6.65 | 12,623,535.98 | 2,681,130.01 | 15,734,710.71 | 5,079,312.56 | 20,814,023.27 |
| — | Lead compounds | 4,360 | 1,254,380.94 | 362,122.44 | 206,138.00 | 6,026,676.69 | 5,986,663.17 | 358,128,545.38 | 371,964,526.62 | 50,213,158.38 | 422,177,685.00 |
| | **Mercury and mercury compounds** | 1,665 | 150,462.84 | 1,805.15 | 1,741.11 | 8,035.04 | 60,008.84 | 4,455,980.78 | 4,678,033.75 | 228,282.95 | 4,906,316.70 |
| 7439-97-6 | Mercury | 537 | 24,698.08 | 341.73 | 460.40 | 0.00 | 19,861.63 | 11,899.27 | 57,261.10 | 76,715.67 | 133,976.77 |
| — | Mercury compounds | 1,128 | 125,764.77 | 1,463.42 | 1,280.71 | 8,035.04 | 40,147.21 | 4,444,081.51 | 4,620,772.66 | 151,567.27 | 4,772,339.93 |
| | **Polycyclic aromatic compounds** | 3,813 | 1,177,581.28 | 17,069.76 | 2.10 | 332.95 | 97,094.05 | 71,292.51 | 1,363,372.65 | 1,622,784.90 | 2,986,157.55 |
| 191-24-2 | Benzo(g,h,i)perylene | 1,509 | 31,455.26 | 685.17 | 1.00 | 1.65 | 3,716.71 | 4,852.90 | 40,712.69 | 86,240.63 | 126,953.32 |
| — | Polycyclic aromatic compounds | 2,304 | 1,146,126.02 | 16,384.58 | 1.10 | 331.30 | 93,377.34 | 66,439.62 | 1,322,659.96 | 1,536,544.27 | 2,859,204.23 |
| 1336-36-3 | Polychlorinated biphenyls (PCBs) | 137 | 1,359.90 | 2.80 | 0.00 | 0.00 | 2,265,476.30 | 225,685.85 | 2,492,524.85 | 12,251.02 | 2,504,775.86 |
| | **Pesticides** | 130 | 6,559.55 | 282.29 | 115.14 | 0.00 | 15,182.57 | 38,277.70 | 60,417.25 | 50,845.10 | 111,262.34 |
| 309-00-2 | Aldrin | 8 | 0.31 | 0.00 | 0.00 | 0.00 | 0.00 | 0.00 | 0.31 | 1.07 | 1.38 |
| 57-74-9 | Chlordane | 20 | 15.49 | 80.00 | 0.00 | 0.00 | 3,630.30 | 0.00 | 3,725.79 | 331.61 | 4,057.40 |
| 76-44-8 | Heptachlor | 15 | 6.04 | 0.00 | 0.00 | 0.00 | 271.69 | 0.00 | 277.73 | 28.24 | 305.97 |
| 465-73-6 | Isodrin | 5 | 0.35 | 0.00 | 0.00 | 0.00 | 19.00 | 0.00 | 19.35 | 441.40 | 460.75 |
| 72-43-5 | Methoxychlor | 15 | 25.19 | 0.00 | 0.00 | 0.00 | 334.69 | 0.00 | 359.88 | 95.93 | 455.81 |
| 40487-42-1 | Pendimethalin | 18 | 3,573.66 | 195.00 | 0.00 | 0.00 | 185.00 | 28,832.00 | 32,785.66 | 46,702.21 | 79,487.87 |
| 8001-35-2 | Toxaphene | 18 | 42.34 | 6.29 | 0.14 | 0.00 | 3,073.89 | 0.00 | 3,122.66 | 854.53 | 3,977.18 |
| 1582-09-8 | Trifluralin | 31 | 2,896.17 | 1.00 | 115.00 | 0.00 | 7,668.00 | 9,445.70 | 20,125.87 | 2,390.11 | 22,515.98 |
| | **Other PBTs** | 168 | 55,273.40 | 463.43 | 23.48 | 0.02 | 19,006.97 | 203,735.50 | 278,502.80 | 637,304.07 | 915,806.87 |
| 118-74-1 | Hexachlorobenzene | 99 | 1,199.39 | 321.61 | 22.00 | 0.02 | 18,586.97 | 4,937.60 | 25,067.59 | 11,107.40 | 36,174.98 |
| 29082-74-4 | Octachlorostyrene | 4 | 0.00 | 0.12 | 0.00 | 0.00 | 0.00 | 193.00 | 193.12 | 508.60 | 701.72 |
| 608-93-5 | Pentachlorobenzene | 17 | 69.10 | 132.70 | 1.48 | 0.00 | 420.00 | 1,929.90 | 2,553.18 | 206.32 | 2,759.50 |
| 79-94-7 | Tetrabromobisphenol A | 48 | 54,004.91 | 9.00 | 0.00 | 0.00 | 0.00 | 196,675.00 | 250,688.91 | 625,481.75 | 876,170.66 |
| | **Total** | 15,794 | 3,024,365.00 | 433,047.31 | 208,019.97 | 6,035,051.53 | 21,066,989.83 | 365,804,742.12 | 396,572,215.77 | 57,844,139.84 | 454,416,355.61 |

Notes: Off-site releases include metals and metal category compounds transferred off-site for solidification/stabilization and for wastewater treatment, including to publicly owned treatment works (POTWs). Off-site releases do not include transfers to disposal sent to other toxics release inventory (TRI) facilities that reported the amount as an on-site release.

*The chemical category dioxin and dioxin-like compounds is reported in grams. Where the category dioxin and dioxin-like compounds is shown on a table with other TRI chemicals, it is presented in pounds. The grams are converted to pounds by multiplying by 0.002205. RCRA=Resource Conservation and Recovery Act.

SOURCE: "Table ES-5: TRI On-site and Off-site Releases, PBT Chemicals, 2001," in *2001 Toxics Release Inventory Public Data Release*, U.S. Environmental Protection Agency, Office of Waste, Washington, DC, 2003 rin, and chlordane, at more than 20 percent of agricultural sites tested. Use of these pesticides has been restricted for several decades, but many of them are "persistent," which means they do not easily degrade. DDT, in particular, can latch tightly onto soil particles, where it can persist for decades. DDT is also bioaccumulative, which means it can work its way from a nonliving medium, such as dirt, to a plant or animal. Since it is not easily broken down by metabolization within living creatures, DDT moves up the food chain as plants and animals containing DDT are consumed by others.

While pesticides have important uses, studies show that some cause serious health problems at certain levels of exposure. For example, pesticide by-products have been linked to breast cancer in humans. Researchers have found that breast tissue from some women with malignant breast tumors contained more than twice as many PCBs and DDE (a component of the pesticide DDT) than are found in the tissue of women who do not have cancer. Scientists indicate that the carcinogen is stored in body fat, making obesity a risk factor for breast cancer.

In general the level of persistent toxins, including pesticides, has declined in humans and wildlife since the 1970s. Newer pesticide compounds are often more toxic than the older types of pesticides but they are generally designed to be less persistent in the environment and tend to cause fewer chronic problems such as birth defects. However, even pesticides originally believed safe are sometimes found to be harmful. In June 2000 researchers announced that recent tests of the pesticide dursban, a very commonly used chemical in residences, found the substance to be harmful, and many applications were withdrawn from the market. Ironically, dursban was often used as a substitute for chlordane, a chemical also withdrawn from use after being discovered to be harmful.

Fear of agricultural toxins has contributed to a rise in the interest in "organic" foods. The federal government's National Organic Program defines organic agriculture as that which excludes the use of synthetic fertilizers and pesticides. More importantly it strives for low environmental impact and enlists natural biological systems—cover crops, crop rotation, and natural predators—to increase fertility and decrease the likelihood of pest infestation.

AGENT ORANGE. Beginning in 1962 approximately 19 million gallons of herbicides were sprayed over South Vietnam, primarily by aircraft, to defoliate vegetation used as cover for enemy troops in the Vietnam War. In 1969 studies linked chemicals in one of these herbicides, Agent Orange (named after the orange band used to mark the drums it was stored in), to birth defects in laboratory animals. Use of the defoliant was subsequently stopped in 1971. However, Vietnam has estimated that more than one million of its citizens were exposed to the spraying, and tens of thousands of Americans who served in the war are also believed to have been exposed to the chemical, which is a form of dioxin. (In 1994 the EPA affirmed the health danger posed by dioxin.)

Title 38 of the *United States Code* prohibits veterans from suing the government for injuries suffered while in the military. However, many Vietnam veterans have sued the manufacturers of Agent Orange for damages because of health problems experienced since their return from Vietnam. In 1979 a number of claimants filed a class action suit, *In re Agent Orange Product Liability Litigation,* which was settled out of court in 1987 for $180 million. The final funds in the case were distributed in 1992. Additional suits against the manufacturers have been attempted but have been prohibited by the courts. The most strongly fought of these legal battles, *Ivy v. Diamond Shamrock,* was supported by the attorneys general of all 50 states. The Supreme Court, however, refused to hear the arguments and the case ended in 1992. The court decreed that the issue was *res judicata* (the matter is settled).

The Department of Veterans Affairs sponsored a study of Agent Orange and announced its findings in 1996. The veterans' illnesses were grouped into four categories: (1) those that have a positive association with the herbicide; (2) those for which there is suggestive, but not conclusive, evidence of a link; (3) those for which there is insufficient evidence to make a determination; and (4) those for which there is little or no evidence of an association. The report also emphasized the need for further study. The illnesses that the study positively linked to herbicide use were soft-tissue sarcomas (a form of cancer), non-Hodgkin's lymphoma, Hodgkin's disease, and the skin disease chloracne. The second category included respiratory cancers, prostate cancer, multiple myeloma, acute peripheral neuropathy (nerve numbness or weakness), and spina bifida (a congenital deformity of the spine) in children of veterans. The federal government has generally agreed to pay medical claims on those illnesses.

## Endocrine Disrupters—Environmental Hormones

Medical and scientific researchers are increasingly linking chemical compounds known as organochlorines to the endocrine systems of humans and wildlife. The endocrine system—also called the hormone system—is made up of glands located throughout the body, hormones that are synthesized and secreted by the glands into the bloodstream, and receptors in the various target organs and tissues. The receptors recognize and respond to the hormones. The function of the system is to regulate the many bodily processes, including control of blood sugar, growth and function of the reproductive systems, regulation of metabolism, brain and nervous system development, and development of the organism from conception through adulthood and old age.

Substances that interfere with these processes are called "endocrine disrupters." Although some occur naturally—for example, plant-derived hormones—most appear to be man-made. Disruption of the endocrine system can produce certain genetic, reproductive, and behavioral abnormalities in humans and wildlife; increases in several types of cancer not related to smoking or age; malformations; and nervous system disorders.

Endocrine disrupters are sometimes referred to as "environmental estrogens" because they are so widely dispersed in the environment that they turn up in rain water, well water, lakes, and oceans, as well as in foods consumed by birds, fish, animals, and humans. Some scientists worry that organochlorines mimic or block the action of natural estrogen, thereby disrupting the endocrine system.

Some effects of certain estrogenic compounds have been well known for some time. Among these are the eggshell thinning and cracking that led to the population decline of the American bald eagle; the reproductive abnormalities of women exposed *in utero* to diethylstilbestrol, a synthetic estrogen prescribed between 1948 and 1971 to prevent miscarriages; and reported declines in the quantity and quality of sperm in humans. In 1996 researchers at the National Biological Service, in a study of Columbia River otters, found a direct correlation between the level of chemicals and pesticides in the otters' livers and the size of the males' genitalia.

Only recently, however, have researchers begun to realize how many compounds in the environment are estrogenic. More than 50 of these endocrine-disrupting chemicals have been observed to disrupt the hormone or reproductive system, but the remainder of the 85,000 chemicals currently in use remain to be studied. Among them are many herbicides, pesticides, insecticides, and industrial cleaning compounds. Many such compounds have been banned in the United States. Nonetheless, they persist in the food chain for many years and accumulate in animal tissue. Moreover, many of these chemicals continue to be used in developing countries.

Researchers in North America and Europe are studying the possibility of a link between environmental estrogens and the occurrence of birth abnormalities, Alzheimer's disease, sterility in both males and females, hyperactivity in children, neurological illnesses, and many cancers. Both men and women appear to be susceptible to endocrine disruption. In men, some studies show that estrogenic compounds affect the development of the Sertoli cells in the testicles. These cells secrete masculinizing hormones that regulate sperm production, the descent of the testicles, and the development of the urethra.

Because of the potentially serious consequences of human exposure to endocrine-disrupting chemicals, Congress included specific language on endocrine disruption in the Food Quality Protection Act of 1996 (PL 104-170) and Safe Drinking Water Act Amendments of 1996 (PL 104-182). The first mandated the EPA to develop an endocrine-disrupter screening program (EDSP), while the latter authorized the EPA to screen endocrine disrupters found in drinking water. In May 2002 the EPA presented its latest report to Congress on the program's progress. The agency plans to implement the EDSP by setting screening priorities for various chemicals and by establishing uniform laboratory tests for determining which chemicals affect humans similarly to naturally occurring hormones.

## Chlorine

Chlorine is a gaseous element first isolated in 1774 by chemist Wilhelm Scheele. The gas has an irritating odor and, in large concentrations, is dangerous. It was the first substance used as a poisonous gas in World War I. It can be liquefied under pressure and is usually transported as a liquid in steel bottles or tank cars.

The use of chlorine to disinfect water supplies is one of the greatest public health success stories of the twentieth century. First used to purify water in the early 1900s, chlorine is, by far, the world's primary water disinfectant and is indisputably valuable in preventing the spread of disease. It is inexpensive, effective, and available nearly everywhere. It is credited with banishing typhoid fever, cholera, and dysentery from the United States and elsewhere. About three-fourths of all U.S. drinking water is chlorinated to kill parasites, viruses, and bacteria, while most of the rest is treated with a combination of chlorine and ammonia.

Researchers, however, are trying to determine if there is any connection between chlorine in drinking water and bladder and rectal cancer in humans. When chlorine is added to water containing organic matter, it produces by-products (such as chloroform) that are suspected of causing harm to humans and other species. Environmentalists contend that chlorine is responsible for a thinning ozone layer and for causing reproductive malformations in aquatic species and cancer in humans.

Many sources claim that reducing or eliminating chlorine in water would cost many lives. Although the body of evidence seems to show a slight increase in cancer risk from chlorine by-products, most scientists consider the risk not "statistically significant" and believe that the risks to public health would be far greater if chlorine were to be reduced or eliminated.

Although researchers are investigating alternatives to chlorination, none have been found as effective and economical. Ozonation is used by some municipal systems, because ozone is an even more powerful disinfectant than chlorine. Ozonation is widely used in Europe. However, ozone is costly to generate, and the effects of ozonation by-products are largely unknown. Use of ultraviolet radiation has been found effective against bacteria in water, but is less effective against viruses and has no effect on cysts and worms. In addition, alternative disinfectants do not provide the residual protection of chlorine-based treatment; they must be used in combination with chlorine or chlorine derivatives to provide a complete disinfection system.

## Lead

Lead is a naturally occurring metal. Exposure to lead in the United States is relatively widespread, because the metal was commonly used in many industries prior to the 1970s. Exposure to even low levels of lead can cause severe health effects in humans.

SOURCES OF LEAD EXPOSURE. Mined along the eastern seaboard since 1621, lead created an important industry, providing bullets, piping, and a base for paint. Because of its malleability, it was valued as a conduit for water. The use of wallpaper steadily declined with the almost universal use of paint, not only for protecting surfaces, but also for interior decorating. "White lead" paint was sold as the best thing to use on interior and exterior surfaces. In cities teeming with millions of new immigrants, the glossy, durable finish of white lead-based paint meant walls could be easily washed. In 1922 a General Motors researcher discovered that the addition of lead to automobile fuel reduced the "knocking" that limited power and efficiency in car engines.

Many structures are still covered by old lead paint. Nearly three-quarters of all U.S. homes constructed before 1980 contain some lead paint. The EPA reports that lead paint poses little danger if stable. But when renovations are made that involve sanding or stripping paint, the old paint may become hazardous. Certain ceramics and crystal ware, especially those made in foreign countries, still contain unacceptable levels of lead. In 1996 researchers announced that ingredients used to manufacture some miniblinds could be toxic to humans because they contain lead. Lead can be found in many other places as well, including weights used for draperies, wheel balances, or fishing lures; seams in stained-glass windows; linoleum; batteries; solder; gun shot; and plumbing. Test kits and laboratories that test for lead can now check questionable items and locations for the presence of the heavy metal.

Most of the lead in water comes from lead pipes and lead solder in plumbing systems. A 1992 EPA report revealed that 20 percent of the nation's large cities exceeded government limits for lead in drinking water. By 1993 all large public water-supply systems were required to add substances such as lime or calcium carbonate to their water lines to reduce the corrosion of older pipes, which releases lead.

REDUCING LEAD EXPOSURE. As early as the late 1890s medical reports concerning problems with lead began to appear. In 1914 the first U.S. case of lead poisoning was reported, although the cause was undetermined. Scientists eventually began to link lead poisoning to lead paints and, as World War II ended, began to address the problem. In the mid-1960s medical reports documented the connection of lead poisoning to both auto emissions and paint.

In 1971 Congress passed the Lead-Based Poisoning Prevention Act (PL 91-695), restricting residential use of lead paint in structures constructed or funded by the federal government. The phasedown of leaded fuel in automobiles began in the 1970s. This effort was not to safeguard health but to protect cars' catalytic converters, which were rendered inoperable by lead.

In an effort to protect families from exposure to the hazards of lead-based paint, Congress amended the Toxic Substances Control Act (TSCA; PL 94-469, 1976) in 1992 to add Title IV, entitled "Lead Exposure Reduction." Title IV directs the EPA to address the general public's exposure to lead-based paint through regulations, education, and other activities. A particular concern of Congress and the EPA is the potential lead exposure risk associated with housing renovation. The law directs the EPA to publish lead hazard information and make it available to the general public, especially to those undertaking renovations.

Also in 1992 Congress passed the Residential Lead-Based Paint Hazard Reduction Act (PL 102-550) to stop the use of lead-based paint in federal structures and to set up a framework to evaluate and remove paint from buildings nationwide. In 1996 Congress once again amended the TSCA, adding section 402a to establish and fund training programs for lead abatement and to set up requirements and training of technicians and lead-abatement professionals.

BLOOD LEAD LEVELS. Lead is highly toxic, causing harm to the brain, kidneys, bone marrow, and central nervous system. Levels as low as ten micrograms of lead per deciliter of blood can have serious health effects in infants, children, and pregnant women. (See Figure 8.1.)

Lead is a cumulative poison. For people who are exposed to it every day, over time, it begins to accumulate in the body. At very high levels of exposure (now rare in the United States), lead can cause mental retardation, convulsions, and even death. More commonly, exposure occurs at very low levels over an extended period of time.

The Centers for Disease Control and Prevention (CDC) monitors blood lead levels (BLLs) of children and adults. Since the 1970s the concentrations of lead measured in blood samples of children aged five and under have declined dramatically. (See Figure 8.2.) However, the *EPA's Draft Report on the Environment 2003* notes that lead poisoning is still a "serious environmental hazard in young children in the U.S." This is particularly true for

FIGURE 8.1

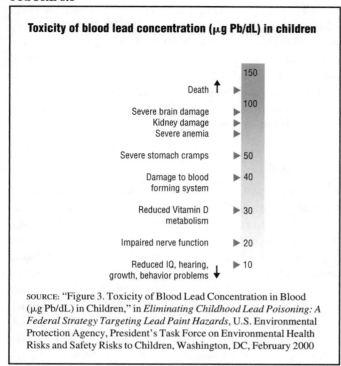

**Toxicity of blood lead concentration (μg Pb/dL) in children**

| | |
|---|---|
| Death ↑ ▶ | 150 |
| | 100 |
| Severe brain damage ▶ | |
| Kidney damage ▶ | |
| Severe anemia ▶ | |
| Severe stomach cramps ▶ | 50 |
| Damage to blood forming system ▶ | 40 |
| Reduced Vitamin D metabolism ▶ | 30 |
| Impaired nerve function ▶ | 20 |
| Reduced IQ, hearing, growth, behavior problems ↓ ▶ | 10 |

SOURCE: "Figure 3. Toxicity of Blood Lead Concentration in Blood (μg Pb/dL) in Children," in *Eliminating Childhood Lead Poisoning: A Federal Strategy Targeting Lead Paint Hazards*, U.S. Environmental Protection Agency, President's Task Force on Environmental Health Risks and Safety Risks to Children, Washington, DC, February 2000

urban areas. The EPA reports that in 2001 slightly more than 10 percent of the children screened for lead in Chicago had elevated blood levels of the metal. The percentage has fallen dramatically since 1996 when it exceeded 25 percent. On a nationwide basis the EPA estimates that approximately 2 to 3 percent of children have elevated BLL.

Many scientists believe that the federal standards for exposure should be lowered, and, in fact, some researchers believe there is no safe level for lead. The CDC believes that effects on the central nervous system of children begin at ten micrograms per deciliter, and the greater the BLL, the higher the risk. The CDC recommends that if many children in a community have BLLs above ten micrograms per deciliter, community-wide lead poisoning prevention activities should begin. The Occupational Safety and Health Administration requires that a worker be removed from a workplace if his or her BLL reaches 50 micrograms per deciliter, although two Harvard School of Public Health studies released in 1996 found kidney damage and hypertension correlated with lead levels in bone below ten micrograms per deciliter.

**DOES LEAD CONTRIBUTE TO DELINQUENCY?** A 1996 University of Pittsburgh Medical Center study of 800 male public school students found that, even after taking into account other predictors of delinquency, such as socioeconomic status, boys with higher lead levels were more likely to engage in antisocial acts. A direct relationship was found between the amount of lead in boys' leg bones and reports from parents, teachers, and the children themselves of criminal or aggressive behavior.

FIGURE 8.2

**Concentrations of lead in blood of children aged 5 and under, selected years 1976–2000**

*[Graph: Blood lead concentrations, micrograms per deciliter (μg/dL), plotted for 1976–1980, 1988–1991, 1992–1994, 1999–2000. The "90th percentile (10 percent of children have this blood lead level or greater)" line begins at approximately 25 in 1976–1980 and declines to about 4.5 in 1999–2000. The "Median value (50 percent of children have this blood lead level or greater)" line begins at approximately 15 in 1976–1980 and declines to under 2 in 1999–2000. A reference marking at 10¹·² is shown on the y-axis.]*

¹10 μg/dL of blood lead has been identified by CDC as elevated, which indicates the need for intervention. (CDC, *Preventing Lead Poisoning in Young Children*, 1991.)
²Recent research suggests that blood levels less than 10 μg/dL may still produce subtle, subclinical health effects in children. (Schmidt, C.W. *Poisoning Young Minds*, 1999.)

SOURCE: "Exhibit 4-8: Concentration of Lead in Blood of Children Age 5 and Under, 1976–1980, 1988–1991, 1992–1994, 1999–2000," in "Chapter 4–Human Health," *EPA's Draft Report on the Environment 2003*, U.S. Environmental Protection Agency, Washington, DC, 2003

## Fertilizers

The United States is the largest producer of fertilizers for domestic use and export, and American farmers are the most productive in the world, producing crops for domestic demand as well as for export to other countries. Agriculture exists in every state, but is concentrated in the Midwest. Although many different crops are farmed in the United States, corn, soybeans, wheat, and hay account for 75 percent of total crops in 2002. (See Figure 8.3.)

The U.S. Department of the Interior defines a fertilizer as any substance applied to soil to enhance its ability to produce plentiful, healthy crops. Fertilizers are natural or manufactured chemicals that contain nutrients known to improve the fertility of soils. Nitrogen, phosphorus, and potassium are the three most important of these nutrients; some scientists also consider sulfur a major nutrient for plant health. The EPA estimates that 90 percent of all fertilizers used contain nitrogen, phosphorus, and to a lesser extent, potassium.

The use of fertilizers sky-rocketed following World War II. Between 1960 and 1980 fertilizer use increased

FIGURE 8.3

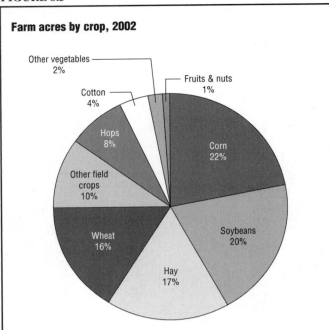

**Farm acres by crop, 2002**

Other vegetables 2%
Fruits & nuts 1%
Cotton 4%
Hops 8%
Other field crops 10%
Corn 22%
Wheat 16%
Soybeans 20%
Hay 17%

SOURCE: Adapted from "Field Crops: Acreage, Yield, Production, Price, and Value" and "Fresh Vegetables: Acreage, Yield, Production, Price, and Value" and "Fruits and Nuts: Noncitrus Fruit Acreage, Utilized Production, Price, and Value" and "Fruits and Nuts: Citrus Fruit Acreage, Utilized Production, Price, and Value" and "Fruits and Nuts: Nut Acreage, Utilized Production, Price, and Value," in *Statistical Highlights of United States Agriculture, 2002/2003,* U.S. Department of Agriculture, National Agricultural Statistics Service, Washington, DC, June 2003

from approximately seven million tons per year to nearly 23 million tons per year. Fertilizer use declined during the 1980s, but then began to increase again, exceeding 20 million tons per year by the year 2000. The USGS estimates that approximately 12 million tons of nitrogen are applied to the land each year from commercial fertilizers. An additional seven million tons of nitrogen are applied as manure. These two sources are blamed for most of the nitrogen entering watersheds across the country.

Overfertilization of crops can cause excess chemicals to leach into surface and groundwater. Runoff of rain and irrigation waters washes nutrients (like nitrogen) into streams, rivers, and lakes. Excessive nutrients are a problem in surface waters because they cause conditions called eutrophication and hypoxia. Eutrophication occurs when nutrients stimulate the rapid growth of algae and aquatic plants. When these plants and algae die, bacteria in the water decompose them. This depletes the amount of dissolved oxygen in the water—a condition called hypoxia. Fish and other aquatic creatures require dissolved oxygen to live and thrive. Hypoxic areas become "dead zones" in the environment.

Nitrate ($NO_3$) is a common form of nitrogen found in water. The EPA has monitored the nitrate load in the country's major rivers since the 1950s. The data show a disturbing trend, particularly in the Midwest, where fertil-

izer use and soil erosion rates are high. The nitrate load in the Mississippi River increased from approximately 250,000 tons per year in the early 1960s to approximately one million tons per year in 1999. The result has been the development of a massive hypoxic area near the mouth of the river in the Gulf of Mexico. Figure 8.4 shows how this area grew in size between 1985 and 2002. The data reflect midsummer measurements, because that is the time of year when hypoxia is at its worst.

## FLUORIDE

Fluoride was first added to drinking water in 1945 to control dental caries and prevent tooth decay. Dental caries is an infectious, communicable disease in which bacteria dissolve the enamel surface of a tooth. Left unchecked, the bacteria may then penetrate the underlying dentin and soft tissue, resulting in tooth loss, discomfort, and even acute infection throughout the body. At the beginning of the twentieth century, dental caries was common in the United States and most developed countries. Failure to meet the minimum standard of having six opposing teeth was a leading cause of rejection from military service in both world wars.

Dental caries declined greatly during the second half of the twentieth century, and many people attribute the decline to fluoridation of the public drinking water supply and the addition of fluoride to toothpastes and mouth washes. By 2000 approximately 57 percent of the U.S. population received drinking water that was fluoridated. (See Figure 8.5.) This is in addition to a small percentage of the population that has access to naturally fluoridated water. Groundwater and surface waters naturally contain about 0.1 to 0.2 parts per million (ppm) of fluoride. The recommended level of fluoride in water to prevent tooth decay is 0.7 to 1.2 ppm.

The widespread use of fluoridated water and dental aids has ensured that virtually everyone in the United States has been exposed to fluoride. The CDC estimates that fluoride has reduced tooth decay in children by as much as 40 to 70 percent and in adults by 40 to 60 percent. Water fluoridation is believed to be especially beneficial in low-income areas where residents often have less access to dental-care services and other sources of fluoride. Consequently, the CDC rates the fluoridation of drinking water as among the top ten greatest public health achievements of the twentieth century.

Critics say that the benefits of fluoride have been grossly overestimated and the hazards largely ignored. They point out that fluoridation of drinking water is not done in much of western Europe, yet those regions have experienced the same declines in dental caries seen in the United States. There have been long-standing concerns about the negative effects of fluoridation. Many communities have rejected flu-

FIGURE 8.4

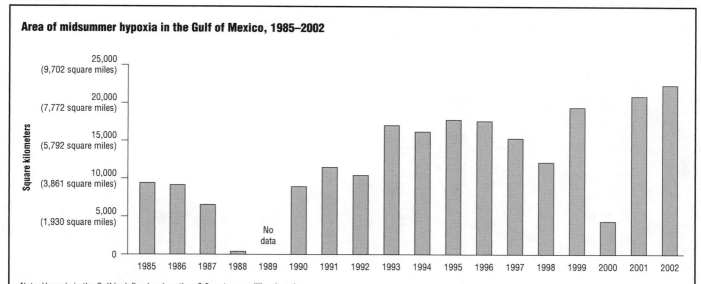

**Area of midsummer hypoxia in the Gulf of Mexico, 1985–2002**

Note: Hypoxia in the Gulf is defined as less than 2.0 parts per million (ppm).
Annual midsummer cruises have been conducted systematically over the past 15 years (with the exception of 1989). Hypoxia in bottom waters covered an average of 8,000–9,000 km² in 1985–92 but increased to 16,000–20,000 km² in 1993–99.

SOURCE: "Exhibit 2-3: Areal Extent of Midsummer Hypoxia in the Gulf of Mexico, 1985–2002," in "Chapter 2–Purer Water," *EPA's Draft Report on the Environment 2003*, U.S. Environmental Protection Agency, Washington, DC, 2003

oridation of their drinking water when the issue was presented for a vote. Worries continue about the possible links between fluoridation and dental fluorosis (mottling of the teeth), skeletal fluorosis (accumulation of excessive fluoride in the bones), kidney disease, birth defects, and cancer. Research as of 2004 has not been conclusive.

## ASBESTOS

Asbestos is the generic name for several fibrous minerals that are found in nature. Very long and thin fibers are bundled together to make asbestos. First used as a coating for candlewicks by the ancient Greeks, asbestos was developed and manufactured in the twentieth century as an excellent thermal and electrical insulator. The physical properties that give asbestos its resistance to heat and decay have long been linked to adverse health effects in humans. Asbestos is found in mostly older homes and buildings, primarily in indoor insulation.

Asbestos tends to break into microscopic fibers. These tiny fibers can remain suspended in the air for long periods of time and can easily penetrate body tissues when inhaled. Because of their durability, these fibers can lodge and remain in the body for many years. No "safe" exposure threshold for asbestos has been established, but the risk of disease generally increases with the length and amount of exposure. Diseases associated with asbestos inhalation include asbestosis (scarring of the lungs), lung and throat cancers, malignant mesothelioma (a tissue cancer in the chest or abdomen), and nonmalignant pleural disease (accumulation of bloody fluid around the lungs).

In May 2003 the CDC released its sixth annual *Work-Related Lung Disease Surveillance Report 2002*. The report notes that 1,265 people died in 2002 from asbestosis. (See Figure 8.6.) This value is up from less than 100 recorded in 1968. In total, 10,914 people have died from asbestosis between 1990 and 1999. The vast majority of the deaths have occurred among white men aged 55 and older. Most were plumbers, pipe fitters, and steamfitters. Construction accounted for, by far, the greatest proportion (24.6 percent) of asbestosis deaths; second was ship/boat building and repairing (6.0 percent). Death from asbestosis usually occurs only after many years of impaired breathing.

Asbestos was one of the first substances regulated under section 112 of the Clean Air Act of 1970 (CAA; PL 91-604) as a hazardous air pollutant. The discovery that asbestos is a strong carcinogen has resulted in the need for its removal or encapsulation (sealing off so that residue cannot escape) from known locations, including schools and public buildings. Many hundreds of millions of dollars have been spent in such cleanups.

Under the CAA, asbestos-containing materials must be removed from demolition and renovation sites without releasing asbestos fibers into the environment. Among other safeguards, workers must wet asbestos insulation before stripping the material from pipes and must seal the asbestos debris in leak-proof containers while still wet to prevent the release of asbestos dust. The laws of most states have specific requirements for asbestos workers.

A number of legal convictions have resulted from improper and illegal asbestos removal. In many cases the

FIGURE 8.5

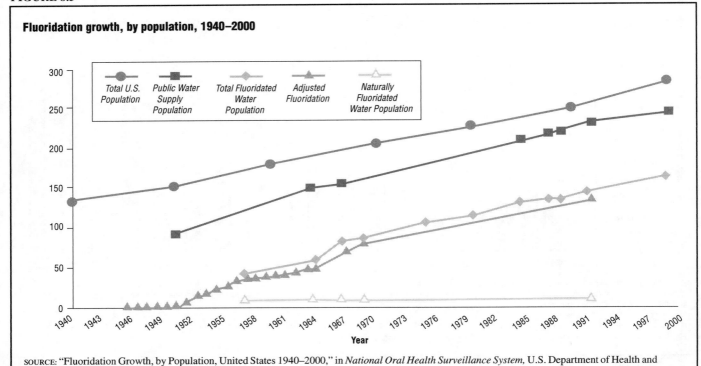

**Fluoridation growth, by population, 1940–2000**

SOURCE: "Fluoridation Growth, by Population, United States 1940–2000," in *National Oral Health Surveillance System,* U.S. Department of Health and Human Services, Centers for Disease Control, Atlanta, GA, March 8, 2002 [Online] http://www.cdc.gov/nohss/FSGrowth.htm [accessed April 12, 2004]

convicted companies had hired homeless people or teenagers to clean up asbestos without advising them that they were dealing with asbestos and without training them in proper handling methods. In response to those cases the Department of Justice and the EPA joined with the National Coalition for the Homeless to issue an advisory to be posted in homeless shelters around the United States. The advisory warns about the dangers of asbestos and cautions workers to be on guard for employers who offer work tearing out old asbestos without providing adequate notice, equipment, and training.

Some observers believe that asbestos poses less risk to humans than previously thought, and suggest that asbestos is less harmful than smoking, drug and alcohol abuse, improper diet, or lack of exercise. They contend that Americans can live safely with asbestos—given careful management procedures—and do not need to spend huge sums of money attempting to remove it completely. In fact, the EPA recommends that asbestos found in good condition be left alone, because "disturbing it may create a health hazard where none existed before."

According to the USGS *Mineral Commodity Summaries, January 2004,* there was no asbestos production in the United States in 2003. The last U.S. asbestos mine closed in 2002. Approximately 6,000 metric tons of asbestos were imported into the country during 2003. Canada supplied 96 percent of the imported asbestos. U.S. demand for asbestos peaked in the 1970s, when it reached 800,000 tons. By 2003 U.S. demand had dropped to 6,600

tons, a level not seen since the 1800s. Most (80 percent) of the asbestos consumed in 2003 was used in roofing products. Gaskets accounted for another 8 percent, followed by friction products with 4 percent.

World production of asbestos in 2003 was around 2 million metric tons, with Russia as the leading producer, followed by China and Canada. World production has dropped since it peaked in 1975 at 5 million metric tons. Growing pressure to ban asbestos around the world is expected to keep pushing markets downward.

## RADIATION

Radiation is energy that travels in waves or particles. Radiation exposure comes from natural and human-made sources. People are exposed to natural radiation from outer space (cosmic radiation), the Earth (terrestrial radiation and radon), and their own bodies. According to the U.S. Nuclear Regulatory Commission, these sources account for about 82 percent of the average person's radiation exposure. (See Figure 8.7.) Radon is, by far, the largest source of radiation exposure, at 55 percent. Human-made sources, mostly medical devices and electromagnetic equipment, account for 18 percent of a person's average radiation exposure.

### Radon

In 1984 a worker in a nuclear plant triggered a radiation contamination alarm as he entered the plant to work. Since the alarms were intended to check for contamina-

## FIGURE 8.6

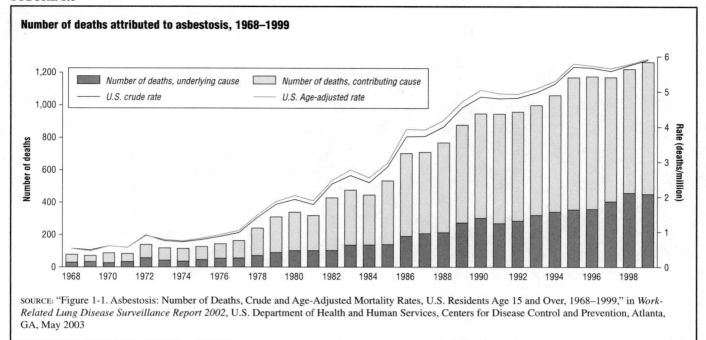

**Number of deaths attributed to asbestosis, 1968–1999**

Legend:
- Number of deaths, underlying cause
- Number of deaths, contributing cause
- U.S. crude rate
- U.S. Age-adjusted rate

SOURCE: "Figure 1-1. Asbestosis: Number of Deaths, Crude and Age-Adjusted Mortality Rates, U.S. Residents Age 15 and Over, 1968–1999," in *Work-Related Lung Disease Surveillance Report 2002*, U.S. Department of Health and Human Services, Centers for Disease Control and Prevention, Atlanta, GA, May 2003

tion as workers left the plant, plant officials were amazed. Investigations discovered the source of the worker's contamination was radon present in extraordinarily high amounts in his home.

Radon is an invisible, odorless radioactive gas formed by the decay of uranium in rocks and soil. This gas seeps from underground rock into the basements and foundations of structures via cracks in foundations, pipes, and sometimes through the water supply. Because it is naturally occurring, it cannot be entirely eliminated from homes.

Although there is no federal regulation addressing radon in indoor air, the EPA recommends that a resident take action if levels reach or exceed four picocuries per liter (pCi/L). The average indoor radon level is 1.25 pCi/L; the average outdoor level is 0.4 pCi/L. The EPA estimates that there are at least six million homes—one in 15 nationwide—with levels greater than four pCi/L, and 100,000 with levels greater than 20 pCi/L. The radon content in most homes can be reduced to 2 pCi/L or less by using devices such as specially designed fans that prevent radon from seeping into a house.

Radon inhaled into the lungs undergoes radioactive decay, releasing particles that damage the DNA in lung tissue. In June 2003 the EPA published its latest estimates of radon-related deaths in *EPA Assessment of Risks from Radon in Homes*. The report estimates that radon causes an estimated 21,000 deaths from lung cancer each year—the second leading cause of lung cancer after smoking. A synergistic effect has been noted: when radon levels are high in a home where a smoker resides, the likelihood of that person contracting lung cancer is greatly increased.

## FIGURE 8.7

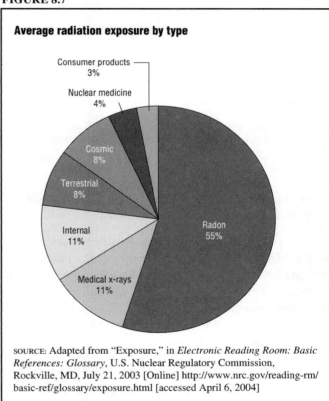

**Average radiation exposure by type**

- Consumer products 3%
- Nuclear medicine 4%
- Cosmic 8%
- Terrestrial 8%
- Internal 11%
- Medical x-rays 11%
- Radon 55%

SOURCE: Adapted from "Exposure," in *Electronic Reading Room: Basic References: Glossary*, U.S. Nuclear Regulatory Commission, Rockville, MD, July 21, 2003 [Online] http://www.nrc.gov/reading-rm/basic-ref/glossary/exposure.html [accessed April 6, 2004]

The EPA has proposed voluntary guidelines calling for builders to install protective measures in hundreds of thousands of new houses across the country to prevent radon from seeping in. The EPA recommends the testing of homes, which is generally quite inexpensive; home test kits are also available to homeowners. Radon contamina-

## FIGURE 8.8

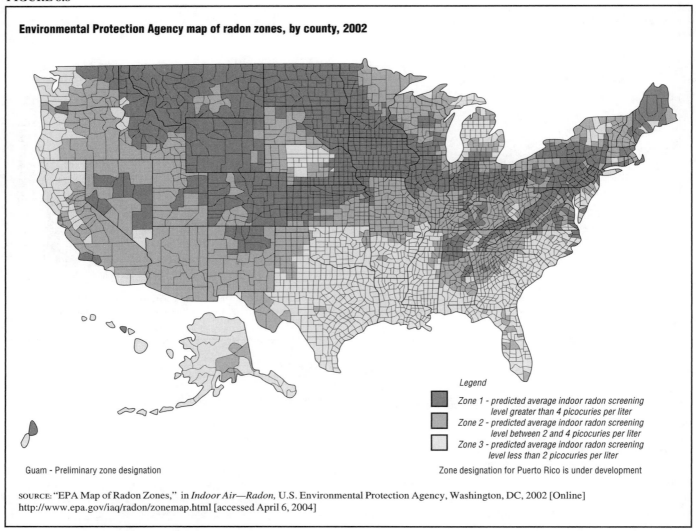

**Environmental Protection Agency map of radon zones, by county, 2002**

Legend

Zone 1 - predicted average indoor radon screening
level greater than 4 picocuries per liter

Zone 2 - predicted average indoor radon screening
level between 2 and 4 picocuries per liter

Zone 3 - predicted average indoor radon screening
level less than 2 picocuries per liter

Guam - Preliminary zone designation

Zone designation for Puerto Rico is under development

SOURCE: "EPA Map of Radon Zones," in *Indoor Air—Radon,* U.S. Environmental Protection Agency, Washington, DC, 2002 [Online] http://www.epa.gov/iaq/radon/zonemap.html [accessed April 6, 2004]

tion is addressed under the Superfund Amendments and Reauthorization Act of 1986 (PL 99-499).

The Indoor Radon Abatement Act of 1988 directed the EPA to identify areas of the country with the potential for elevated levels of indoor radon. The EPA assessed 3,141 counties in terms of geology, aerial radioactivity, soil permeability, foundation type, and indoor radon measurements. Each county was assigned to one of three zones based on its predicted average indoor radon screening level:

- Zone 1—predicted level greater than four pCi/L

- Zone 2—predicted level of two to four pCi/L

- Zone 3—predicted level less than two pCi/L

The EPA map of radon zones is shown in Figure 8.8. The map is not intended to indicate which homeowners should test their homes for radon, but to provide a general guide to state and local organizations dealing with radon abatement. The EPA cautions that homes with elevated levels of radon have been found in all three zones.

**RADON IN DRINKING WATER.** In addition to exposure through the air, humans can also contact radon in drinking water. Radon gas can dissolve and accumulate in groundwater such as that found in wells. The EPA estimates that only about 1 to 2 percent of radon comes from drinking water. However, ingesting radon-contaminated drinking water can lead to the development of internal-organ cancers, primarily stomach cancer—although the risk is smaller than that of developing lung cancer from radon released to the air from tap water. The National Academy of Sciences report *Risk Assessment of Radon in Drinking Water* (1999), a study mandated by the Safe Drinking Water Act Amendments, estimates that radon in drinking water causes about 169 cancer deaths a year; 89 percent of those are caused by breathing radon released into the air from tap water while 11 percent are caused by consuming water that contains radon.

Not all water contains radon. Surface waters—such as rivers, lakes, or reservoirs—usually do not carry radon because the radon tends to evaporate before it has a

FIGURE 8.9

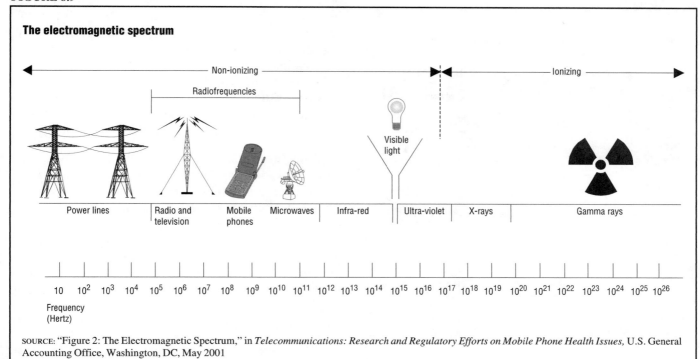

**The electromagnetic spectrum**

SOURCE: "Figure 2: The Electromagnetic Spectrum," in *Telecommunications: Research and Regulatory Efforts on Mobile Phone Health Issues*, U.S. General Accounting Office, Washington, DC, May 2001

chance to reach residences. Underground sources such as groundwater, however, may contain radon. Those who get water from a public water system that serves 25 or more year-round residents may obtain an annual water quality report that tells where the water comes from, what is in that water, and whether or not radon was found.

As of April 2004 there is no federally enforced drinking water standard for radon. The EPA has proposed two alternatives to the states and water agencies. The first is a maximum contaminant level of 300 pCi/L for radon in drinking water. The second alternative is a standard of 4,000 pCi/L if the state or water agency also implements a program to reduce indoor air radon. Neither proposal is popular with drinking water agencies, and the EPA continues to review the radon rule. A final ruling is expected by December 2004.

For homeowners who get their water from private wells—which would not be regulated—radon in the water can be reduced by using a carbon filter or aeration devices that bubble air through the water and carry radon gas out through an exhaust fan.

### Electromagnetic Fields

Earth produces electromagnetic fields (EMFs) naturally, such as during thunderstorms and deep within the planet's molten core. Electricity also occurs naturally in the human body where it can be measured by electroencephalograms of brain wave activity, or electrocardiograms of heart rhythms.

Electric power is a fact of life in America and the developed world. Many consumer and industrial products use some form of electromagnetic energy. While the danger of electric shock is well known, another concern has arisen about electric power—does it cause certain types of cancer, particularly leukemia and cancer of the central nervous system?

One type of electromagnetic energy that is of increasing importance worldwide is radio frequency (RF) energy, which includes radio waves and microwaves. RF energy is used for a wide array of applications including television broadcasting, cellular telephones, pagers, cordless phones, radio communications for police and fire departments, amateur radio, and satellite transmissions. In the United States the Federal Communications Commission (FCC) authorizes or licenses most RF telecommunications activities. Noncommunications uses of RF energy include microwave ovens, radar, and diathermy (the delivery of heat to body tissues for medical or surgical purposes).

The spectrum of electromagnetic waves ranges from extremely low-frequency energy to X-rays and gamma rays, which have very high frequencies. Frequency is the number of waves passing a given point in one second. (See Figure 8.9.) All humans are exposed to EMFs at some time during the course of a day, either from natural sources or sources produced by humans. Human-produced sources include electrical items found in the typical home or office. Most people are exposed to some level of EMF daily and that level can differ at different times of the day. (See Figure 8.10.)

**FIGURE 8.10**

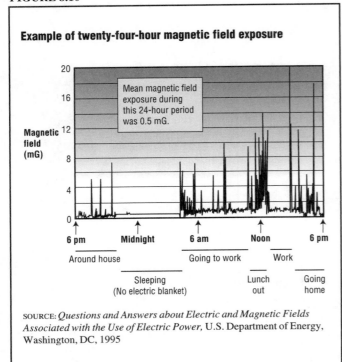

**Example of twenty-four-hour magnetic field exposure**

Mean magnetic field exposure during this 24-hour period was 0.5 mG.

Magnetic field (mG)

20
16
12
8
4
0

6 pm    Midnight    6 am    Noon    6 pm

Around house    Going to work    Work

Sleeping (No electric blanket)    Lunch out    Going home

SOURCE: *Questions and Answers about Electric and Magnetic Fields Associated with the Use of Electric Power,* U.S. Department of Energy, Washington, DC, 1995

Scientists have known for years that exposure to high levels of RF radiation can be harmful because of the ability of RF energy to heat biological tissue rapidly (this is the principle by which microwave ovens work). Two areas of the body, the eyes and the testes, are known to be particularly vulnerable to RF heating. RF radiation, even in low levels, has caused cataracts and sterility in laboratory rabbits.

In 1996 the U.S. Public Health Service reviewed the scientific data on EMFs and determined the following:

> In total, the epidemiological data on both residential and occupational exposures show a moderate risk of cancer, generally between 1.1 and 3.0 times, for adults and children exposed to magnetic fields. This is not extremely large relative to other known risks (for example, the smoking-related risk of approximately 10 or the asbestos-related risk of 5). However, what is unusual about magnetic field exposures is that they are universal. Virtually all of us are exposed.... This means that (a) the observed risk may be underestimated because we cannot identify a truly unexposed comparison group; (b) because of the widespread exposure, even a small risk may result in a large number of individual cancers.

The Public Health Service concluded that "the cost of mitigation [cleanup] already instituted far exceeds the health protection offered, and mitigation of other environmental risks is more important. From a cost-benefit view only limited, low-cost mitigation should be considered."

In its 1999 *Questions and Answers about Biological Effects and Potential Hazards of Radiofrequency Electro-magnetic Fields,* the FCC reported: "Environmental levels of RF energy routinely encountered by the general public are far below levels necessary to produce significant heating and increased body temperature. However, there may be situations, particularly workplace environments near high-powered RF sources, where recommended limits for safe exposure of human beings to RF energy could be exceeded. In such cases, restrictive measures or actions may be necessary to ensure the safe use of RF energy."

Nonetheless, some scientists have continued to suggest that exposure to electric and magnetic fields generated by electric power is responsible for certain cancers (particularly among children), reproductive dysfunction, birth defects, neurological disorders, and Alzheimer's disease. Some activist groups allege the hazards to be so great that they have called for the closure of schools and other public facilities near power lines, and restructuring of the entire power delivery system. EMFs are even cited as causing decreases in property values. Some utilities, with equally strong beliefs, claim there is no proof of risk.

In response to concern about the risks of EMFs, in 1992, under the Energy Policy Act of 1992 (PL 102-486), Congress authorized the Electric and Magnetic Fields Research and Public Information Dissemination Program. The program mandated that the National Institute of Environmental Health Sciences (NIEHS), the National Institutes of Health (NIH), and the U.S. Department of Energy (DOE) research any possible link between EMFs and health. In 1998 an international panel of 30 scientists met to consider the evidence. In 1999 the panel issued its findings in *Health Effects from Exposure to Power-Line Frequency Electric and Magnetic Fields.*

The report found that the evidence was lacking to prove EMFs were a "known human carcinogen" or "probable human carcinogen," and thus EMFs did not warrant aggressive regulatory action. However, a majority of the members concluded that exposure to power-line frequencies is a "possible human carcinogen." The NIEHS concluded that EMF exposure cannot be recognized as entirely safe because of weak scientific evidence that exposure may pose a leukemia hazard. However, because virtually everyone in the United States uses electricity and is thus exposed, the panel recommended research into ways to reduce exposure. The researchers found no indication of increased cancer incidence in experimental animal studies. The strongest evidence of any connection between EMF exposure and cancer came from observations of human populations, although the scientists noted that some other factor could explain the findings, especially since the correlation was weak, at best.

In 1998 the FCC studied EMF emission levels of cellular phones, vehicle-mounted antennas, and fixed transmitting antennas used for such devices. In *Information on*

*Human Exposure to Radiofrequency Fields from Cellular and PCS Radio Transmitters,* the FCC reported that a human would have to remain within a few feet of a main transmitting beam for extremely long periods of time to be exposed to levels of RF energy in excess of recognized safety levels. In addition, the study noted that, regarding vehicle antennas, vehicle occupants are effectively shielded by the metal auto body, especially when the antennas are mounted in the center of the roof or trunk. The study yielded similar results for handheld cellular devices—exposures vary greatly depending on use. Nevertheless researchers continue a number of programs to study EMFs.

As reported in "Exposure to Power-Frequency Magnetic Fields and the Risk of Childhood Cancer," (*Journal of the American Medical Association,* February 2000) the American Medical Association (AMA), after conducting a study of the possible link between EMFs and childhood leukemia, found no evidence that exposure to EMFs increased the risk for childhood cancer.

In June 2002 researchers at Finland's Radiation and Nuclear Safety Authority announced that exposure to cellular phone radiation had been shown to increase activity in human cell proteins grown in the laboratory. More research was recommended to determine the seriousness of these changes and their possible health effects, particularly in the brain.

The huge growth in cellular phone use has brought renewed attention to the potential dangers of EMF exposure, particularly since the phone is held so close to the brain. More than 140 million Americans were cellular phone subscribers in 2003, compared to only 16 million in 1994. Worldwide subscribership is expected to reach two billion by 2006. In May 2001 the U.S. General Accounting Office (GAO) recommended that the FCC improve cellular phone testing and urged both the FCC and the Food and Drug Administration (FDA) to improve consumer information regarding EMF exposure and related health issues.

The FCC requires wireless phones sold in the United States to demonstrate compliance with human exposure limits for radiofrequency energy. The value used to describe the relative amount of this energy absorbed by a person's head while using a wireless phone is called the Specific Absorption Rate (SAR). The FCC safety limit for wireless phones is 1.6 watts per kilogram. Consumers can look up the SAR for a particular phone model using an FCC identification number printed on the case of the phone. Instructions for using the SAR database are available at http://www.fcc.gov/oet/rfsafety/sar.html.

## TOBACCO

*Tobacco use remains the leading preventable cause of death in the United States, causing more than 440,000 deaths each year and resulting in an annual cost of more than $75 billion in direct medical costs. Nationally, smoking results in more than 5.6 million years of potential life lost each year.*

— Centers for Disease Control and Prevention, Atlanta, GA [Online] http://www.cdc.gov/tobacco/issue.htm [accessed May 7, 2004]

The federal government has been warning people about the dangers of smoking for decades. In 1964 the first *Surgeon General's Report on Smoking and Health* was issued. Tobacco use, which had skyrocketed since 1900, dropped dramatically.

The latest smoking statistics by the CDC were published in *Health, United States, 2003.* In 2001 approximately 46.2 million American adults smoked cigarettes, representing 22.7 percent of the adult population (18 years of age and older). (See Table 8.4.) This percentage has declined significantly since 1965 when 41.9 percent of the adult population smoked. In 2001 slightly more adult males (24.7 percent) than females (20.8 percent) were smokers. African-American males aged 45-64 years (34.3 percent) and white males aged 18–24 years (32.5 percent) were most likely to smoke. Approximately one-third of both groups were current smokers in 2001. Education level and smoking are inversely related in adults; people with limited education are much more likely to smoke than are those with advanced degrees.

Table 8.5 shows the prevalence of smoking among high school seniors and eighth- and tenth-graders for various years between 1980 and 2002. In 2002 more than one quarter of high school seniors were smokers, as were nearly 18 percent of tenth-graders and nearly 11 percent of eighth-graders. These values have all declined in recent years.

In April 2002 the CDC issued its latest report on mortality associated with cigarette smoking. Based on data for 1995–99, the report states that smoking kills more than 440,000 people annually. Smoking is blamed for the deaths of 264,087 men and 178,311 women each year from 1995 to 1999. Smoking resulted in the average loss of 13.2 years of life for men and 14.5 years of life for women. Smoking during pregnancy is blamed for 1,000 infant deaths annually. The CDC noted that each pack of cigarettes sold in the United States costs the country $7.18 in medical care costs.

The CDC summarized its findings on the link between cigarette smoking and disease in the 2003 publication of "Cigarette Smoking—Attributable Morbidity—United States, 2000." The report estimates that 8.6 million people in the United States suffered from serious illnesses in 2000 due to smoking. Chronic bronchitis and emphysema accounted for 59 percent of all smoking-attributable diseases. The data are presented in Table 8.6.

In addition to nicotine, which is very addictive, cigarette smoke contains hundreds of mutagens, carcinogens,

**TABLE 8.4**

## Cigarette smoking among adults, 1965–2001

[Data are based on household interviews of a sample of the civilian noninstitutionalized population]

| Sex, race, and age | 1965 | 1974 | 1979 | 1983 | 1985 | 1990 | 1995 | 1997[1] | 1998[1] | 1999[1] | 2000[1] | 2001[1] |
|---|---|---|---|---|---|---|---|---|---|---|---|---|
| **18 years and over, age adjusted[2]** | | | | | Percent of persons who are current cigarette smokers[3] | | | | | | | |
| All persons | 41.9 | 37.0 | 33.3 | 31.9 | 29.9 | 25.3 | 24.6 | 24.6 | 24.0 | 23.3 | 23.1 | 22.7 |
| Male | 51.2 | 42.8 | 37.0 | 34.8 | 32.2 | 28.0 | 26.5 | 27.1 | 25.9 | 25.2 | 25.2 | 24.7 |
| Female | 33.7 | 32.2 | 30.1 | 29.4 | 27.9 | 22.9 | 22.7 | 22.2 | 22.1 | 21.6 | 21.1 | 20.8 |
| White male[4] | 50.4 | 41.7 | 36.4 | 34.2 | 31.3 | 27.6 | 26.2 | 26.8 | 26.0 | 25.0 | 25.5 | 24.9 |
| Black or African American male[4] | 58.8 | 53.6 | 43.9 | 41.7 | 40.2 | 32.8 | 29.4 | 32.4 | 29.0 | 28.4 | 25.7 | 27.6 |
| White female[4] | 33.9 | 32.0 | 30.3 | 29.6 | 27.9 | 23.5 | 23.4 | 22.8 | 23.0 | 22.5 | 22.0 | 22.1 |
| Black or African American female[4] | 31.8 | 35.6 | 30.5 | 31.3 | 30.9 | 20.8 | 23.5 | 22.5 | 21.1 | 20.5 | 20.7 | 17.9 |
| **18 years and over, crude** | | | | | | | | | | | | |
| All persons | 42.4 | 37.1 | 33.5 | 32.1 | 30.1 | 25.5 | 24.7 | 24.7 | 24.1 | 23.5 | 23.3 | 22.8 |
| Male | 51.9 | 43.1 | 37.5 | 35.1 | 32.6 | 28.4 | 27.0 | 27.6 | 26.4 | 25.7 | 25.7 | 25.2 |
| Female | 33.9 | 32.1 | 29.9 | 29.5 | 27.9 | 22.8 | 22.6 | 22.1 | 22.0 | 21.5 | 21.0 | 20.7 |
| White male[4] | 51.1 | 41.9 | 36.8 | 34.5 | 31.7 | 28.0 | 26.6 | 27.2 | 26.3 | 25.3 | 25.8 | 25.1 |
| Black or African American male[4] | 60.4 | 54.3 | 44.1 | 40.6 | 39.9 | 32.5 | 28.5 | 32.2 | 29.0 | 28.6 | 26.1 | 27.6 |
| White female[4] | 34.0 | 31.7 | 30.1 | 29.4 | 27.7 | 23.4 | 23.1 | 22.5 | 22.6 | 22.1 | 21.6 | 21.7 |
| Black or African American female[4] | 33.7 | 36.4 | 31.1 | 32.2 | 31.0 | 21.2 | 23.5 | 22.5 | 21.1 | 20.6 | 20.8 | 18.0 |
| **All males** | | | | | | | | | | | | |
| 18–24 years | 54.1 | 42.1 | 35.0 | 32.9 | 28.0 | 26.6 | 27.8 | 31.7 | 31.3 | 29.5 | 28.5 | 30.4 |
| 25–34 years | 60.7 | 50.5 | 43.9 | 38.8 | 38.2 | 31.6 | 29.5 | 30.3 | 28.5 | 29.1 | 29.0 | 27.2 |
| 35–44 years | 58.2 | 51.0 | 41.8 | 41.0 | 37.6 | 34.5 | 31.5 | 32.1 | 30.2 | 30.0 | 30.2 | 27.4 |
| 45–64 years | 51.9 | 42.6 | 39.3 | 35.9 | 33.4 | 29.3 | 27.1 | 27.6 | 27.7 | 25.8 | 26.4 | 26.4 |
| 65 years and over | 28.5 | 24.8 | 20.9 | 22.0 | 19.6 | 14.6 | 14.9 | 12.8 | 10.4 | 10.5 | 10.2 | 11.5 |
| **White male[4]** | | | | | | | | | | | | |
| 18–24 years | 53.0 | 40.8 | 34.3 | 32.5 | 28.4 | 27.4 | 28.4 | 34.0 | 34.1 | 30.5 | 30.9 | 32.5 |
| 25–34 years | 60.1 | 49.5 | 43.6 | 38.6 | 37.3 | 31.6 | 29.9 | 30.4 | 29.2 | 30.8 | 29.9 | 29.0 |
| 35–44 years | 57.3 | 50.1 | 41.3 | 40.8 | 36.6 | 33.5 | 31.2 | 32.1 | 29.6 | 29.5 | 30.6 | 27.8 |
| 45–64 years | 51.3 | 41.2 | 38.3 | 35.0 | 32.1 | 28.7 | 26.3 | 26.5 | 27.0 | 24.5 | 25.8 | 25.1 |
| 65 years and over | 27.7 | 24.3 | 20.5 | 20.6 | 18.9 | 13.7 | 14.1 | 11.5 | 10.0 | 10.0 | 9.8 | 10.7 |
| **Black or African American male[4]** | | | | | | | | | | | | |
| 18–24 years | 62.8 | 54.9 | 40.2 | 34.2 | 27.2 | 21.3 | *14.6 | 23.5 | 19.7 | 23.6 | 20.8 | 21.6 |
| 25–34 years | 68.4 | 58.5 | 47.5 | 39.9 | 45.6 | 33.8 | 25.1 | 31.6 | 25.2 | 22.7 | 23.3 | 23.8 |
| 35–44 years | 67.3 | 61.5 | 48.6 | 45.5 | 45.0 | 42.0 | 36.3 | 33.9 | 36.1 | 34.8 | 30.8 | 29.9 |
| 45–64 years | 57.9 | 57.8 | 50.0 | 44.8 | 46.1 | 36.7 | 33.9 | 39.4 | 37.3 | 35.7 | 32.2 | 34.3 |
| 65 years and over | 36.4 | 29.7 | 26.2 | 38.9 | 27.7 | 21.5 | 28.5 | 26.0 | 16.3 | 17.3 | 14.2 | 21.1 |
| **All females** | | | | | | | | | | | | |
| 18–24 years | 38.1 | 34.1 | 33.8 | 35.5 | 30.4 | 22.5 | 21.8 | 25.7 | 24.5 | 26.3 | 25.1 | 23.4 |
| 25–34 years | 43.7 | 38.8 | 33.7 | 32.6 | 32.0 | 28.2 | 26.4 | 24.8 | 24.6 | 23.5 | 22.5 | 23.0 |
| 35–44 years | 43.7 | 39.8 | 37.0 | 33.8 | 31.5 | 24.8 | 27.1 | 27.2 | 26.4 | 26.5 | 26.2 | 25.7 |
| 45–64 years | 32.0 | 33.4 | 30.7 | 31.0 | 29.9 | 24.8 | 24.0 | 21.5 | 22.5 | 21.0 | 21.6 | 21.4 |
| 65 years and over | 9.6 | 12.0 | 13.2 | 13.1 | 13.5 | 11.5 | 11.5 | 11.5 | 11.2 | 10.7 | 9.3 | 9.2 |
| **White female[4]** | | | | | | | | | | | | |
| 18–24 years | 38.4 | 34.0 | 34.5 | 36.5 | 31.8 | 25.4 | 24.9 | 29.4 | 28.1 | 29.6 | 28.7 | 27.2 |
| 25–34 years | 43.4 | 38.6 | 34.1 | 32.2 | 32.0 | 28.5 | 27.3 | 26.1 | 26.9 | 25.5 | 25.1 | 25.5 |
| 35–44 years | 43.9 | 39.3 | 37.2 | 34.8 | 31.0 | 25.0 | 27.0 | 27.5 | 26.6 | 26.9 | 26.6 | 27.0 |
| 45–64 years | 32.7 | 33.0 | 30.6 | 30.6 | 29.7 | 25.4 | 24.3 | 20.9 | 22.5 | 21.2 | 21.4 | 21.6 |
| 65 years and over | 9.8 | 12.3 | 13.8 | 13.2 | 13.3 | 11.5 | 11.7 | 11.7 | 11.2 | 10.5 | 9.1 | 9.4 |
| **Black or African American female[4]** | | | | | | | | | | | | |
| 18–24 years | 37.1 | 35.6 | 31.8 | 32.0 | 23.7 | 10.0 | *8.8 | 11.5 | *8.1 | 14.8 | 14.2 | 10.0 |
| 25–34 years | 47.8 | 42.2 | 35.2 | 38.0 | 36.2 | 29.1 | 26.7 | 22.5 | 21.5 | 18.2 | 15.5 | 16.8 |
| 35–44 years | 42.8 | 46.4 | 37.7 | 32.7 | 40.2 | 25.5 | 31.9 | 30.1 | 30.0 | 28.8 | 30.2 | 24.0 |
| 45–64 years | 25.7 | 38.9 | 34.2 | 36.3 | 33.4 | 22.6 | 27.5 | 28.4 | 25.4 | 22.3 | 25.6 | 22.6 |
| 65 years and over | 7.1 | *8.9 | *8.5 | *13.1 | 14.5 | 11.1 | 13.3 | 10.7 | 11.5 | 13.5 | 10.2 | 9.3 |

*Estimates are considered unreliable.

[1]Data starting in 1997 are not strictly comparable with data for earlier years due to the 1997 questionnaire redesign. Cigarette smoking data were not collected in 1996.

[2]Estimates are age adjusted to the year 2000 standard population using five age groups: 18–24 years, 25–34 years, 35–44 years, 45–64 years, 65 years and over.

[3]Beginning in 1993 current cigarette smokers reported ever smoking 100 cigarettes in their lifetime and smoking now on every day or some days.

[4]The race groups, white and black, include persons of Hispanic and non-Hispanic origin.

SOURCE: "Table 59. Current Cigarette Smoking by Persons 18 Years of Age and Over According to Sex, Race, and Age: United States, Selected Years 1965–2001," in *Health, United States, 2003*, U.S. Department of Health and Human Services, Centers for Disease Control and Prevention, National Center for Health Statistics, Hyattsville, MD, 2003

and some 4,000 other chemical compounds including carbon monoxide and radioactive polonium. These chemicals not only enter the lungs, but also the bloodstream where they circulate into internal organs. Smoking can be responsible for, or contribute to, asthma; heart disease; cancer of the lungs, esophagus, mouth, bladder, pancreas,

TABLE 8.5

## Cigarette smoking among teens, 1980–2002

[Data are based on a survey of high school seniors and eighth-graders in the coterminous United States]

| Substance, sex, race, and grade in school | 1980 | 1990 | 1991 | 1995 | 1998 | 1999 | 2000 | 2001 | 2002 |
|---|---|---|---|---|---|---|---|---|---|
| **Cigarettes** | | | | Percent using substance in the past month | | | | | |
| All seniors | 30.5 | 29.4 | 28.3 | 33.5 | 35.1 | 34.6 | 31.4 | 29.5 | 26.7 |
| Male | 26.8 | 29.1 | 29.0 | 34.5 | 36.3 | 35.4 | 32.8 | 29.7 | 27.4 |
| Female | 33.4 | 29.2 | 27.5 | 32.0 | 33.3 | 33.5 | 29.7 | 28.7 | 25.5 |
| White | 31.0 | 32.5 | 31.8 | 37.3 | 41.0 | 39.1 | 36.6 | 34.1 | 30.9 |
| Black or African American | 25.2 | 12.0 | 9.4 | 15.0 | 14.9 | 14.9 | 13.6 | 12.9 | 11.3 |
| All tenth-graders | - - - | - - - | 20.8 | 27.9 | 27.6 | 25.7 | 23.9 | 21.3 | 17.7 |
| Male | - - - | - - - | 20.8 | 27.7 | 26.2 | 25.2 | 23.8 | 20.9 | 16.7 |
| Female | - - - | - - - | 20.7 | 27.9 | 29.1 | 25.8 | 23.6 | 21.5 | 18.6 |
| White | - - - | - - - | 23.9 | 31.2 | 32.4 | 29.1 | 27.3 | 24.0 | 20.8 |
| Black or African American | - - - | - - - | 6.4 | 12.2 | 13.8 | 11.0 | 11.3 | 10.9 | 9.1 |
| All eighth-graders | - - - | - - - | 14.3 | 19.1 | 19.1 | 17.5 | 14.6 | 12.2 | 10.7 |
| Male | - - - | - - - | 15.5 | 18.8 | 18.0 | 16.7 | 14.3 | 12.2 | 11.0 |
| Female | - - - | - - - | 13.1 | 19.0 | 19.8 | 17.7 | 14.7 | 12.0 | 10.4 |
| White | - - - | - - - | 15.0 | 21.7 | 21.1 | 19.0 | 16.4 | 12.8 | 11.1 |
| Black or African American | - - - | - - - | 5.3 | 8.2 | 10.8 | 10.7 | 8.4 | 8.0 | 7.3 |

SOURCE: "Table 63. Use of Selected Substances by High School Seniors, Eighth-, and Tenth-Graders, According to Sex and Race: United States, Selected Years 1980–2002," in *Health, United States, 2003*, U.S. Department of Health and Human Services, Centers for Disease Control and Prevention, National Center for Health Statistics, Hyattsville, MD, 2003

and pharynx; bronchitis; emphysema; and low birth weight babies. One ironic result of the attempt to reduce smoking has been the marked increase in the use of smokeless tobacco, which can cause oral cancers. Tobacco use is known to increase the risks for nine cancers. According to the U.S. National Cancer Institute, it is the leading cause of 30 percent of all cancer deaths and 87 percent of lung cancer deaths.

### Environmental Tobacco Smoke

While the dangers of smoking to smokers have been known for quite some time, the risks to nonsmokers have only recently attracted attention. In May 2000 the NIH released its ninth *Report on Carcinogens* in which substances, such as metals, pesticides, and other chemicals, are identified as "known" or "reasonably anticipated" to cause cancer and to which a significant number of Americans are exposed. For the first time the report included environmental tobacco (secondhand) smoke (ETS) as a "known" human carcinogen.

ETS is classified by the EPA as a Group A carcinogen, because it is known to cause cancer in humans. According to the CDC, exposure to ETS causes 3,000 lung cancer deaths annually among nonsmokers and increases the risk for heart disease. As many as 62,000 deaths every year from coronary heart disease are estimated as attributable to ETS. Secondhand smoke is also blamed for aggravating respiratory problems and ear infections in children and increasing the risk of sudden infant death syndrome. Passive smokers (those who inhale the smoke from others' cigarettes) have a 30 percent greater risk of dying of lung cancer than those who are not passive smokers.

ETS contains cotinine, a chemical that results from the breakdown of nicotine in the body. Cotinine levels in urine, saliva, hair, and blood can be measured. Active smokers have cotinine levels in excess of 15 nanograms per milliliter (ng/mL). Nonsmokers with average exposure to ETS have cotinine levels less than 1 ng/mL. The CDC reports that nearly 90 percent of the U.S. population had measurable levels of serum cotinine in their blood in 1991. Among nonsmokers the median cotinine level was 0.20 ng/mL. By 1999 this value had dropped to less than 0.050 ng/mL, a 75 percent decrease. Figure 8.11 shows that the concentration of cotinine measured in children's blood declined by more than 50 percent between 1988 and 2000.

Until the late 1990s nonsmokers generally had no choice about breathing tobacco smoke in many public buildings, including hospitals. This is no longer the case and, in fact, smokers may find themselves ostracized or, at the very least, required to smoke in designated areas. Many American companies are banning smoking from the workplace; some even refuse to hire smokers. Also, some restaurants and social clubs—places where smoking has historically been common—have begun to ban, or severely curtail, smoking on their premises.

### Tobacco and the Legal System

In June 2002 a 67-year-old man from Kansas was awarded $15 million in punitive damages from R. J.

TABLE 8.6

**Relationship between cigarette smoking and serious medical conditions, 2000**

| Condition | Current smokers | | Former smokers | | Overall | |
|---|---|---|---|---|---|---|
| | No. | (%) | No. | (%) | No. | (%) |
| Chronic bronchitis | 2,633,000 | (49) | 1,872,000 | (26) | 4,505,000 | (35) |
| Emphysema | 1,273,000 | (24) | 1,743,000 | (24) | 3,016,000 | (24) |
| Heart attack | 719,000 | (13) | 1,755,000 | (24) | 2,474,000 | (19) |
| All cancer except lung cancer | 358,000 | (7) | 1,154,000 | (16) | 1,512,000 | (12) |
| Stroke | 384,000 | (7) | 637,000 | (9) | 1,021,000 | (8) |
| Lung cancer | 46,000 | (1) | 138,000 | (2) | 184,000 | (1) |
| Total[4] | 5,412,000 | (100) | 7,299,000 | (100) | 12,711,000 | (100) |

Notes: Cigarette smoking-attributable conditions considered are stroke, heart attack, emphysema, chronic bronchitis, and cancer of the lung, bladder, mouth/pharynx, esophagus, cervix, kidney, larynx, and pancreas. Current smokers were defined as persons who reported smoking ≥100 cigarettes during their lifetime and who now smoke some days or every day. Former smokers were defined as persons who reported having smoked ≥100 cigarettes during their lifetime but did not smoke at the time of interview. Results are adjusted for age, race, sex, and state/area of residence and rounded to the nearest 1,000. Numbers might not add to total because of rounding.

SOURCE: "Cigarette Smoking—Attributable Morbidity—United States, 2000," in *Morbidity and Mortality Weekly Report*, vol. 52, no. 35, U.S. Department of Health and Human Services, Centers for Disease Control and Prevention, Atlanta, GA, September 5, 2003

FIGURE 8.11

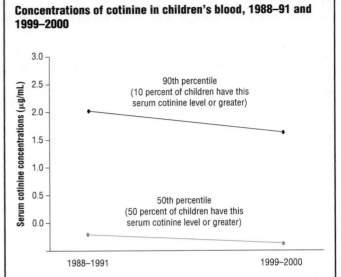

**Concentrations of cotinine in children's blood, 1988–91 and 1999–2000**

SOURCE: "Exhibit 4-11: Concentrations of Cotinine in Children's Blood, 1988–1991, 1999–2000," in "Chapter 4–Human Health," *EPA's Draft Report on the Environment 2003*, U.S. Environmental Protection Agency, Washington, DC, 2003

Reynolds (RJR) tobacco company. The man, who smoked for more than 40 years, had lost both legs to a circulatory disease that he blamed on smoking. It was the first time that a federal judge awarded punitive damages against a tobacco company. The judge ruled that the company had concealed the addictiveness of cigarettes. In the same week a man who had lost his tongue to cancer was awarded $37.5 million from Philip Morris, Brown and Williamson, and the Liggett Group by a Miami jury.

When individuals sue tobacco companies, they claim that cigarette smoking contributed to their ill health in a variety of ways. Some of the issues raised in these suits include the following:

- The cost to society for medical expenses related to tobacco use

- Advertising and selling tobacco products to youth

- The location and placement of tobacco products

- Taxes on tobacco products

- The harmful effects of secondhand smoke

- The addictive nature of nicotine

- The failure of tobacco companies to divulge information about the harmful and addictive nature of smoking, and their concealment of evidence to that fact

- Smoking in public buildings

Tobacco companies argue that smokers choose to smoke and that they must assume the risk for doing so. In the landmark 1994 case *Cipollone v. Liggett Group Inc.*

(60 LW 4703), the Supreme Court ruled that smokers may sue cigarette companies for concealing facts about smoking and that the Liggett Group was at least partially liable in the death of Rose Cipollone. However, the court also determined that Cipollone herself was partially responsible for her cigarette use and subsequent death.

In 1991 the AMA publicly charged RJR with targeting children through its Joe Camel advertising campaign. Later that year Janet Mangini, a California attorney, brought suit to end the Joe Camel campaign, becoming the first person to legally challenge the tobacco industry for targeting minors in its advertising. Six years of "discovery," that is, taking depositions and obtaining records, followed. As the trial date neared in May 1997, RJR offered to halt the ad campaign in order to stop the trial and also agreed to provide for public release of its documents about youth marketing and the Joe Camel campaign.

In the 1997 case *Broin v. Philip Morris,* a flight attendant sued claiming that secondhand smoke in airplanes had harmed her health. The suit was settled out of court with Philip Morris agreeing to pay some $300 million to establish a medical foundation.

Nevertheless most legal action against "Big Tobacco" prior to 1998 was unsuccessful because most of the suits were filed as class action suits, where large numbers of complainants unite to sue. The 1995 case *Castono v. The American Tobacco Company Inc.* (85 F 3rd 734, 5th Cir. 1996), filed in Louisiana, included people who had purchased and smoked cigarettes and had become nicotine dependent. The U.S. District Appeals Court in Louisiana "decertified" the suit, ruling that individual issues predominated over common issues, thus making the case not a proper candidate for a class action.

That ruling—that individual issues were greater than common issues in suits against the tobacco industry—has been handed down both by state and federal courts in at least 30 suits. In 1996, however, in *Howard A. Engle, MD v. R. J. Reynolds Tobacco, Philip Morris, Brown and Williams, Lorillard Tobacco, the American Tobacco Company, et al.,* filed in Dade County, Florida, the state appellate court ruled that the suit could proceed to trial, although it was reduced to include only Florida residents. In 1999 the jury ruled for the plaintiffs and, in July 2000, assessed penalties of $145 billion in punitive damages against the country's five largest tobacco companies. The money was to be split among Florida residents who could prove they became ill from smoking. Philip Morris was ordered to pay $73.96 billion, RJR $36.28 billion, Brown and Williamson $17.59 billion, Lorillard Tobacco $16.25 billion, and the Liggett Group $790 million.

The trial, which lasted for two years, was the longest civil trial in the history of tobacco litigation, and the penalty was the largest ever levied in any case. Rather than building their case around the dangers of smoking, the plaintiffs focused on the negligent conduct of the tobacco companies that they claimed had covered up smoking risks for more than four decades. While some observers suggested that the jury's decision would encourage other class actions, tobacco industry executives contend that the ruling will be reversed on appeal. They claim that the case will be overturned because it should not have been allowed as a class action and that the fine is outrageous and will put the tobacco companies out of business. In any case, legal experts predicted the verdict was many years away from being final.

In the mid-1990s a number of state governments initiated suits against tobacco companies to recoup state Medicaid spending on tobacco-related illnesses. With the pressure of state and private lawsuits building up, the tobacco industry has begun to seek settlements with the states. In November 1998, in what was termed a "Master Settlement Agreement" between the major tobacco manufacturers and 46 state attorneys general (Texas, Florida, Minnesota, and Mississippi settled independently), the tobacco companies agreed to accept a number of limitations on how they marketed and sold their products. These included: ceasing youth-targeted advertising, marketing, and promotion by stopping the use of cartoon characters in advertising; limiting brand-name sponsorship of events with significant youth audiences; terminating outdoor advertising; banning youth access to free samples; and setting the minimum cigarette package size at 20. In addition to these limits on their business practices, the tobacco industry agreed to pay more than $200 billion to the states.

In the 1990s testimony before Congress claimed that the tobacco industry deliberately manipulated the amount of nicotine in its cigarettes, and that cigarettes were nothing more than a delivery system for nicotine, a drug now widely recognized as physiologically addictive. The industry feared that if cigarettes were perceived as a delivery system for nicotine, tobacco products might fall under the control of the FDA. In 1995 the FDA ruled that nicotine was indeed a drug and liable to its regulation—the first time the tobacco industry had been regulated. In March 2000, however, the Supreme Court ruled 5–4 that the FDA does not have jurisdiction to regulate tobacco products or cigarette-company marketing practices under existing law.

In his 1999 State of the Union address, President Bill Clinton announced his intent to sue the tobacco industry to recover money spent by the federal government to treat illnesses caused by smoking. Accordingly, the U.S. Department of Justice filed suit in September 1999 (*United States of America v. Philip Morris Inc., et al.*) in the U.S. District Court for the District of Columbia. The government accused the tobacco companies of misleading and defrauding the public about the dangers of smoking. In September 2000 a federal judge dismissed part of the lawsuit in which the government was seeking to recover billions of dollars spent in health care costs related to smoking. As of April 2004 the government is seeking $289 billion in its case against the tobacco companies. The case is expected to go to trial in late 2004. It will likely be the most complex and lengthy case ever tried in U.S. court. Government attorneys plan to submit more than one million pages of exhibits as evidence.

## INDOOR AIR TOXINS

Indoor pollution has become a serious problem in America. Although most people think of outdoor air when they think of air pollution, studies now reveal that indoor environments are not safe havens from air pollution. As discussed earlier, lead, asbestos, radon, and ETS all pose serious and costly problems in indoor settings. Modern indoor environments contain a variety of pollution sources, including synthetic building materials, consumer products, and dust mites (minute insects that live on house dust and human skin residue). People, pets, and indoor plants also contribute to airborne pollution. Improvements in home and building insulation and the widespread use of central air conditioning and heating systems have largely ensured that any contaminant present indoors will not be diluted by outside air and, therefore, will become more concentrated.

Indoor pollution, however, is not just a product of modern, well-insulated homes and buildings. In 1995 doctors in the Bronx, New York, began observing an emerging epidemic of asthma, citing hospitalization rates as high as 17.3 per 1,000 people and death rates as high as 11 per 100,000. Both rates were eight times the national average and the rate of incidence among children was twice the national rate. The Bronx, an area with many dilapidated and neglected buildings, is among the worst places in the country for asthma. Among the causes cited by area physicians were many factors associated with indoor pollution: dust mites, cockroaches, smoking, dust, and respiratory viruses and bacteria (which spread easily in crowded quarters). Scientists suspected that stress and the start-up of a waste incinerator in the Bronx three years earlier were factors as well.

The EPA and other sources identify indoor pollution as one of the most serious environmental risks to human health because:

- concentrations of pollutants in indoor air can be more than two to five times higher than those in outdoor air;

- people spend an average of 80 to 90 percent of their time indoors, even more for certain groups, such as the elderly, the chronically ill, and the very young; and

- the indoor environment is unique in that it contains materials and surfaces that act as emitters and reservoirs of pollutants.

Reports of illness and allergy among building occupants have become commonplace. Scientific evidence suggests that respiratory diseases, allergies, mucous membrane irritation, nervous system defects, cardiovascular symptoms, reproductive problems, and lung cancer may be linked to exposure to indoor air pollution.

Poorly ventilated buildings sometimes become the cause of "sick-building syndrome." The term, first employed in the 1970s, describes a spectrum of specific and nonspecific complaints reported by building occupants. Such symptoms might include headaches, fatigue, or difficulty breathing, which begin soon after entering a building and subside after leaving that building. When 20 percent of a building's occupants report complaints, the World Health Organization (WHO) calls that building "sick." Experts generally believe there are many people who do not complain, even when they experience symptoms, and the WHO and the EPA estimate that 30 percent of all buildings worldwide are unfit for human occupation. Surveys and assessments by private corporations and federal agencies estimate that sick-building syndrome goes undetected in another 10 to 15 percent of structures. Many billions of dollars in income and productivity are lost annually because of employees falling ill from problems linked to sick-building syndrome.

The first legislation to deal specifically with indoor air quality was Title IV of the Superfund Amendments and Reauthorization Act of 1986, which called for the EPA to establish an advisory committee to conduct research and disseminate information. In October 1991 the GAO reported on the progress of the legislation in *Indoor Air Pollution: Federal Efforts Are Not Effectively Addressing a Growing Problem*. The GAO concluded not only that the EPA's emphasis on indoor pollution was not commensurate with the health risks posed by the problem, but also that research had been, and would likely continue to be, constrained by a lack of funding. Accordingly, the proposed Indoor Air Quality Act of 1991 was not enacted by Congress.

Federal agencies reported that they spent a total of almost $1.1 billion on indoor pollution-related research from 1987 to 1999. Most of that amount went toward indoor air research, followed by studies of lead, asbestos, and radon. The NIEHS spent the most (almost $400 million), with the National Heart, Lung, and Blood Institute spending $175.2 million, the EPA $140.4 million, the DOE $136.5 million, and the National Institute of Allergy and Infectious Diseases $93.7 million.

In 1999 the GAO once again reviewed the status of indoor air quality activities. In *Indoor Pollution: Status of Federal Research Activities,* the GAO found that significant strides have been made in understanding the risks posed by chemicals and other contaminants commonly found in homes, offices, and schools. Nonetheless it concluded that "many gaps and uncertainties remain in the assessment of exposures to known indoor pollutants." These gaps include specific sources of exposures; the magnitude of exposures; the relative role of specific exposures such as inhalation, ingestion, and skin contact; the nature, duration, and frequency of human activities that contribute to exposures; and the geographic distribution of exposures to certain pollutants for the U.S. population as a whole.

## NOISE POLLUTION

Noise is unwanted sound. The word is derived from the Latin word *nausea,* meaning seasickness. Experts agree that noise pollution is bad and getting worse in America. Noise from road traffic, airplanes, jet skis, garbage trucks, construction equipment, manufacturing processes, lawn mowers, subways, and leaf blowers are just a few of the unwanted sounds that are routinely broadcast into the air. Physicists, audiologists, engineers, architects, and physicians report that permanent hearing loss caused by amplified music is a widespread affliction in the United States. Although hearing loss is the most dramatic effect of noise pollution, even smaller amounts of noise can negatively affect health and well-being. Besides hearing loss, some other problems related to noise include the following:

- stress

- high blood pressure

- sleep loss

- distraction and lost worker productivity

- a general decline in quality of life

Noise levels are measured in decibels (db). A noise level of less than 65 db is considered acceptable from an environmental standpoint, although several studies have found that levels of 60 to 65 db are annoying to 9 percent of the public. Soft whispers have a decibel level of 30. An air conditioner at 20 feet measures 60 db. The noise level of a vacuum cleaner or a crowded restaurant is about 70 db. Average city traffic, garbage disposals, or alarm clocks at 2 feet could be 80 db. The subway, a motorcycle, or a lawn mower would be approximately 90 db; a basketball arena 108 db. A rock concert or thunderclap (120 db), a gunshot blast or jet plane (140 db), or a rocket launching pad (180 db) can be dangerous to those under constant exposure. Researchers for the EPA have found that 20 percent of the population is "highly annoyed" if sound levels reach 55 db.

Many cities have pressed the Federal Aviation Administration to steer airplane flight paths around metropolitan areas in order to reduce noise over residential areas. The airline industry has responded by beginning to build jet engines with noise levels in mind.

### Legislation against Noise

The air into which noise is emitted is a "commons," or a public space. It belongs to no one person but to everyone. People, organizations, and businesses, therefore, do not have unlimited rights to broadcast noise. The United States has been slow to confront the issue of noise. At a time when European nations were addressing the issue of noise abatement, in 1981 Congress eliminated the EPA's previously allocated funds for noise programs.

The Office of Noise Abatement and Control (ONAC) of the EPA was established by the Noise Control Act of 1972 (NCA; PL 92-574). During President Ronald Reagan's administration, noise pollution came to be viewed as a local problem because noise pollution does not travel very far and quickly dissipates. Some legislators believed that state and local regulation was more efficient than federal regulation since local governments could more easily gauge and respond to noise situations in their area. Consequently, in 1981 Congress eliminated all funding for ONAC, although it did not repeal the NCA. And while many of the provisions of that original law have become outdated and obsolete, others still—technically—could be invoked, although they generally have not been.

Thus, noise pollution has fallen to state and local governments to define and regulate. Much like the federal government, most states have been slow to do so. Increas-ingly, however, citizens are filing lawsuits based on noise issues. People regularly file noise complaints against airports and road builders, and police often respond to noise-related neighborhood conflicts.

## TOXINS IN FOOD

### Chemicals and Pesticides

Noncommercially caught fish and wildlife are sometimes contaminated with chemicals, such as mercury, PCBs, and DDT. In order to protect consumers from health risks associated with consuming such pollutants, the EPA and the states issue consumption advisories to inform the public that high concentrations of contaminants have been found in local specimens. According to the EPA in *Update: National Listing of Fish and Wildlife Advisories* (May 2003), in 2002 (the latest year for which data are available), 2,800 advisories were in effect.

The total number of lake areas and river miles under advisory for various pollutants between 1993 and 2003 is shown in Figure 8.12 and Figure 8.13. All five of the specific contaminants listed are bioaccumulative, meaning that they accumulate in the tissues of aquatic organisms at much higher concentrations than are found in the water. These contaminants also persist in the environment for a relatively long time.

The GAO report *Information on EPA's Draft Reassessment of Dioxins* (April 2002) provided the EPA's estimates of the average U.S. adult's exposure to dioxins in food on a daily basis. (See Table 8.7.) Beef and freshwater fish and shellfish are the major sources of exposure. These levels are associated with adverse health effects, but are below the levels associated with cancer. Levels of exposure are thought to be even greater in people who have diets high in fat content.

### Pathogens

According to federal officials, the U.S. food supply is among the safest in the world. Nevertheless, episodes of food poisoning and diseases occur in the United States. Based on "Food-Related Illness and Death in the United States" by Paul S. Mead et al. (*Emerging Infectious Diseases,* vol. 5, no. 5, 1999), the CDC estimates that as many as 76 million illnesses, 325,000 hospitalizations, and 5,000 deaths annually are caused by foodborne hazards. Unfortunately, most of the incidences cannot be traced to a particular pathogen, but are attributed to "unknown" agents. Known pathogens are associated with 14 million illnesses, 60,000 hospitalizations, and 1,800 deaths annually. Three pathogens, *Salmonella, Listeria,* and *Toxoplasma,* are blamed for more than 75 percent of the deaths.

Foodborne illnesses became the object of intense public scrutiny following an outbreak of *Escherichia*

FIGURE 8.12

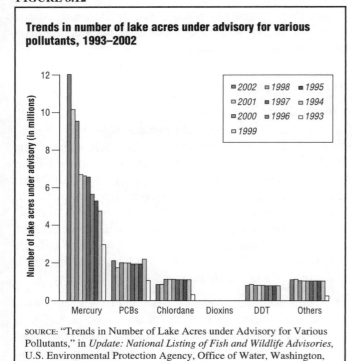

**Trends in number of lake acres under advisory for various pollutants, 1993–2002**

SOURCE: "Trends in Number of Lake Acres under Advisory for Various Pollutants," in *Update: National Listing of Fish and Wildlife Advisories*, U.S. Environmental Protection Agency, Office of Water, Washington, DC, May 2003

FIGURE 8.13

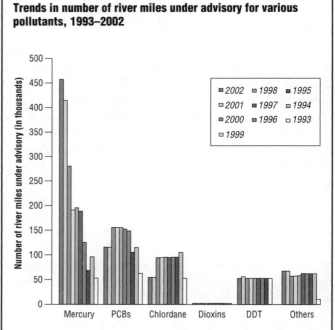

**Trends in number of river miles under advisory for various pollutants, 1993–2002**

SOURCE: "Trends in Number of River Miles under Advisory for Various Pollutants," in *Update: National Listing of Fish and Wildlife Advisories*, U.S. Environmental Protection Agency, Office of Water, Washington, DC, May 2003

*coli* (*E.coli*) in 1993 that killed four people and sickened hundreds. The illness was attributed to undercooked hamburgers from fast-food restaurants. The FDA responded by raising the recommended internal temperature for cooked hamburgers to 155 degrees Fahrenheit. A sampling program was begun to test for *E. coli* in raw ground beef. New labels containing food handling instructions were required on consumer packages of raw meats and poultry.

In 1996 more illnesses were attributed to *E. coli,* this time in unpasteurized apple juice. The FDA proposed new regulations to improve the safety of fresh and processed juices. In that same year several federal and state agencies established a surveillance program called FoodNet to monitor laboratory-identified foodborne diseases related to seven pathogens in parts of five states. By 2002 the program had grown to monitor 12 pathogens and syndromes in nine states, encompassing 37.4 million people (13 percent of the U.S. population). FoodNet data for 2002 are presented in Table 8.8.

FoodNet identified 16,580 cases of foodborne illnesses related to monitored pathogens in 2002. *Salmonella* accounted for 36 percent of cases, followed by *Campylobacter* (30 percent), and *Shigella* (23 percent).

The incidence of diseases attributed to *Listeria, Yersinia,* and *Campylobacter* decreased dramatically between 1996 and 2002. The CDC attributes the decline to several factors including increased public awareness about foodborne diseases and food safety, new pathogen

reduction measures implemented by the U.S. Department of Agriculture (USDA) at meat and poultry slaughterhouses and processing plants, egg quality assurance programs, better agricultural practices that ensure produce safety, increased regulation of imported foods and fruit and vegetable juices, and the introduction of hazard reduction measures in the seafood industry.

## Eating Habits and Food Preparation

Americans have changed their eating habits, and with such changes have come additional risks. Some explanations for the high rate of foodborne illnesses include the following:

- Americans are eating out more than in the past.

- More and more foods are being imported from foreign countries.

- New pathogens and strains of organisms are turning up in the food supply.

- Many people are careless about food preparation in the home.

In 1997 the CDC found that, among the factors that contributed to the transmission of foodborne disease, improper handling temperatures caused the most cases, followed by poor personal hygiene of handlers, contaminated equipment, inadequate cooking, and food from unsafe sources.

TABLE 8.7

**Environmental Protection Agency estimates of the average adult's daily exposure to dioxins -from dietary intake, picograms per day**

| Food type | Dietary exposure to CDDs and CDFs | Dietary exposure to PCBs | Total dietary exposure to dioxins |
|---|---|---|---|
| Beef | 9.0 | 4.2 | 13.2 |
| Freshwater fish and shellfish | 5.9 | 7.1 | 13.0 |
| Dairy products (cheese, yogurt, etc.) | 6.6 | 3.2 | 9.8 |
| Other meats (lamb, baloney, etc.) | 4.5 | 1.0 | 5.5 |
| Marine fish and shellfish | 2.5 | 2.4 | 4.9 |
| Milk | 3.2 | 1.5 | 4.7 |
| Pork | 4.2 | 0.2 | 4.4 |
| Poultry | 2.4 | 0.9 | 3.3 |
| Eggs | 1.4 | 1.7 | 3.1 |
| Vegetable fat (oils, margarine, etc.) | 1.0 | 0.6 | 1.6 |
| **Total** | **40.7** | **22.8** | **63.5** |

Note: The average adult is assumed to weigh 70 kilograms (154 pounds). A picogram is one-trillionth of a gram.
CDDs = Chlorinated Dibenzo-p-dioxins
CDFs = Chlorinated Dibenzofurans
PCBs = Polychlorinated Biphenyls

SOURCE: "Table 1. EPA's Estimates of the Average U.S. Adult's Daily Exposure to Dioxins from Dietary Intake, Picograms Per Day," in *Information on EPA's Draft Reassessment of Dioxins*, GAO-02-515, U.S. General Accounting Office, Washington, DC, April 2002

## Contamination from Produce

The per capita consumption of fresh produce has increased in the United States in recent years. Some Americans may be eating more produce for health reasons and, because of growing commerce between nations all over the globe, a wider variety of fruits and vegetables are available today.

Researchers Larry R. Beuchat and Jee-Hoon Ryu conducted a study for the CDC of factors associated with produce contamination. As reported in "Produce Handling and Processing Practices" (*Emerging Infectious Diseases*, vol. 3, no. 4, 1997), they determined that contamination of produce can occur in the field or orchard, during harvesting or processing, in transport or marketing, or in the home or restaurant. The scientists categorized sources of contamination as preharvest or postharvest.

Preharvest sources include feces in soil or fertilizer, organisms in the soil, pollution of water used to irrigate or spray crops, dust and air, animals (including birds), insects, and human handling. Postharvest factors include feces, handling by workers or consumers, farm equipment, transport containers, insects, and animals. Other possible postharvest sources are air and dust, wash water, processing equipment, ice, transport vehicles, improper storage or packaging, inappropriate temperatures, cross-contamination from other foods, and incorrect handling. Beuchat and Ryu concluded: "Control or elimination of pathogenic microorganisms from fresh fruit and vegetables can be achieved only by addressing the entire system—from the field, orchard, or vineyard to the point of consumption."

## Organic Foods—A Booming Industry

Some consumers are concerned about certain technologies being applied to the food supply, including irradiation, the use of hormones in milk production, and genetically engineered crops. Irradiation, in particular, is a highly controversial subject. About 40 countries worldwide use irradiation, in which food is briefly exposed to radiant energy, such as gamma rays or high-energy electrons, as a means of controlling pathogens. Some foods in the United States have been irradiated since the 1960s. (See Table 8.9.) A 2000 report on food irradiation by the GAO reported that 95 million pounds of food products were irradiated in 1999, representing about 10 percent of their total consumption. The GAO concluded that the benefits of irradiation in terms of reduced mortality, illnesses, and associated costs outweighed the minimal risks. Risks commonly attributed to food irradiation include possible creation of chemical by-products and loss of nutritional value.

The USDA defines organic agriculture as an "ecological production management system that promotes and enhances biodiversity, biological cycles, and soil biological activity. It is based on minimal use of off-farm inputs and on management practices that restore, maintain, and enhance ecological harmony." Organic agriculture is both an approach to food production based on biological methods that avoid the use of synthetic crops or livestock products, and a broadly defined philosophical approach to farming that puts value on ecology, conservation, and nonintensive animal breeding practices. Some conventional practices not accepted in organic agriculture include the following:

- Synthetic fertilizer and pesticides
- Confinement livestock operations such as feed lots or cages where animals are fattened before slaughter
- Routine use of growth-enhancing animal drugs such as hormones or antibiotics
- Genetically modified crops
- Irradiation of foods for preservation or decontamination

While organic methods of farming emerged in the United States and Europe in the early 1900s, it was not until the late 1980s that research groups and consumers began to express widespread interest in such practices. Beginning in 1989 sales of organically produced products began to climb, growing, on average, 20 percent per year. According to the 2003 USDA report *U.S. Organic Farming in 2000-2001: Adoption of Certified Systems* by Catherine Greene and Amy Kremen, in 2001 sales of organic products reached approximately $9 billion in the United States and $21 billion worldwide. Organic food sales accounted for 1–2 percent of total food sales in all major world markets.

In the same report the USDA notes that there were nearly 7,000 certified organic operations in the United

# TABLE 8.8

**Cases of infection[1] with nine pathogens and of one syndrome under surveillance in the Foodborne Diseases Active Surveillance Network (FoodNet) by selected characteristics, 2002**

| Syndrome | California | Colorado | Connecticut | Georgia | Maryland | Minnesota | New York | Oregon | Tennessee | Overall | Object |
|---|---|---|---|---|---|---|---|---|---|---|---|
| Campylobacter | 31.67 | 13.99 | 15.79 | 7.58 | 6.72 | 18.95 | 12.97 | 16.01 | 6.37 | 13.37 | 12.30 |
| Escherichia coli O157 | 0.99 | 2.12 | 1.37 | 0.67 | 0.48 | 3.62 | 1.69 | 5.13 | 0.70 | 1.73 | 1.00 |
| Listeria | 0.37 | 0.12 | 0.47 | 0.18 | 0.43 | 0.10 | 0.45 | 0.26 | 0.11 | 0.27 | 0.25 |
| Salmonella | 15.85 | 13.46 | 13.28 | 21.43 | 17.13 | 11.87 | 16.22 | 9.53 | 19.61 | 16.10 | 6.80 |
| Shigella | 11.45 | 5.53 | 3.04 | 19.06 | 21.53 | 4.46 | 1.60 | 2.71 | 5.07 | 10.34 | NA[2] |
| Vibrio | 0.37 | 0.12 | 0.32 | 0.32 | 0.35 | 0.10 | 0.12 | 0.43 | 0.25 | 0.27 | NA |
| Yersinia | 0.50 | 0.12 | 0.47 | 0.51 | 0.26 | 0.38 | 0.63 | 0.49 | 0.60 | 0.44 | NA |
| Cryptosporidium | 0.99 | 0.88 | 0.55 | 1.43 | 0.52 | 4.10 | 1.75 | 1.15 | 0.46 | 1.42 | NA |
| Cyclospora | 0.05 | NR[3] | 0.20 | 0.27 | 0.05 | NR | 0.33 | NR | 0.04 | 0.11 | NA |
| HUS[4] | 1.01 | 2.62 | 1.79 | 0.84 | NR | 2.42 | 0.49 | 8.96 | NR | 1.78 | NA |
| Population in surveillance (millions)[5] | 3.20 | 2.46 | 3.43 | 8.38 | 5.38 | 4.97 | 3.32 | 3.47 | 2.84 | – | – |

[1]Per 100,000 persons.
[2]Not applicable.
[3]None reported.
[4]Hemolytic uremic syndrome. Incidence per 100,000 children aged < 5 years.
[5]Population for some sites is entire state, for other sites, selected countries. For some sites, the catchment area for *Cryptosporidium* and *Cyclospora* is larger than for bacterial pathogens.

SOURCE: "Incidence of Cases of Infection with Nine Pathogens and of One Syndrome Under Surveillance in the Foodborne Diseases Active Surveillance Network, by Site, Compared with National Health Objectives for 2010–United States, 2002," in *Morbidity and Mortality Weekly Report*, vol. 52, no. 15, April 18, 2003

States in 2001, up from 5,000 operations in 1997. The acres of farmland managed under certified systems increased by 74 percent during those years, from 1.4 million acres in 1997 to 2.2 million acres in 2001.

For most of its history the organic food industry established its own organizations and standards, with approximately 33 private certification operations. However, there was no consistency in labeling and no guarantee that foods labeled as organic are actually grown and processed in a purely organic fashion.

In 1990 Congress passed the Organic Foods Production Act (Title 21 of PL 10-624) to regulate the organic food industry. The act authorized the National Organic Program (NOP) to be administered by the USDA. The program would define standard practices and certify that operations meet those standards. It would be illegal for anyone to use the word "organic" on a product if it does not meet the defined criteria.

## CHEMICAL AND BIOLOGICAL WEAPONS

More and more the world faces threats from toxins intentionally and maliciously introduced into the environment with the sole purpose of harming people. The anthrax letters that followed the terrorist attacks of September 11, 2001, have brought new focus to the dangers of toxins that can be used for chemical and biological weapons. While some of these toxins can be identified, such as anthrax, others remain of mysterious origin, making it much more difficult to treat those who are affected.

### Anthrax

Anthrax is a disease caused by the spore-forming bacterium *Bacillus anthracis*. Anthrax is found in nature mostly in agricultural areas, where it is associated with livestock, such as cattle, sheep, and goats. The spores can survive in soils for many years. Human exposure to anthrax is usually through infected animals or animal products, such as meat, hides, wool, or leather. Anthrax can enter the body either through inhalation, via the skin (cutaneous), or ingestion. Inhaled anthrax is the most lethal. Anthrax is not contagious and is treatable by antibiotics if detected early enough in the disease.

Between 1955 and 1999 there were only 236 reported cases of anthrax in the United States, most of them cutaneous. An inhalation case of anthrax had not occurred since 1976, when a man working with infected imported yarns died of the disease. In October and November 2001 there were ten confirmed cases of inhalation anthrax in Florida, New York, New Jersey, and the District of Columbia. All resulted from intentional release of the spores through mailed letters or packages. Five of those infected died.

### Gulf War Syndrome

Many veterans of the 1991 Persian Gulf War returned with physical complaints that have baffled medical experts. "Gulf War syndrome" is the name given to an array of symptoms—fatigue, skin rashes, memory loss, and headaches—experienced by men and women who served in the war. Some sources have claimed these conditions may have resulted from exposure to poison gas that was inadvertently released when American troops destroyed caches of the gas following the actual conflict, others contend the troops were heavily exposed to fumes of burning oil during the war, while still others suggest that the immunizations given to the soldiers may have caused such responses.

**TABLE 8.9**

**Food products approved for irradiation, August 21, 1963–July 21, 2000**

| Food product | Agency and approval date | Purpose for irradiation | Maximum permitted dosage (kiloGray) |
|---|---|---|---|
| Wheat and wheat powder | FDA - August 21,1963 | Insect deinfestation | 0.20 to 0.50 |
| White potatoes | FDA - July 8, 1964 | Inhibit sprout development | 0.05 to 0.15[1] |
| Spices and dry vegetables | FDA - July 5, 1983 | Microbial disinfection and insect deinfestation[2] | 10.0 |
| Dry or dehydrated enzyme preparations | FDA - June 10, 1985 | Microbial disinfection | 10.0 |
| Pork carcasses or fresh nonheated processed cuts | FDA - July 22, 1985 | Control *Trichinella spiralis* | 0.30 to 1.00 |
| Fresh foods | FDA - April 18, 1986 | Delay maturation | 1.0 |
| Dry or dehydrated aromatic vegetable substances[3] | FDA - April 18, 1986 | Microbial disinfection | 30.0 |
| Fresh, frozen uncooked poultry | FDA - May 2, 1990 USDA - October 21, 1992 | Control foodborne pathogens | 3.0 |
| Refrigerated and frozen uncooked beef, lamb, goat, and pork | FDA - December 3, 1997 USDA - February 22, 2000 | Control foodborne pathogens and extend shelf life | 4.5 (refrigerated) 7.0 (frozen) |
| Fresh shell eggs | FDA - July 21, 2000 | Control *salmonella* | 3.0 |

[1]Maximum dose increased from 0.10 to 0.15 on November 9, 1965.
[2]Insect deinfestation approved June 1984.
[3]Refers to substances used as ingredients for flavoring or aroma (e.g., culinary herbs, seeds, spices, and vegetable seasonings). Includes turmeric and paprika when used as color additives.

SOURCE: "Appendix III. Food Products Approved for Irradiation in the United States," in *Food Irradiation: Available Research Indicates that Benefits Outweigh Risks: Report to Congressional Requesters,* (GAO/RCED-00-217), U.S. General Accounting Office, Washington, DC, August 2000

The Department of Veterans Affairs (VA) offers Gulf War veterans eligibility for medical treatment. More than 67,000 service members have responded to the VA's Gulf War Registry program, which offers physical examinations to all eligible service members. Most veterans are diagnosed and treated; for some, however, symptoms have been chronic. Data from about 10,000 claimants' health exams found unexplained illness in approximately 12 percent of the cases. If a veteran's symptoms defy diagnosis, the veteran can be referred to one of the nation's four Gulf War Referral Centers for treatment.

In 1995 the federal government began devoting $115 million for 121 research projects related to Gulf War ill-nesses. The VA has provided compensation payments to chronically disabled Gulf War veterans with undiagnosed illnesses. (A disability is considered chronic if it lasts more than six months.)

The federal response to the health consequences of Gulf War service is led by the Persian Gulf Veterans Coordinating Board, composed of the VA and the U.S. Departments of Defense and Health and Human Services. The Presidential Advisory Committee on Gulf War Veterans' Illnesses was formed by President Bill Clinton in 1995. In 1999 he signed into law the Veterans Millennium Health Care and Benefits Act (PL 106-117), which extended medical care to veterans, their spouses, and their children until December 31, 2003.

# CHAPTER 9
# DEPLETION AND CONSERVATION OF NATURAL RESOURCES

Human activity on Earth has always altered the land. When populations were small enough, and productive and accessible land was abundant, people could abandon land that had been damaged by overuse and move on. While some countries still have excess land available, if population growth continues at the expected rate, virtually all arable (fit for cultivation) land will be in use.

## THE ECONOMIC VALUE OF THE WORLD'S ECOSYSTEMS—HOW MUCH IS NATURE WORTH?

Nature performs valuable, practical, measurable functions, without which the human economy could not exist. Many experts contend that, as human activity gradually consumes or destroys this natural capital, the monetary value of the ecosystem to the economy must be calculated and considered. Thirteen economists, ecologists, and geographers studied 16 different biomes (ecological areas such as lakes, urban areas, and grasslands) to estimate the economic value of 17 ecosystem services. To do this they assigned dollar values to services performed by nature that are considered necessary to the human economy. Their report, published in the journal *Nature* (May 1997), estimated that ecosystems perform at least $33 trillion worth of services annually. Marine systems contribute about 63 percent of the value, mainly from coastal systems ($10.6 trillion). Terrestrial systems account for 37 percent of the value, mainly from forests ($4.7 trillion) and wetlands ($4.9 trillion). This total was 1.8 times the 1997 global gross national product of approximately $18 trillion. In other words the services performed by nature were 180 percent as valuable as all of humankind's economic activities.

Most experts, including the authors of the study, recognize the figure is a crude, conservative, "starting point" for an estimate of what the environment does for humans. Others contend nature's value is incalculable. Virtually everyone agrees that without nature's ecological contribu-tion, human life could not exist. Furthermore, as ecosystems become more stressed and scarce in the future, their value will increase. If significant, irreversible thresholds are passed, the valuable services of these ecosystems may become irreplaceable.

## THE ROLE OF FORESTS AND HABITAT

For millennia humans have left their mark on the world's forests, although it was difficult to see. By the twenty-first century, however, forests that humans once thought were endless are shrinking before their eyes. Forests are not only a source of timber; they perform a wide range of social and ecological functions. They provide a livelihood for forest dwellers, protect and enrich soils, regulate the hydrologic cycle, affect local and regional climate through evaporation, and help stabilize the global climate. Through the process of photosynthesis they absorb carbon dioxide ($CO_2$) and release the oxygen humans and animals breathe. They provide habitat for half of all known plant and animal species, are the main source of wood for industrial and domestic heating, and are widely used for recreation.

Forests are attractive and accessible sources of natural wealth. They are not, however, unlimited. Deforestation is caused by farmers, ranchers, logging and mining companies, and fuel wood collectors. Governments have often encouraged the settlement of land through cheap credit, land grants, and the building of roads and infrastructure. Much of these activities led to the destruction of forests, causing some governments to reverse their policies.

Forests play a particularly crucial role in the global cycling of carbon. The Earth's vegetation contains two trillion tons of carbon, roughly triple the amount stored in the atmosphere. When trees are cleared, the carbon they contain is oxidized and released into the air, adding to the atmospheric store of carbon dioxide. Many scientists believe that carbon dioxide contributes to global warming.

This release happens slowly if the trees are used to manufacture lumber or are allowed to decay naturally. If they are burned as fuel, however, or in order to clear forestland for farming, almost all of their carbon is released rapidly. The clearing for agriculture in North America and Europe has largely stopped, but the burning of tropical forests has taken over the role of producing the bulk of carbon dioxide added to the atmosphere by land use changes.

A 1995 National Oceanic and Atmospheric Administration (NOAA) study found that about half the carbon dioxide emitted by burning fossil fuels is absorbed by plants in the Northern Hemisphere. The finding showed that plants play a role about equal to that of oceans, to which most of the absorption had previously been attributed. The study showed that plants absorb carbon dioxide that is rich in the carbon isotope 12, or C12, while ocean water absorbs C12 and C13 equally. By determining the ratios of the isotopes, scientists can determine the relative effects of oceans and plants on the atmosphere's carbon dioxide concentration.

## Rising Pressures on Forests

According to Janet N. Abramovitz in *Taking a Stand: Cultivating a New Relationship with the World's Forests* (Washington, DC: Worldwatch Institute, 1998), the Worldwatch Institute, an independent, nonprofit environmental research organization, reported that between 1980 and 1995 alone at least 494 million acres of forests vanished—an area larger than Mexico.

According to the United Nations Food and Agriculture Organization (FAO), approximately half the wood cut worldwide is used for fuel and charcoal. Most fuel wood is used in developing countries. In dry countries such as India, the majority of trees cut are for fuel; in moist tropical areas such as Malaysia, most trees are cut for industrial timber.

## Tropical Rain Forests

Tropical forests are the most "alive" places on Earth. Although they cover less than 2 percent of the Earth's surface, they are home to as many as 30 million species of plants and animals—more than half of all life forms on the planet. A single acre of tropical rain forest supports 60 to 80 tree species and an enormous number of vines and mosses.

Rain forests also play an essential role in the weather. They absorb solar energy, which affects wind and rainfall worldwide. Regionally, they reduce erosion and act as buffers against flooding. Tropical trees contain huge amounts of carbon which, when they are destroyed, is released into the atmosphere as carbon dioxide.

Of roughly 3,000 plants identified as having cancer-fighting properties, 70 percent grow in the rain forests. One of every three species of birds in the world nest there. In addition to wildlife, more than 1,000 indigenous tribes

still survive in tropical forests, just as they have for thousands of years.

The Natural Resources Defense Council (NRDC), a private organization that supports environmental health, reports that nearly half of the forests on the planet have been decimated and every year more than 30 million acres of tropical forests and woodlands are destroyed for agriculture or logging. Exacerbating this problem, global wood consumption is set to double over the next 30 years, according to the NRDC, further stressing the survival of global forests..

In *Combined Summaries: Technologies to Sustain Tropical Forest Resources and Biological Diversity* (Washington, DC: U.S. Government Printing Office, 1992) the now-defunct U.S. Office of Technology Assessment concluded that the major underlying cause of deforestation and species extinction is the lack of alternative employment for the growing populations of tropical countries. Logging and the conversion of forestland to short-term, usually unsustainable, agricultural use results in destruction of the land, declining fisheries, erosion, and flooding.

Species in tropical rain forests possess a high degree of mutuality, in which two species are completely dependent on one another for survival; for example, a species of wasp and a species of fig tree. Such relationships are believed to evolve as a result of the relatively constant conditions in the tropics. Any species dependent on trees therefore becomes imperiled when a tree is cut down.

THE AMAZON—AN EXAMPLE. The Amazon rain forests, located in South America, are the most famous of the Earth's tropical forests. They serve as a good example of the controversies surrounding rain forests worldwide. This controversy generally centers on the interest of environmentalists (often from developed countries) in stabilizing the environment and the developing world's basic need to cut down its forests for fuel and livelihood. Most developing nations claim that these needs are too great to be set aside for the sake of the environment. They also resent the industrialized world's disdain of practices the developed countries once followed themselves in building their own nations. These poorer, developing countries also wonder why they are expected to pay for the cleanup of a world that they did not contaminate.

There are also international incentives for continuing to cut down the rain forests. Foreign countries, especially Asian nations, are increasingly eyeing the Amazon forests as a source of ancient trees to make plywood, ornamental moldings, and furniture. Granting logging rights to these nations may seem an appealing option for those South American countries desperate for money.

Based on studies of satellite photographs taken over the Amazon, researchers believe that as much as 10 percent of the original Amazon forest has been destroyed,

FIGURE 9.1

**Location of protected forest and other forest in the United States, 2001**

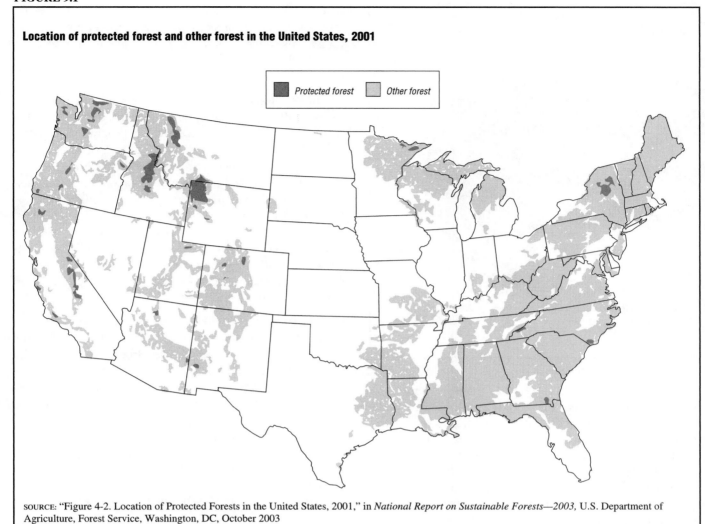

SOURCE: "Figure 4-2. Location of Protected Forests in the United States, 2001," in *National Report on Sustainable Forests—2003,* U.S. Department of Agriculture, Forest Service, Washington, DC, October 2003

mainly through "slash and burn" methods of clearing land that are used to convert the land to farming use. The cleared land's productivity usually decreases within a few years, and farmers often have to abandon the fields and move on—slashing and burning a new area.

## U.S. FORESTS UNDER STRESS

The United States has 747 million acres of forested lands; they comprise roughly one-third of the nation's total land area. (See Figure 9.1.) Forests are valued for a variety of ecological and economical reasons. In their natural state they provide vital habitat for wildlife and play an important role in the carbon cycle. (See Figure 2.10 in Chapter 2.) Forests are also a source of recreation for humans and provide wood for fuel and lumber. Human uses combined with natural environmental stresses (such as disease and drought) pose a constant threat to the health and vitality of the nation's forests.

As shown in Figure 9.2 almost half of America's forested lands are in the hands of private owners with no ties to

industry. Another 20 percent are part of the National Forest System overseen by the U.S. Forest Service, an agency of the U.S. Department of Agriculture. Other federal agencies control 13 percent of the country's forest lands. Industrial entities (such as timber companies) own 10 percent of America's forests, while the remaining 8 percent are under state control.

### Forest Health—The Latest Government Assessment

In May 2003 the U.S. Forest Service published its latest assessment on the health and well-being of the nation's forests. According to *America's Forests: 2003 Health Update,* there are five key areas of concern:

- Wildfires

- Outbreaks of native insects

- Nonnative invasive insects and pathogens (diseases)

- Invasive plant species

- Ecologically damaging changes in forest type

FIGURE 9.2

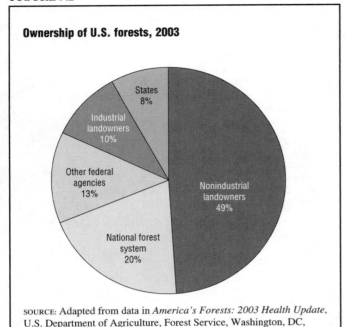

**Ownership of U.S. forests, 2003**

States 8%

Industrial landowners 10%

Other federal agencies 13%

Nonindustrial landowners 49%

National forest system 20%

SOURCE: Adapted from data in *America's Forests: 2003 Health Update,* U.S. Department of Agriculture, Forest Service, Washington, DC, May 2003

**WILDFIRES.** The Forest Service manages about 155 national forests across the country. (See Figure 9.3.) About 70 percent of these lands are located in the dry, interior areas of the western United States. Management practices in the past called for the Forest Service to put out all wildfires in the national forests. Scientists have recently put forward the idea that wildfires are necessary for forest health. They point out that wildfires are natural occurrences that serve to remove flammable undergrowth without greatly damaging larger trees.

Before pioneers settled the West, fires occurred about every five to 30 years. Those frequent fires kept the forest clear of undergrowth, fuels seldom accumulated, and the fires were generally of low intensity, consuming undergrowth but not igniting the tops of large trees. Disrupting this normal cycle of fire has produced an accumulation of vegetation capable of feeding an increasing number of large, uncontrollable, and catastrophic wildfires. Thus, the number of large wildfires has increased over the past decade, as have the costs of attempting to put them out.

Because the national forests are attractive for recreation and enjoyment, human population has grown rapidly in recent years along the boundaries scientists refer to as the "wildland/urban interface." According to a 1999 U.S. General Accounting Office (GAO) study, *Western National Forests—A Cohesive Strategy Is Needed to Address Catastrophic Wildfire Threats,* this combination of rising population and increased fire risk poses a catastrophic threat to human health and life along the wildland/urban interface areas. In addition to the risk fires pose to nearby inhabitants, smoke from such fires contains substantial amounts of particulate matter that contaminates the air for many hundreds of miles. In addition, forest soils become subject to erosion and mud slides after fires, further threatening the ecosystem and those who live near the forests.

In 1997 the Forest Service began an attempt to improve forest health by reducing, through "controlled burns," the amount of accumulated vegetation, a program to be completed by 2015. The GAO found that lack of funding and inadequate preparedness may render the program "too little, too late." The National Commission on Wildfire Disasters concluded: "Uncontrollable wildfire should be seen as a failure of land management and public policy, not as an unpredictable act of nature. The size, intensity, destructiveness and cost of ... wildfires ... is no accident. It is an outcome of our attitudes and priorities.... The fire situation will become worse rather than better unless there are changes in land management priority at all levels."

The summer of 2000 was considered the worst fire season in 50 years in the United States. Nearly 123,000 fires burned more than 8.4 million acres. Ironically, one of these fires resulted when a "controlled burn" near Los Alamos, New Mexico, raged out of control, sweeping across hundreds of acres of land and destroying homes and businesses for miles. In June 2002 a massive fire swept through Arizona destroying hundreds of homes and businesses and causing 30,000 people to flee. The fire burned 375,000 acres in only a week and was called a "tidal wave" by fire fighters trying to contain it. Numerous other fires roared through the American West during the summer of 2002.

According to *America's Forests: 2003 Health Update,* catastrophic fires are due to decades of fire suppression that have allowed forests to become overcrowded with highly combustible undergrowth. The situation is aggravated by a lingering drought in the West and trees stressed by pests and disease. The report warns that "the fire risk in many forested areas remains high."

Figure 9.4 shows the number of acres burned by wildland fires between 1960 and 2002. In 2002 federal agencies spent $1.66 billion putting out destructive wildfires.

Following the disastrous 2000 fire season the Forest Service collaborated with other agencies to develop *The National Fire Plan,* a long-term strategy for more effectively dealing with fire threats and preventing future wildfires. In August 2002 the Bush administration presented its plan for wildfire management in *Healthy Forests: An Initiative for Wildfire Prevention and Stronger Communities.* The so-called Healthy Forests Initiative implements core strategies of the National Fire Plan.

**OUTBREAKS OF NATIVE INSECTS.** Native insects of concern in American forests include bark beetles and southern pine beetles. Under certain conditions these

FIGURE 9.3

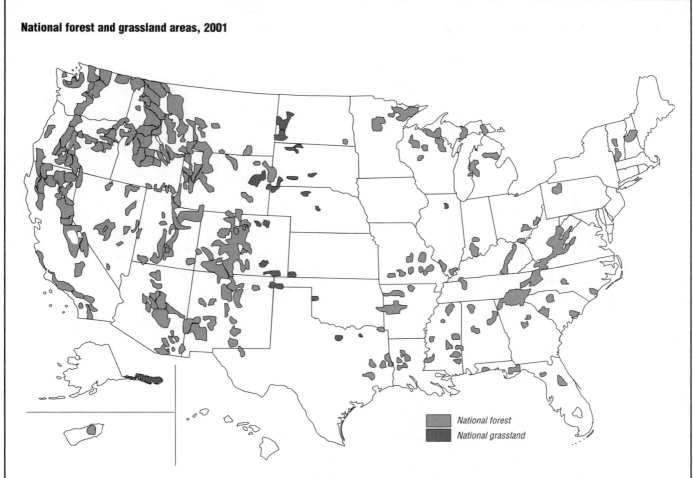

**National forest and grassland areas, 2001**

National forest
National grassland

SOURCE: "Map," in *Recreation, Heritage & Wilderness Resources,* U.S. Department of Agriculture, U.S. Forest Service, Washington, DC, March 13, 2002 [Online] http://www.fs.fed.us/recreation/map/finder.shtml [accessed May 6, 2004]

insects can infest huge areas of forests and kill thousands of trees. This is damaging by itself and exaggerates other threats to forests, such as wildfires. Wildfires are more likely to spread quickly and burn hotter when forests contain large amounts of trees that have been weakened or killed by insect damage.

The Forest Service estimates that southern pine beetles pose a moderate to high risk to more than 90 million forested acres across the Southeast. In 2001 beetle outbreaks affected tens of thousands of acres in the South resulting in $200 million of damages. The Forest Service spent $10 million that year alone fighting the beetle outbreak. In the western United States the bark beetle known as the Mountain Pine Beetle is a major killer of pine trees. Thousands of acres of pine forest across the West are considered at risk. The Forest Service focuses its resources on tracking, suppressing, and preventing beetle outbreaks and on replanting forests decimated by the pests.

**NONNATIVE INVASIVE INSECTS AND PATHOGENS (DISEASES).** Another major threat to America's forests is the spread of nonnative invasive insects and pathogens. Nonnative (or exotic) species can be very harmful, because they do not have natural predators in their new environment. This allows them to "invade" their new territory and spread very quickly.

Species of major concern to forest health are as follows:

• Gypsy Moths—An insect that arrived in the United States during the 1800s from Europe and Asia. In the springtime they devour newly emerged leaves on hundreds of tree species (primarily oaks). They are concentrated in eastern forests where they are blamed for defoliating more than 80 million acres of trees.

• Hemlock Woolly Adelgid—An insect that arrived in the United States during the 1920s from China and Japan. The pest eats the leaves off of eastern hemlock trees. It has infested hemlock forests across the Northeast and South Central states from Maine to northern Georgia. Trees die within only a few years of being infested.

FIGURE 9.4

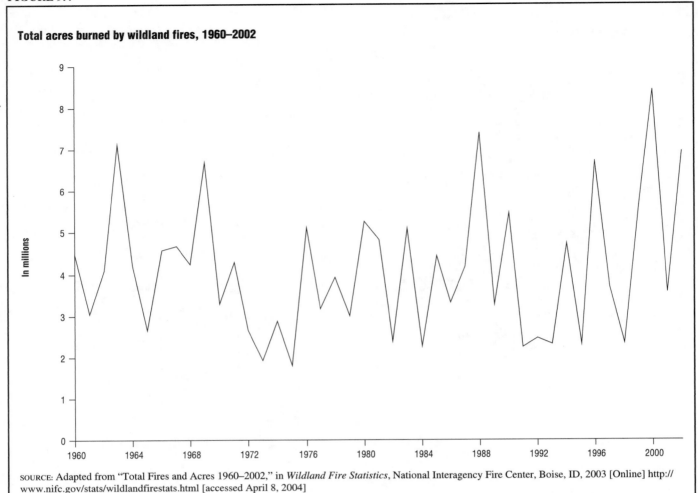

**Total acres burned by wildland fires, 1960–2002**

SOURCE: Adapted from "Total Fires and Acres 1960–2002," in *Wildland Fire Statistics*, National Interagency Fire Center, Boise, ID, 2003 [Online] http://www.nifc.gov/stats/wildlandfirestats.html [accessed April 8, 2004]

- White Pine Blister Rust—A fungus native to Europe introduced to western Canada around 1910. It migrated quickly southward across the mountainous states where it is particularly lethal to high-altitude pine forests. Once firmly entrenched in an area the fungus can kill more than 95 percent of the trees it infects.

- Sudden Oak Death—A disease caused by the pathogen *Phytophthora ramorum*. Its origin is unknown, but it was introduced to the United States within the past few decades. So far, it has been found in forests in California and southern Oregon where it has killed thousands of trees (primarily oak) and ornamental and wild shrubs. Scientists fear that it could spread eastward and cause enormous damage to the country's massive oak forests.

- Emerald Ash Borer—An exotic wood-boring beetle from Asia that targets ash trees. Believed to have entered the United States in cargo packing materials, the beetle was discovered in Southeastern Michigan in the summer of 2002 and has so far killed millions of

trees. Ashes are killed when the beetles' larvae bore tunnels within the wood, cutting off the tree's water and nutrients. Infested ash trees, which are predominantly in the Northeastern United States and Canada, die within two to three years of infestation.

The Forest Service employs a variety of measures to combat nonnative invasive pests, including application of insecticides and release of biological control agents. These agents include insects and pathogens found to prey upon the nonnative invasive pests. For example, since 1999 the Forest Service has raised and released more than 500,000 ladybird beetles into forests infested with Hemlock Woolly Adelgids. The beetles, which are native to the United States, feed on the Adelgids and their eggs. Experts hope this measure will wipe out nearly all of the Adelgid population in the forests treated.

**INVASIVE PLANT SPECIES.** Insects and pathogens are not the only invaders causing damage to America's forests. Certain plants (native and nonnative) become a threat when they grow out of control and overpower regular forest vegetation. Invasive plants of major concern include leafy spurge in northern states (particularly in the West), kudzu in the South, and mile-a-minute weed in the Northeast and mid-Atlantic states.

Although these plants are most often a problem in rangelands they are increasingly affecting forests. Invasive plants strangle and smother young seedlings and gobble up resources, such as water and nutrients needed by other plants. They also contribute to buildup of highly combustible undergrowth, making forests more susceptible to hot-burning wildfires. This is one of the reasons that the National Fire Plan targets invasive plants for reduction. In addition, there is the Federal Interagency Committee for the Management of Noxious and Exotic Weeds. This is a collaboration of 17 agencies working to develop control techniques for invasive plants across the country. The Forest Service is working on a variety of measures, primarily biological agents (such as insects or fungi known to attack invasive plants).

ECOLOGICALLY DAMAGING CHANGES IN FOREST TYPE. America's forests have been changed over the years by many human and natural factors. This has led to ecological changes in entire forest types. For example, prior to the 1900s the forests of the Appalachian Mountains were dominated by the American Chestnut. A fungus introduced from Europe virtually wiped out the chestnut population by the 1950s. Other species of trees soon became predominant. Today scientists consider this type of forest change to be harmful from an ecological standpoint. Major changes in forest type have profound effects on the overall health of a forest, wildlife habitat, and even soil conditions. The Forest Service worries that a combination of stressors, including fire, drought, destructive pests, and human activities, pose major dangers to forests. Human activities that can negatively affect forests include agriculture and residential development.

## Forest Health—Other Problems

Environmentalists fear that the forests of the northwest United States are being depleted by "clear-cutting" practices—the method of logging in which all trees in an area are cut—as opposed to "selective management" techniques, in which only certain trees are removed from an area. The lumber industry continually battles with environmentalists and the U.S. Forest Service over the right to clear-cut ancient forests. Experts believe that North American "old growth" forests (stands of old, large trees) may store more carbon than any of the world's other sinks (repositories).

National Aeronautics and Space Administration (NASA) scientists report that satellite pictures show a high level of damage to the evergreen forests of the Pacific Northwest. They attribute the damage to clear-cutting and claim the region has been so fragmented by clear-cutting that the overall health of the forest is at risk. Many observers believe that the biggest threat from this logging technique is the loss of diversity of species in the area. The logging industry contends that restrictions on logging devastate rural communities by causing the loss of thousands of jobs and leading to an increase in retail prices for lumber nationwide.

Logging roads are increasingly blamed for contributing to landslides, floods, and changes in rivers and streams. The Roadless Area Conservation Rule was adopted in January 2001 to protect nearly 60 million acres of national forests from further road building and logging, while keeping them open for recreational uses. The rule had both environmental and economic goals. The Forest Service oversees approximately 386,000 miles of roads, and their upkeep costs billions of dollars. The high cost of building and maintaining these roads is often cited as a reason many national forests lose money on timber sales. The rule was immediately challenged in court by a variety of groups, but a federal appeals court upheld the rule in December 2002.

REPLANTING. In an effort to counteract tree loss, forests are often "replanted" or replaced. Most experts contend that, when a natural forest (that has been replanted after clear-cutting) is replanted with commercially valuable trees, the plot becomes a tree farm, not a forest, and the biological interaction is damaged. Primary forests represent centuries, perhaps a millennium, of undisturbed growth. Trees will rebound after clear-cutting within 70 to 150 years but, researchers have found, the plants and herbs of the understory (growth under the canopy of the trees) never regain the richness of species diversity and complexity of their predecessors.

## The Effects of Pollution

Many biologists believe that regional air pollution is a serious anthropogenic (made by humans) threat to temperate forest ecosystems. The most dangerous impact on forests comes from ozone, heavy metals, and acid deposition. Ozone exposure reduces forest yields by stunting the growth of seedlings and increasing stresses on trees. Such damage can take years to become evident. Numerous studies suggest that both photosynthesis and growth decline significantly after one or two weeks of ozone at levels of 50 to 70 parts per billion (ppb), more than twice the normal background level of 20 to 30 ppb. During growing seasons average ozone levels are highest in the West (California, Nevada, Utah, and Arizona) and on the East Coast south of Pennsylvania.

In 1995, in a four-year study of a widespread timber species called loblolly pine, researchers at the Oak Ridge National Laboratory in Tennessee determined that ground-level ozone levels that frequently occur in the eastern United States caused growth to slow, especially under drier soil conditions. The loblolly pine, covering approximately 60 million acres, contributes billions of dollars to the economy of the South.

In 2001 the Hubbard Brook Research Foundation reported that more than half of large-canopy red spruce trees in the Adirondack Mountains and the Green Mountains had died since the 1960s. Acid rain was considered the primary cause. Along with acid rain, the Environmental Protection Agency (EPA) also blames other pollutants and natural stress factors for the increased death and decline of northeastern red spruce at high altitudes (in the Adirondacks, for example) as well as the decreased growth of red spruce in the southern Appalachians. Acid rain is also closely linked to the decline of sugar maple trees in Pennsylvania.

In the March 1999 report *Soil Calcium Depletion Linked to Acid Rain and Forest Growth in the Eastern United States,* the U.S. Geological Survey (USGS) stated that calcium levels in forest soils had declined at locations in ten states in the eastern United States. Calcium is necessary to neutralize acid rain and is an essential nutrient for tree growth.

## THE BATTLE OVER PUBLIC LANDS

The GAO reported in 2001 that the federal government managed just over 680 million acres or about 29 percent of the nation's total land surface. Of these lands, 96 percent are managed by four agencies—the National Park Service, the Fish and Wildlife Service, the Bureau of Land Management, and the Forest Service. Most public lands are located in western states.

In much of the West ranchers have petitioned Congress to loosen restrictions on grazing on thousands of acres of federally owned ranch land. Environmental groups strongly oppose the proposal, claiming that grazing imperils land conservation, wildlife, and recreation. Grazing, they charge, is especially destructive to stream banks and sensitive wildlife habitat. Such concern for the soil and wildlife also lies at the heart of the dispute between oil companies and environmentalists over control of public lands such as the Arctic National Wildlife Refuge in Alaska.

### A Land Grab

Unfortunately, love of the land has led Americans and developers into isolated, undeveloped areas in record numbers, threatening to destroy the very beauty they enjoy. Increasingly, developers are trying to build in choice, remote locations, such as the Sonoran Desert in Arizona.

In Phoenix and Scottsdale, some government leaders are concerned that developers will transform the desert into a sea of asphalt. Residents have indicated a willingness to invest tax money to protect mountain preserves

from encroaching development and have declared several mountain areas off limits to developers. The state of Arizona and those cities involved have undertaken to purchase as many as 700,000 acres of land with tax revenues with the purpose of doing nothing with it and simply allowing it to remain in a natural state.

In 1995 South Carolina's Office of Ocean and Coastal Resource Management denied permission to the owners of 7,000 acres of Sandy Island to build a bridge connecting the island to the mainland. Although the owners claimed the bridge would be used to transport harvested timber from there to the mainland, opponents believed the bridge would, in fact, lead to the construction of homes, condominiums, and golf courses on the island.

In Texas, in an effort to balance development with wildlife preservation, the city of Austin invited the Nature Conservancy, a nonprofit environmental group, to develop a plan to protect the environment while enabling building. The Endangered Species Act of 1973 (ESA; PL 93-205) allows such regional arrangements. The result is the Balcones Canyonlands Conservation Plan, a 70,000- to 75,000-acre preserve in the Texas Hill Country, home to a number of endangered species.

In 1997 the Nature Conservancy finalized one of its many negotiations to keep family ranches in environmentally sensitive areas of the country—especially the West—from being broken up. The organization purchased the Dugout Ranch in Utah, which consists of 5,167 acres of privately owned pastureland and 250,000 acres of grassland leased from the government for grazing, with $4.6 million donated by individuals, foundations, and corporations. The ranch was a favored setting for movies and commercials. The Nature Conservancy will maintain it as a working ranch, ecological preserve, and model for how cattle grazing and conservation—at odds throughout the West—can work hand in hand. As of 2003 the Nature Conservancy owned and managed about 15 million acres in the United States, in addition to assisting with conservation efforts around the world.

In May 2002 the North Carolina chapter of the Nature Conservancy purchased 38,000 acres of isolated woodlands and wetlands from a paper company for $24 million. The purchase is part of a massive 100,000-acre area owned by the organization that will supply protected habitat for black bears and other wildlife.

## WETLANDS—FRAGILE ECOSYSTEMS

Marshes, swamps, bogs, estuaries, and bottomlands comprise about 5 to 9 percent of the 48 contiguous states and about 40 percent of Alaska. Although these terms refer to specific biosystems with sometimes very distinctive characteristics, they are commonly grouped together under the name "wetlands." Wetlands provide a vivid example of the dynamic, yet fragile interactions that create, maintain, and repair the world's ecological system. Unfortunately, the fate of many wetlands can also offer concrete evidence of the harmful consequences of human activities that are carried out without regard for, and often without knowledge of, the relationship of each part of the ecosystem to the whole.

Once regarded as useless swamps, good only for breeding mosquitoes and taking up otherwise valuable space, wetlands have become the subject of increasingly heated debate. Many people want to use them for commercial purposes such as agricultural and residential development. Others want them left in their natural state because they believe that wetlands and their inhabitants are indispensable parts of the natural cycle of life on Earth.

## What Are Wetlands?

"Wetlands" is a general term used to describe areas that are always or often saturated by enough surface or groundwater to sustain vegetation that is typically adapted to saturated soil conditions, such as cattails, bulrushes, red maples, wild rice, blackberries, cranberries, and peat moss. The Florida Everglades and the coastal Alaskan salt marshes are examples of wetlands, as are the sphagnum-heath bogs of Maine. Because some varieties of wetlands are rich in minerals and nutrients and provide many of the advantages of both land and water environments, they are often dynamic systems that teem with a diversity of species, including many insects—a basic link in the food chain.

Wetlands are generally located along sloping areas between uplands and deep-water basins such as rivers, although they may also form in basins far from large bodies of water. Of the 90 million acres of wetlands in the lower 48 states, almost all (95 percent) are inland, freshwater areas; the remaining 5 percent are coastal saltwater wetlands. Alaska is estimated to have more than 200 million acres of wetlands.

There are several distinct forms of wetlands, each with its own unique characteristics. The main factors that distinguish each type of wetland are location (coastal or inland), source of water (precipitation, rivers and streams, groundwater), salinity (freshwater or saltwater), and the dominant type of vegetation (peat mosses, soft-stemmed, or woody plants). Wetlands are a continuum in which plant life changes gradually from predominantly aquatic to predominantly upland species. The difficulty in defining the exact point at which a wetland ends and upland begins results in much of the confusion as to how wetlands should be regulated.

## The Many Roles of Wetlands

Experts have understood some of the functions of wetlands for many years. Other purposes have come to light more recently.

**FOOD AND HABITAT.** Wetlands are a source of food and habitat for numerous game and nongame animals. For some species of waterfowl and freshwater and saltwater fish, wetlands are essential for nesting and breeding. About one in five plant and animal species listed as endangered by the U.S. government depend on wetlands for their survival. Two-thirds of the species of Atlantic fish and shellfish that humans consume depend on wetlands for some part of their life cycle, as do nearly half of all species listed as endangered or threatened.

Coastal marshes and some inland freshwater wetlands boast some of the highest rates of plant productivity of any natural ecosystem, thus supporting abundant animal populations within the food chain. After a plant dies nearly 70 percent of it breaks down and is flushed into adjacent waters where it can be consumed by fish and shellfish.

Inland wetlands also serve as way stations for migrating birds. The 30,000-acre region in the north central United States and south central Canada, for example, provides a resting place and nourishment for nearly half of the more than 800 species of protected migratory birds (which individually number in the millions) during the migration season. Without this stopover the flight to their Arctic breeding grounds would be impossible.

**IMPROVING WATER QUALITY.** Wetlands can temporarily or permanently trap pollutants such as excess nutrients, toxic chemicals, suspended materials, and disease-causing microorganisms—thus cleansing the water that flows over and through them. Some pollutants that become trapped in wetlands are biochemically converted to less harmful forms; other pollutants remain buried there; still others are absorbed by wetland plants and either recycled through the wetland or carried away from it. (See Figure 9.5.)

**COMMERCIAL FISHING.** Between 60 and 90 percent of the United States' commercial fish species spawn in coastal wetlands. More than one-half of the country's seafood catch depends on wetlands during some part of their life cycle.

**FLOODWATER REDUCTION.** Isolated and floodplain wetlands can reduce the frequency of flooding in downstream areas by temporarily storing runoff water. For example, the Cache River watershed in southern Illinois retains about 8.4 percent of the watershed's total runoff during flooding.

**SHORELINE STABILIZATION.** Because of their density of plant life, wetlands can dramatically lessen shoreline erosion caused by large waves and major flooding along rivers and coasts.

**RECREATION.** Many popular recreational activities, including fishing, hunting, and canoeing, occur in wet-

**FIGURE 9.5**

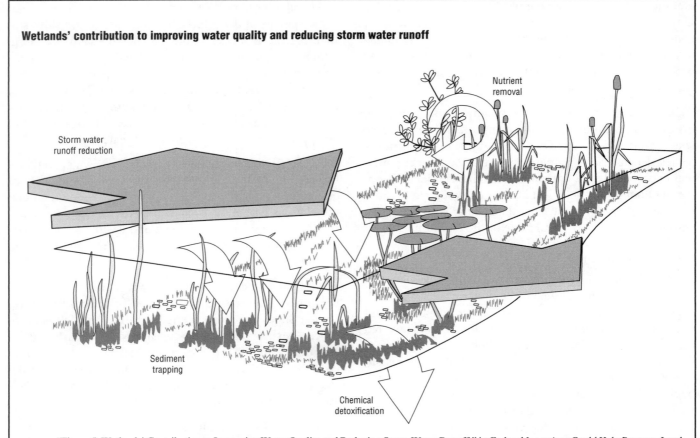

**Wetlands' contribution to improving water quality and reducing storm water runoff**

Nutrient removal

Storm water runoff reduction

Sediment trapping

Chemical detoxification

SOURCE: "Figure 5. Wetlands' Contribution to Improving Water Quality and Reducing Storm Water Runoff," in *Federal Incentives Could Help Promote Land Use That Protects Air and Water Quality*, GAO-02-12, U.S. General Accounting Office, Washington, DC, October 2001

lands. In addition wetland areas provide open space, an important but increasingly scarce commodity.

### History of Wetlands Use

Early Americans considered wetlands nature's failure, a waste in nature's economy. They sought not to preserve nature in its original form but to increase the efficiency of natural processes. In an agricultural economy, land unable to produce crops or timber was considered worthless. Many Americans began to think of draining these lands, an undertaking requiring government funds and resources.

In the nineteenth century state after state passed laws to drain (reclaim) wetlands by the formation of drainage districts and statutes. Coupled with an agricultural boom and technological improvements, reclamation projects multiplied in the late nineteenth and early twentieth centuries. The farmland under drainage doubled between 1905 and 1910 and again between 1910 and 1920. By 1920 state drainage districts in the United States encompassed an area larger than Missouri.

After the Great Depression of the 1920s and 1930s, programs such as the Works Progress Administration and

the Reconstruction Finance Corporation encouraged wetland conversion to form land for urban development. In 1945, at the end of World War II, the total area of drained farmland increased sharply.

In the final few decades of the twentieth century, however, conservationists and the courts have challenged reclamation. Where drainage was once thought to improve the look of the land, today it is more likely to be seen as degrading it. Wetlands turned out not to be wastelands but the conservationist's ideal—systems efficient in harnessing the Sun's rays to feed the food chain. Studies have shown that wetlands have far greater value for flood protection than for their potential agricultural use.

**A LOSS IN RECENT YEARS.** When the first Europeans arrived in America, there were an estimated 215 million acres of wetlands. By the beginning of the twenty-first century only about 90 million or so acres remained. In the 200 years since the birth of the United States, more than 50 percent of the wetlands in the 48 contiguous states have been taken over for agriculture, mining, forestry, oil and gas extraction, and urbanization. Some loss resulted from natural causes such as erosion, sedimentation (the buildup of soil by the settling of fine particles over a long period of time), subsidence (the sinking of land because of diminishing underground water supplies), and a rise in the sea level. However, 95 percent of the losses since the 1970s have been caused by humans, especially by the conversion of wetlands to agricultural land. (See Figure 9.6.)

Eighty percent of the wetland conversions were for agricultural purposes; 8 percent for the construction of impoundments (water confined within an enclosure) and large reservoirs; 6 percent for urbanization; and 6 percent for other purposes such as mining, forestry, and road construction.

More than half (56 percent) the losses of coastal wetlands resulted from dredging for marinas, canals, port development, and, to some extent, from natural shoreline erosion. Urbanization accounted for 22 percent, 14 percent was from creating beaches, 6 percent from natural or human-made transitions of saltwater wetlands to freshwater wetlands, and only 2 percent from agriculture.

California, Ohio, Iowa, Indiana, and Missouri have lost almost all their wetlands. The Fish and Wildlife Service estimates that the United States is losing more than 250,000 acres each year—about 30 acres every hour. Developers have discovered that many of the most tempting sites for new housing or shopping centers are wetlands.

The conversion of wetlands causes the loss of natural pollutant sinks (repositories). As water floods into wetlands from rivers and streams, the loss in velocity causes sediments and their absorbed pollutants to settle out in the wetland before they can enter other water bodies. In the United States artificial wetlands have been proposed as a means of controlling pollution from nonpoint sources. The dramatic decline in wetlands globally suggests not only loss of habitat but also diminished water quality.

One of the largest wetlands in the United States is the Everglades of south Florida. The fresh water that used to flow naturally into this ecosystem has been diverted by decades of canal building. According to Nicole Duplaix in "South Florida Water: Paying the Price" (*National Geographic,* July 1990), drainage changes implemented since 1920 to create farmland or housing had dried out half of the Everglades National Park, and left the rest heavily polluted. Scientists are studying the complex water flow problems in this area and are working to restore some natural drainage characteristics to the Everglades. According to the Comprehensive Everglades Restoration Plan (CERP) Web site (http://www.evergladesplan.org/index.cfm), as of 2002 some limited progress had been made and environmental improvements were expected by 2010.

**Concern over Property Rights**

The dispute over wetlands regulation reflects Americans' ambivalence when private property and public rights intersect, especially since three-fourths of the nation's

FIGURE 9.6

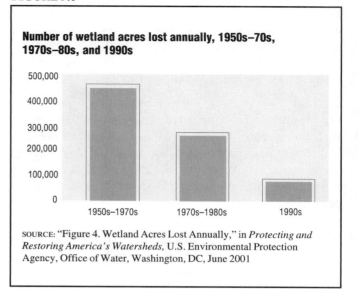

**Number of wetland acres lost annually, 1950s–70s, 1970s–80s, and 1990s**

SOURCE: "Figure 4. Wetland Acres Lost Annually," in *Protecting and Restoring America's Watersheds,* U.S. Environmental Protection Agency, Office of Water, Washington, DC, June 2001

wetlands are owned by private citizens. In recent years many landowners have complained that wetland regulation devalued their property by blocking its development. They argue that efforts to preserve the wetlands have gone too far, citing instances where a small wetland precludes the use of much larger surrounding areas. Some large landowners have long opposed any federal (and state and local) powers to protect resources such as wetlands that might limit their land use options.

Another policy question of concern to the public is the right of the federal government to take property without compensation. The "takings" clause of the Constitution (the Fifth Amendment) provides that, when private property is taken for public use, just compensation must be paid to the owner. Owners claim that when the government—through its laws—eliminates some uses for their land, the value is decreased and they should, therefore, be paid for the loss.

**MOUNTAINS**

Mountains are one of Earth's most important features. They span one-fifth of the landscape and house one-tenth of Earth's population. Roughly two billion people live downstream from mountains and depend on the water, hydropower, grassland, timber, and mineral resources generated by those mountains.

**What Is a Mountain?**

A mountain is a landmass that projects conspicuously above its surroundings and is higher than a hill, generally at least 985 feet (300 meters) in height. An additional criterion is that a mountain's rise creates climates, soils, and vegetation distinct from those in surrounding lowlands. Mountains share common physical attributes of steepness, instability, and ecology that create natural hazards, microclimates, niches of biodiversity, and inaccessibility. The collision of tectonic plates produces mountain uplift and numerous physical hazards, such as earthquakes, volcanic eruptions, landslides, avalanches, and floods. The slope and altitude of mountains create variations in climate—temperature, radiation, wind, and moisture—over very short distances.

Mountains function as the Earth's water towers by attracting much of its precipitation—they are the predominant and most dependable source of fresh water for humankind. A diversity of wild plants makes mountain ecosystems vital sources of food and pharmacological benefits for humans.

## A Naturally Vulnerable Resource

A distinguishing feature of mountains is vulnerability to disturbance, largely due to the vertical dimension (height and slope). Because a doubling in water speed magnifies the size of materials that water can transport, the erosive power of rapid runoff from mountains is immense. Unlike lowland environments, mountain ecosystems are typically less able to recuperate from disruptions such as soil erosion or loss of vegetation; soils are usually thin and poorly anchored, and gravity-powered erosion speeds silt and sediment movement. Also, many mountains are still growing and are less geographically stable than flatter landmasses. Seven of the world's 14 tropical "hotspots" for plants threatened by destruction have at least half their area in mountains, and 131 of the world's 247 bird habitats are in tropical mountains.

As studies seem to confirm the warming of Earth's environment, experts predict that such warming will proceed too fast for many ecosystems to adapt. Austrian researchers have found that nine plant species typical of the nival zone (above alpine grasslands) are migrating to higher altitudes at one meter per decade, but will have to move eight to ten meters per decade to keep up with the current rate of warming.

## Impacts on Mountain Ecology

Mountains face threats from poor land use patterns, resource extraction, and mass tourism and recreation. Half of U.S. rangeland, most of it in the mountainous West, is now considered severely degraded, with its livestock-supporting ability reduced by 50 percent. The FAO reports that hill and mountain forests are more susceptible to ecological damage from excessive population densities than are lowland forests. Because their slopes permit gravity to increase the power of flowing waters, mountains attract most of the Earth's hydroelectric projects and irrigation reservoirs.

Of all the economic activities in the world's mountains, nothing rivals the destructive power of mining. Environmental impacts include habitat destruction, erosion, air pollution, acid drainage, and metal contamination of water bodies. The result is often denuded forests, eroded hillsides, and dammed or polluted rivers.

**TOURISM AND RECREATION.** Many lowlanders feel that mountains are a refuge from modern life, but problems—environmental ones—are following them up the slopes. In slick ads and commercials, images of pristine mountain wilderness lure multitudes for respite and sport. In industrial countries mass tourism and recreation are fast becoming the largest threats to mountain environments.

Since 1945 visits to the ten most popular mountainous national parks in the United States have increased 12-fold. Infrastructure for leisure and recreation in the mountains can be exorbitant. In 1990 there were 100 Alpine golf courses; by the end of 1996, 500 existed. Thundering helicopters bring skiers to untracked slopes not only in the American Rockies but worldwide. The populations of many small ski towns, like Vail, Colorado, have more than doubled since 1980, causing new home construction and retail sales to grow at double or triple the national average. White-water rafters and mountain climbers prowl the slopes in numbers some consider dangerous. A generation ago conquering Mount Everest was considered an unimaginable feat; today, dozens of climbers reach the peak every month, leaving behind trash—even the bodies of climbers that die on the dangerous climb are sometimes left behind. The interference of trash and excess traffic has become a serious problem on the mountain.

## EROSION

Erosion is the process in which the materials of the Earth's crust are worn and carried away by wind, water, and other natural forces. The destruction of forests and native grasses has allowed water and wind greater opportunity to erode the soil. Changes in river flow and seepage from human technology have shifted the runoff patterns of water and the sediment load of rivers that, in turn, deposit into lakes and oceans. Erosion has become a problem in much of the world in areas that are overfarmed or where topsoil cannot be protected, such as on coasts, which are often overdeveloped.

### Coastal Erosion

In *Evaluation of Erosion Hazards,* a study prepared for the Federal Emergency Management Agency in April 2000, the H. John Heinz III Center for Science, Economics, and the Environment, a nonprofit research organization, found that approximately 25 percent of structures within 500 feet of the U.S. coastline will suffer the effects of coastal erosion within 60 years. Especially hard hit will be areas along the Atlantic and Gulf of Mexico coasts, which are expected to suffer 60 percent of nationwide losses.

The nation's highest average erosion rates—up to six feet or more per year—occur along the Gulf of Mexico. The average erosion rate on the Atlantic coast is two to three feet per year. A major storm can erode 100 feet of coastline in a day. The Heinz Center estimates that roughly 10,000 structures are within the estimated ten-year erosion zone closest to the shore. This does not include structures in the densest areas of large coastal cities, such as New York, Chicago, Los Angeles, and Miami, which are heavily protected against erosion.

The powerful effects of erosion were dramatized by the predicament of the Cape Hatteras lighthouse in North Carolina. When it was constructed in 1870 the lighthouse was 1,500 feet from the shore. By 1987 the lighthouse stood only 160 feet from the sea and was in danger of collapsing. In 1999 the National Park Service, at a cost of $9.8 million, successfully moved the lighthouse back 2,900 feet.

Erosion of beaches on the East Coast is becoming a more serious problem as development inches closer to the ocean. The Army Corps of Engineers has been rebuilding eroded beaches since the 1950s. The federal government pays 65 percent of the cost of beach rebuilding, with states and local governments paying the remaining 35 percent. Many experts, however, believe that beach replenishment is a futile effort and that funds could be better spent elsewhere.

## Soil Erosion and Agriculture

Agricultural lands are the principal source of eroded soil. According to the U.S. Department of Agriculture (USDA), approximately 20 percent of the nation's land is set aside for cropland. Three-quarters of this land is actively used to grow crops for harvesting. The remainder is used for pasture or is idled for various reasons. This would include cropland enrolled in the Federal Conservation Reserve Program (CRP). (See Figure 9.7.)

Demands on the Earth to feed growing populations and changes in the Earth's landscape caused by human activities have speeded up soil erosion. Soil erosion has increased to the point where it far exceeds the natural formation of new soil, and experts consider the problem to be of epidemic proportions. According to *Excessive Erosion on Cropland, 1997* (2000) the USDA's Natural Resources Conservation Service found that 108 million acres were eroding excessively.

FARMING PRACTICES. Agriculture depends primarily on the top six to eight inches of topsoil. Fields planted in rows, such as corn, are most susceptible to soil runoff. In 2002 corn comprised 22 percent of total acres used for crops in the United States. Cover crops, such as hay, provide more soil cover to hold the land. Hay crops accounted for 17 percent of total acres used for crops in 2002. (See Figure 8.3 in Chapter 8.)

FIGURE 9.7

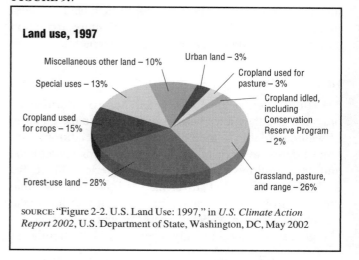

Land use, 1997

Miscellaneous other land – 10%
Urban land – 3%
Special uses – 13%
Cropland used for pasture – 3%
Cropland used for crops – 15%
Cropland idled, including Conservation Reserve Program – 2%
Forest-use land – 28%
Grassland, pasture, and range – 26%

SOURCE: "Figure 2-2. U.S. Land Use: 1997," in *U.S. Climate Action Report 2002*, U.S. Department of State, Washington, DC, May 2002

Historically, when most of the topsoil was lost farmers would abandon the land. Now, however, farmers continue to plow the soil, even when it consists of as much subsoil as topsoil. It costs more money to produce food on such land than on land where topsoil is present. Farmers often use more fertilizer to make up for the decreasing productivity of the soil, and that, in turn, adds to environmental pollution.

The GAO estimates that about 28 percent of the nation's cropland is highly erodible. The states with the highest percentage of highly erodible cropland are New Mexico (90 percent), Arizona (81 percent), and Colorado (77 percent). In absolute terms Texas and Montana have the most erodible land. The amount of erosion has declined in past decades. The USDA attributes the decline to the CRP, which pays farmers to take land out of production for ten years, and to the Conservation Compliance Program. As part of the 1985 Farm Act (PL 99-198), the Conservation Compliance Program was initiated as a major policy tool. To be eligible for agricultural program benefits, farmers must meet minimum levels of conservation on highly erodible land.

ENVIRONMENTAL QUALITY INCENTIVES PROGRAM. Congress, under the 1996 Farm Bill (PL 104-127), authorized the Environmental Quality Incentives Program (EQIP) to address agriculture's natural resource and environmental problems. It is a flexible, voluntary, and effective conservation program that allocates millions of dollars each year to farmers' conservation efforts. EQIP was reauthorized in the Farm Security and Rural Investment Act of 2002 (Farm Bill).

Under EQIP the USDA provides assistance to family-sized farms and ranches for up to 75 percent of the cost of certain environmental protection practices, such as grassed waterways, filter strips, manure management facilities, capping abandoned wells, and wildlife habitat enhancement. The USDA may also offer incentive payments to encourage producers to apply such land management practices to the use of nutrients, manure, irrigation water, wildlife, and integrated pest management.

## IRRIGATION

Human beings have survived in deserts or arid areas only because they have been able to increase the quantity of water available to meet their needs. An elaborate system of dams, reservoirs, irrigation pipelines, aqueducts, and canals allows residents of the American West, for example—especially California—to ignore the fact that they live in a naturally dry climate. That fact has made California's Central Valley the most productive agricultural region in the world on only 3 percent of U.S. farmland.

Constant irrigation, however, is not a miracle solution as once was thought. According to the United Nations Environment Program (UNEP), 90 percent of the land in Egypt, 68 percent in Pakistan, 50 percent in Iraq, 38 percent in Peru, 30 percent in the United States, and 20 percent each in India, Russia, and Australia is suffering salinization (saltiness) caused by irrigation. Sodium in the soil or irrigation water accumulates at the root level of soils or turns into a sterile, rock-hard crust. An estimated 5 million acres of irrigated land are pulled from production each year because of waterlogging and salinization, the result of poor land management. In addition, irrigated land is often paved over for housing, factories, and roads, especially in the United States and Asia, further reducing the productive use of land for agriculture.

## BIODIVERSITY

Biological diversity, or biodiversity, refers to the full range of plant, animal, and microbial life and the ecosystems that house them. Environmentalists began using the term during the 1980s when biologists increasingly warned that human activities were causing a loss of plant and animal species.

Studies of deforestation have supported the concerns about declining biodiversity, showing that tropical rain forests have dwindled from 3.5 billion acres before the industrial era to fewer than two billion acres. Deforestation has meant extinction for hundreds of species of plants and animals each year. The exact number of species in the remote forests is unknown, although it is generally accepted that they house the greatest number of species on the planet.

### Extinction Rates

No one knows how many species of plants and animals exist in the world. By the beginning of the twenty-first century scientists had named and documented 1.4 million species. Educated guesses of the total number of different species range from five million to 100 million. Just as the health of a nation is promoted by a diverse economy, so the health of the biosphere is promoted by a diverse ecology.

Widespread extinctions have occurred infrequently in Earth's history and are generally believed to have been due to major geological and astronomical events. Scientists call the disappearance of only a few species over the period of a million years a "background rate." When that background rate doubles for many different groups of plants and animals at the same time, a mass extinction is taking place.

THE SIXTH EXTINCTION? At least five times in the last 600 million years planet-wide cataclysms, such as drastic climate change or colliding asteroids, have wiped out whole families of organisms. Because of these losses scientists believe that more than 95 percent of all species that have ever existed are extinct. Researchers predict that, as tropical ecosystems are converted to farms and pasture, the extinction rate will approach several hundred extinctions per day before the mid-twenty-first century—millions of times higher than the background rate. The Worldwatch Institute believes that more species of flora and fauna may disappear in our lifetime than were lost in the mass extinction that included the disappearance of the dinosaur 65 million years ago.

The loss of diversity leads to problems beyond the simple loss of animal and plant variety. When local populations of species are wiped out, the genetic diversity within that species that enables it to adapt to environmental change is diminished, resulting in a situation of "biotic impoverishment." Those organisms that do survive are likely to be hardy, "opportunistic" organisms tolerating a wide variety of conditions—characteristics often associated with pests. Experts suggest that, as some species dwindle, their places may be taken by a disproportionate number of pest or weed species that, while a natural part of life, will be less beneficial to human beings.

Most living species have never been identified. Mammals, including humans, make up barely three-tenths of 1 percent of all known organisms. There were 1,263 species in the United States listed as endangered or threatened as of April 2004. Another 558 species are listed for foreign countries. (See Table 9.1.) The number of endangered or threatened species listed in the Unites States has increased dramatically since 1980 when less than 300 species were listed.

Scientists participating in the "Global Biodiversity Strategy," an international team of 500 researchers, believe that the extinction of species will deprive future generations of new medicines and new strains of food crops. With as many as 50 plant species disappearing daily, the researchers calculate that the planet's diversity

TABLE 9.1

**Summary of listed endangered and threatened species, as of April 1, 2004**

| Group | Endangered U.S. | Endangered Foreign | Threatened U.S. | Threatened Foreign | Total species |
|---|---|---|---|---|---|
| Mammals | 69 | 251 | 9 | 17 | 346 |
| Birds | 77 | 175 | 14 | 6 | 272 |
| Reptiles | 14 | 64 | 22 | 15 | 115 |
| Amphibians | 12 | 8 | 9 | 1 | 30 |
| Fishes | 71 | 11 | 43 | 0 | 125 |
| Clams | 62 | 2 | 8 | 0 | 72 |
| Snails | 21 | 1 | 11 | 0 | 33 |
| Insects | 35 | 4 | 9 | 0 | 48 |
| Arachnids | 12 | 0 | 0 | 0 | 12 |
| Crustaceans | 18 | 0 | 3 | 0 | 21 |
| Animal subtotal | 391 | 516 | 128 | 39 | 1074 |
| Flowering plants | 569 | 1 | 144 | 0 | 714 |
| Conifers and cycads | 2 | 0 | 1 | 2 | 5 |
| Ferns and allies | 24 | 0 | 2 | 0 | 26 |
| Lichens | 2 | 0 | 0 | 0 | 2 |
| Plant subtotal | 597 | 1 | 147 | 2 | 747 |
| **Grand total** | **988** | **517** | **275** | **41** | **1821** |

SOURCE: "Summary of Listed Species—Species and Recovery Plans as of 4/1/2004," in *Threatened and Endangered Species System (TESS)—Listed Species Summary (Boxscore),* U.S. Department of the Interior, U.S. Fish and Wildlife Service, Washington, DC, April 12, 2004 [Online] http://ecos .fws.gov/tess_public/TESSBoxscore?format=display&type=archive&sysdate =4/01/2004 [accessed April 12, 2004]

could be reduced by 10 percent by 2015. One-fourth of all medical prescriptions in the United States contain active ingredients from plants. Among the medically useful species are some used in the treatment of cancer, Human Immunodeficiency Virus (HIV) and Acquired Immune Deficiency Syndrome (AIDS), circulatory disorders, bacterial infections, anxiety, inflammatory diseases, and for the prevention of organ rejection in transplants.

## Species Loss—Crisis or False Alarm?

As with most environmental questions, not all experts agree about the threat to species diversity. Some observers believe that extensive damage to species diversity has not been proven and claim that, while wild habitats are disappearing because of human expansion, the seriousness of the extinction has been exaggerated and is unsupported by scientific evidence. They point to the fact that the total number of species and their geographic distribution are unknown. How, they ask, can forecasts be made based on such sketchy data?

Other observers contend that extinctions, even mass ones, are inevitable and occur as a result of great geological and astronomical events that humans cannot affect. They do not believe that disruptions caused by human activity are enough to create the mega-extinction prophesied by people they consider "alarmists."

Furthermore, some critics of the environmental movement believe that the needs of humans are being made

secondary to those of wildlife. They contend that the Endangered Species Act protects wildlife regardless of the economic cost to human beings. Sometimes, as in the case of the spotted owl of the Pacific Northwest forests, that cost is the loss of jobs for people. The owl's presence halted logging there—following protests by environmental groups—at considerable economic loss to communities and families in the area. Furthermore, critics contend that halting development because it threatens a species whose whole population occupies only a few acres and numbers only in the hundreds is simply nonsense.

According to a 1998 poll conducted for the American Museum of Natural History by Louis Harris and Associates, biologists overwhelmingly view the loss of biodiversity as a serious problem. According to the survey, most scientists agreed that, if trends continue, the loss of species will have a very negative effect on the Earth's ability to recover from both natural and human-made disasters.

## Earth Summit Biodiversity Treaty

At the 1992 Earth Summit in Rio de Janeiro, 156 nations signed a pact to conserve species, habitats, and ecosystems. This Biodiversity Treaty is regarded as one of two main achievements of the United Nations Conference on Environment and Development, the other being a treaty on global warming. The Biodiversity Treaty makes nations responsible for any environmental harm in other countries produced by companies headquartered in their country.

One provision of the treaty concerns "biotechnology," a term referring to the ownership of genetic material. Plants, seeds, and germ plasm have historically been in the public domain (belonging to the general public), rather than belonging to any particular government. Therefore, anyone could exploit or use them without compensation to the country of origin. For example, the rosy or Madagascar periwinkle, a plant found only in the tropical rain forests of Madagascar, is used as a base for medication to treat Hodgkin's disease and childhood leukemia. Madagascar receives no compensation for use of the plant. The biotechnology treaty drafted in Rio called for compensation to be paid for the use of those genetic materials.

The United States did not sign the treaty at the time. The administration of George H. W. Bush, while agreeing with many provisions of the pact, believed the economic requirements for accomplishing those goals were unacceptable to American businesses because they would be forced to compensate for the use of these species. President Bill Clinton signed the treaty in 1994. However, as of April 2004 the treaty had not been ratified by the U.S. Senate.

## Endangered Species Act

The 1973 Endangered Species Act (ESA), passed into law during the administration of President Richard Nixon, was originally intended to protect creatures like grizzly bears and whales with whose plight Americans found it easy to identify. In the words of its critics, however, it has become the "pit bull of environmental laws," policing the behavior of entire industries. In three decades the ESA has gone from being one of the least controversial laws passed by Congress, to one of the most contentious.

The ESA regulates industries that can cause fish and wildlife populations to decline. It also determines the criteria to decide which species are endangered. Since the act was first passed, the pendulum has periodically swung between increased protection and the need to soften the law's economic impact.

The U.S. Supreme Court, in *Manuel Lujan, Jr., Secretary of the Interior v. Defenders of Wildlife et al.* (504 US 55 1992), determined that groups and individuals cannot sue the government solely on behalf of the public interest or on behalf of the flora and fauna they seek to protect. Instead, they must demonstrate harm to themselves. The ruling has been generally regarded as a victory for business interests and a defeat for environmentalists in their efforts to protect endangered species, since immediate harm is often difficult to show in environmental issues.

## WILDLIFE

The loss of habitats, the contamination of water and food supplies, poaching, and indiscriminate hunting and fishing have depleted the population of many species. Most scientists agree that prospects for the survival of many species of wildlife, and hence biodiversity, are worsening. The expansion of human development into wildlife habitats has resulted in some animals being squeezed into cities and suburbs where encounters between humans and wildlife have become increasingly common.

Species loss and habitat loss are related. Scientists have recognized for some 150 years the connection between the size of an area and the number of species it contains—as large tracts of land are lost, so are some species that make their homes there. Other major causes of animal extinction are hunting and invasive (nonnative) species.

### Invasive Species

An invasive species is one that is not native to a particular ecosystem and whose presence there causes environmental or economic harm or harm to human health. This includes species purposely introduced (such as the plant saltcedar, which was brought to the United States to control erosion) and unintentionally introduced (such as zebra mussels, which are thought to have arrived in the ballast water of ships). Invasive species often have high reproductive rates and lack predators in their new environments. They can choke out or "out-compete" native species.

Many scientists consider invasive species to be one of the most serious issues threatening the environment. In response to this threat, the National Invasive Species Council was established by the U.S. government in 1997. The council includes members from a variety of agencies including the EPA, the USDA, and the U.S. Department of the Interior. In 2001 the council issued its management plan for dealing with the invasive species problem in *Meeting the Invasive Species Challenge.*

The report states that invasive plants infest approximately 100 million acres in the United States and cost around $137 billion annually for prevention and control. Zebra mussels, which are believed to have arrived in the ballast of ships in the Great Lakes, are one invasive species that has spread rapidly. By 1999 zebra mussel populations extended all the way to the Gulf Coast. The zebra mussel is considered so permanently entrenched that wholesale eradication would be virtually impossible. Instead, authorities are concentrating on limiting further spread of the pests, which clog water intake pipes.

In addition, authorities are increasingly concerned about the West Nile virus, an invasive pathogen that is thought to have originated in Africa. The virus was first detected in the United States in 1999 in New York. It infected animals and birds throughout the East and spread west quickly, carried by migratory birds. The virus can be transmitted to humans by mosquitoes that have bitten infected animals and birds. In 1999 there were 62 human cases reported to the Centers for Disease Control and Prevention (CDC), and 7 people died from the virus. The number of human cases and deaths grew steadily each following year. In 2003 there were 9,858 cases reported to the CDC and 262 deaths. Figure 9.8 shows the counties as of March 24, 2004 that have experienced positive test results for West Nile virus in humans.

### Sharing the Planet

In the nineteenth century, miners took parakeets with them into the mines. If a bird died, they knew they were in danger from noxious gases. While more scientific and humane procedures now exist to determine how dangerous the situation is, some scientists believe that plants and animals may still serve as indicators of the safety of the world. When biologists discover toxic amounts of poisons in wildlife, they ask whether human beings are also ingesting these poisons.

Some observers believe that animals should be protected out of an intrinsic respect for life, aside from any market value or use to humans. Others contend that humankind must manage wildlife correctly because biodiversity makes good economic and survival sense. Still others believe that there is no species-loss "problem," that species loss is a natural part of evolution. All of these issues are being deliberated as the people of the world

FIGURE 9.8

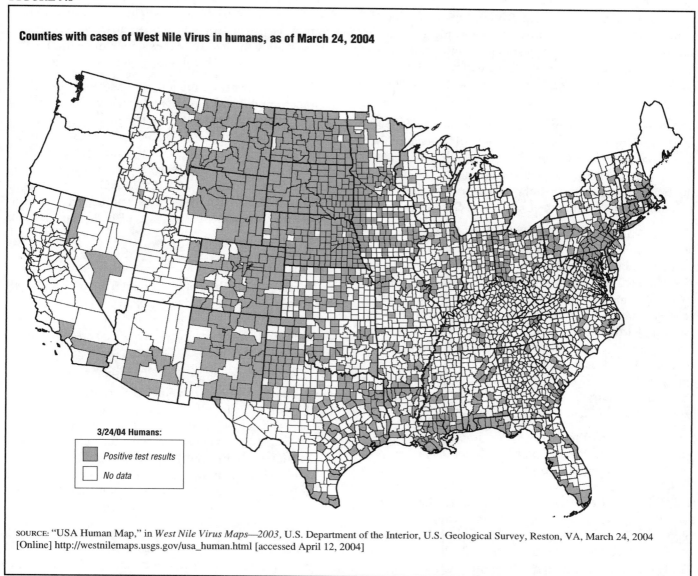

**Counties with cases of West Nile Virus in humans, as of March 24, 2004**

3/24/04 Humans:

☐ Positive test results

☐ No data

SOURCE: "USA Human Map," in *West Nile Virus Maps—2003*, U.S. Department of the Interior, U.S. Geological Survey, Reston, VA, March 24, 2004 [Online] http://westnilemaps.usgs.gov/usa_human.html [accessed April 12, 2004]

struggle to decide how best to live with the other animals and plants that populate the Earth.

A 1996 study by the Nature Conservancy on more than 20,000 American plant and animal species found that about one-third of species were rare or imperiled, a larger fraction than some scientists had expected. The study, the most comprehensive assessment to date of the state of American plants and animal species, found that mammals and birds were doing relatively well compared to other groups, but that a high proportion of flowering plants and freshwater marine species, like mussels, crayfish, and fish, were in trouble. Of the 20,481 species examined, about two-thirds were secure, 1.3 percent were extinct or possibly extinct, 6.5 percent were critically imperiled, 8.9 percent were imperiled, and 15 percent were considered vulnerable. The destruction or degradation of habitat was considered to be the main threat.

## City Life Collides with Wilderness

The growth of urban areas has resulted in a collision between city life and wildlife. Increasingly, humans are encountering wild animals in their communities. The 600,000-acre Angeles National Park in the Los Angeles outback of the San Gabriel Mountains has been the site of numerous attacks on visitors by snakes and wild animals. In addition humans are using some areas designed for wildlife for undesirable purposes; poachers and some hunters shoot deer out of season and prey on the endangered Nelson bighorn sheep. During Christmas season trees are cut down. Crowds of picnickers and hikers often swell to music-festival size and clog roads. Toxic-waste outlaws heave garbage and poisons into creeks and abandoned mine shafts. Criminals sell drugs in the forest and even have small marijuana plantations located there. The Angeles National Park has also become a well-known dumping ground for homicide victims—eight bodies were found in the forest in 1995.

## Some Cases of Threatened Species

Almost daily the decline or threat to some plant or animal is reported. Scientists attribute the decline of salmon on the West Coast to spoiled habitat and disruption of river flow. Erosion of the coastline in Florida has left no place for sea turtles to dig their nests, and they are dying off.

Peregrine falcons, one of the first species to be listed on the Endangered Species List, were dying because they were consuming DDT in the food chain. Following their listing under the ESA, and the banning of DDT in 1972, the falcon population has rebounded. In 1999 they were officially removed from the Endangered Species List.

The ivory tusks of African elephants are very valuable as they can be fashioned into jewelry and artwork. In the mid-twentieth century African elephants were so extensively hunted for their ivory that their population dropped to dangerous levels. The international community responded in 1990 by banning trade in African elephant ivory under the Convention on the International Trade in Endangered Species. Poaching of elephants continued but their population began to rebound. By 1999 there were so many elephants in Zimbabwe, Namibia, and Botswana that those countries (unsuccessfully) requested permission to resume limited trade in ivory.

Dolphins tend to swim with schools of tuna in the Pacific Ocean and nets used by commercial fisheries to catch tuna also entrap dolphins. Since netting began in 1958 an estimated seven million dolphins have been killed. In 1972 Congress passed the Marine Mammal Protection Act (PL 92-522) to reduce the deaths of the dolphins. The law was amended in 1985 and 1988 to regulate tuna imported from other countries. Trade groups have challenged these regulations by pointing to the economic losses of companies and nations that abide by the law. Some companies ignore the law, while other companies have printed a "Dolphin Safe" label on their tuna products to show that they obey the law.

Environmentalists have long argued with government and industry over the question of logging in the Pacific Northwest forests. Environmentalists claim that the biological health of the ecosystem is in decline and more than 100 species of plants and animals are threatened with extinction, while the timber industry responds that the forest provides jobs for thousands of Americans and lumber for millions of people.

The argument came to a head in 1990 when the spotted owl—which lived only in this particular region—was added to the list of endangered species. Logging was halted and a succession of lawsuits was filed against the Forest Service and the Department of the Interior. In 1992 President Bush grudgingly restricted logging in that area but, at the same time, moved to amend the law to allow economic considerations to be taken into account. In 1994 President Clinton worked out what was claimed to be a compromise between environmentalists and business interests, allowing logging to resume with restrictions on the size, number, and distribution of trees to be cut.

Scientists investigating a worldwide decline in frogs and other amphibians have found evidence identifying a number of factors that have contributed to the decline: ultraviolet radiation caused by the thinning of the ozone layer, chemical pollution, and a human taste for frog meat. These species are considered indicator species because their sensitivity makes them early indicators of environmental damage. Butterflies, another creature considered an indicator species, are also disappearing in many areas.

## Roads and Wildlife

Almost four million miles of public roads cross the 48 contiguous states. As roadways reach further and further into undeveloped areas, encounters with wildlife are inevitable. In *Critter Crossing: Linking Habitats and Reducing Roadkill* (2000) the U.S. Department of Transportation reported that roads impact wildlife in several ways:

- Roadkill—Vehicles traveling U.S. roads kill millions of birds, mammals, reptiles, and amphibians every year. The ocelot, an already endangered cat, is in further jeopardy due to highway kills. In addition humans are sometimes killed or injured in animal-vehicle collisions. The insurance industry estimates the cost of these fatalities and injuries is about $200 million; motorists pay at least $2,000 in vehicle repair when they hit a large animal.

- Habitat loss—When humans build highways and develop areas, they destroy habitat. This forces animals into smaller and smaller areas and into areas inhabited by humans. Some species cannot migrate, and therefore die; others are forced to compete for fewer resources to live and breed.

- Habitat fragmentation—When roads cut through wild areas, they divide wildlife populations into smaller, more isolated, and less stable groups. These animals become more vulnerable to predators and are given to inbreeding with its resulting genetic defects.

Under the 1998 Transportation Equity Act for the Twenty-First Century (PL 105-178), the Federal Highway Administration can provide wildlife crossings—"habitat connectivity measures"—for new and existing roads. Among the strategies used to counteract habitat loss and roadkill are overpasses and underpasses, tunnels, and culverts.

## Deep-Sea Harvesting

Worldwide, after centuries of steady growth, the total catch of wild fish peaked in the early 1990s and has declined ever since. In *State of World Fisheries and Aquaculture, 2002,* the FAO reported that approximately 47 percent of all commercial fish stocks have been fully exploited. A further 18 percent of stocks were reported to be overexploited.

A result of the declining catches of fish in shallow fisheries is the recent scouring of the deep seas for other varieties of fish such as the nine-inch-long royal red shrimp, rattails, skates, squid, red crabs, orange roughy, oreos, hoki, blue ling, southern blue whiting, and spiny dogfish. Although limited commercial fishing of the deep has been practiced for decades, new sciences and technologies are making it more practical and efficient. As stocks of better-known fish shrink and international quotas tighten, experts say the deep ocean waters will increasingly be targeted as a source of seafood. Scientists worry that the rush for deep-sea food will upset the ecology of the ocean.

### A Blue Revolution?

The decline in the availability of fish has produced the growth of aquaculture, or farmed fish. Worldwide, one in every five fish eaten is raised on a farm, a share that is expected to rise in the years ahead. Aquaculture is one of the fastest growing sectors in world food production.

As a source of animal protein, farmed fish are an economical alternative to beef or pork and are on par with chicken. While 4 kilograms of grain are required to produce each kilogram of pork and 7 for each kilogram of beef, only 2 kilograms of grain are needed for a kilogram of chicken or fish. In addition only 40 percent of the weight of sheep is eaten and 50 percent of pigs and chicken, while 65 percent of finfish (as opposed to shellfish) is actually consumed. (Because fish are supported by water they have little bone structure; therefore, more of their weight is edible.) Fish are also low in fat and cholesterol, an advantage over other meats.

### Contamination of Fish

Noncommercially caught fish and wildlife are sometimes contaminated with chemicals, such as mercury, polychlorinated biphenyls (PCBs), and DDT. In order to protect consumers from health risks associated with consuming such pollutants, the EPA and the states issue consumption advisories to inform the public that high concentrations of contaminants have been found in local specimens. In May 2003 the EPA published its annual publication *Update: National Listing of Fish and Wildlife Advisories.* During 2002 there were 2,800 fish consumption advisories in effect around the country. (See Figure 6.11 in Chapter 6.)

### MINERALS AND OIL

Materials extracted from the Earth are needed to provide humans with food, clothing, and housing and to continually upgrade the standard of living. Some of the materials needed are renewable resources, such as agricultural and forestry products, while others are nonrenewable, such as minerals.

The USGS reported in *Materials Flow and Sustainability* (1998) that a significant trend is the decreasing use of renewable resources and the increasing demand for nonrenewable resources. Since 1900 the use of construction materials such as stone, sand, and gravel, has soared.

The large-scale exploitation of minerals began in earnest with the Industrial Revolution around 1760 in England and has grown rapidly ever since. In a world economy based on fossil fuels, minerals and oil are valuable. The value increases in proportion to demand—which is increasing—and supply—which is decreasing. The result is that the search for minerals and fuel sources has become very aggressive and may be detrimental to the environment.

Mining has always been a dirty industry. As early as 1550, German mineralogist and scholar Georgius Agricola wrote: "The fields are devastated by mining operations ... the woods and groves cut down ... then are exterminated the beasts and birds.... Further, when the ores are washed, the water that has been used poisons the streams, and either destroys the fish or drives them away."

Centuries later mining still pollutes the environment, only on a larger scale. The Clean Air Act (PL 101-576), the Clean Water Act (PL 92-500), and the Resource Conservation and Recovery Act of 1976 (PL 94-580) regulate certain aspects of mining but, in general, the states are primarily responsible for regulation, which varies widely from state to state.

### Gold Mining

An extraction process in which cyanide is used for the retrieval of gold from tailings or residues left over from other mining operations has become quite controversial. A series of mishaps around the world have left some communities with polluted lakes, rivers, and streams. A major disaster occurred in Romania on January 30, 2000, when an overflow of polluted mud and wastewater from the Aurul Gold smelter dam sent 100,000 cubic meters of cyanide-tainted wastewater into the adjacent Lapus and Somes Rivers. The cyanide was subsequently carried downstream to the Danube River in Yugoslavia. The spill contaminated drinking water for 2.5 million people and killed more than 100 tons of fish.

### Oil in the Arctic

The search for oil has led to the exploration of the Alaskan wilderness. Since the oil supply from the existing North Slope Reserve will steadily decline and then eventually disappear, exploratory oil drillers are focusing their

**FIGURE 9.9**

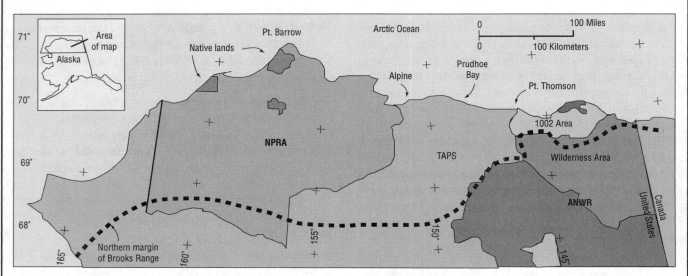

Northern Alaska, showing locations and relative sizes of the National Petroleum Reserve in Alaska (NPRA) and the Arctic National Wildlife Refuge (ANWR)

SOURCE: "Figure 1. Map of Northern Alaska Showing Locations and Relative Sizes of the National Petroleum Reserve in Alaska (NPRA) and the Arctic National Wildlife Refuge (ANWR)," in *U.S. Geological Survey 2002 Petroleum Resource Assessment of the National Petroleum Reserve in Alaska (NPRA)*, U.S. Department of the Interior, Washington, DC, 2002

attention on the National Petroleum Reserve in Alaska (NPRA) in the Arctic wilderness. The NPRA is a 23-million-acre area in northwestern Alaska. (See Figure 9.9.) Geologists consider northern Alaska to be the last, great untapped oil field in North America. Environmental experts fear that oil and gas development will seriously harm the area.

In 2002 the USGS assessed the NPRA and found a significantly greater supply of petroleum (5.9 to 13.2 billion barrels) than previously estimated. Only up to 5.6 billion barrels of this petroleum are technically and economically recoverable at existing market prices. The USGS suspects that there may be as much as 83.2 trillion cubic feet of undiscovered natural gas in the same area. Transportation of this gas to markets would require a new pipeline. There is already a pipeline system in place for oil—the Trans-Alaska Pipeline System (TAPS), which lies between the NPRA and the Arctic National Wildlife Refuge (ANWR) as shown in Figure 9.9. The ANWR is a 19-million-acre area of pristine wilderness along the Alaska-Canada border. It, too, is being considered for oil exploration, a move strongly opposed by environmentalists.

The future of the refuge lies in the hands of the federal government. The administration of George H. W. Bush made drilling there a major foundation of the national energy policy. Under the Clinton administration oil and mineral development was prohibited within the wildlife refuge. In April 2002, following heated debate, the U.S. Senate killed a proposal by the administration of George

W. Bush to let oil companies drill in ANWR. Republicans raised the issue again in the fall of 2003, citing the need for the nation to reduce its dependence on oil imported from the Middle East. As of April 2004 the U.S. Congress has not yet approved drilling in ANWR.

## Antarctic Resources—Regulate or Prohibit?

**THERE ARE MINERALS THERE** ... Dispute is ongoing over another polar area—Antarctica—as the southern polar region attracts new interest as a source of petroleum and minerals. Antarctica covers an area of 5.4 million miles—one-tenth of the Earth's land surface—and is larger than the United States and Mexico combined. Geologists believe that considerable quantities of mineral deposits probably exist there, as in all large landmasses. Based on the geology of the region, geologists believe they can find base metal (copper, lead, and zinc) and precious metal (gold and silver) deposits. There are already some known mineral deposits in Antarctica. The huge mass of ice would make recovery difficult, especially in some areas and seasons.

**... BUT INTERNATIONAL AGREEMENT PROHIBITS THEIR MINING.** In 1959, 12 countries (Argentina, Australia, Norway, South Africa, Chile, the United Kingdom, Sweden, France, New Zealand, Belgium, Japan, and the United States) agreed to preserve the region south of 60 degrees south latitude, which includes Antarctica, as an area for scientific research and as a zone of peace. They concluded the Antarctic Treaty, giving equal participation in governance to the signing countries "in the interests of all mankind." The treaty established provisions for new member nations; 39 countries representing more than three-fourths of the world's population are party to the treaty.

Seven nations claim territorial sovereignty in Antarctica—Argentina, Australia, Chile, France, Great Britain, New Zealand, and Norway. A 1991 agreement prohibits all mining exploration and development for 50 years, protects wildlife, regulates waste disposal and marine pollution, and provides for increased scientific study of the continent.

Environmentalists want to ban all mining in Antarctica indefinitely. Critics of mining believe the ultimate solution to the problem of mining's destruction of the environment lies in changes in mineral use and a shift from fossil fuels to renewable energy sources. These changes, however, would represent huge transformations in the way people live. Whether these changes are justified, and whether many people are prepared to make them, will be a matter of debate for years to come.

## SEEKING GLOBAL SOLUTIONS

The indebtedness and poverty of many developing countries reduce opportunities for conservation. Local, national, and international efforts must be linked to deal effectively with the underlying pressures on the ecosystems that support biological diversity. Some international efforts include "debt-for-nature" programs and ecotourism (travel oriented around natural sites, native species, and traditional cultural practices).

Since the 1980s a number of debt forgiveness programs have involved debt-for-nature swaps, where governments or conservation groups buy back, or forgive, a portion of a country's debt, usually at a discounted market price, in exchange for the commitment to fund conservation programs.

Another approach to stemming biodiversity-depleting exploitation is to support alternative, less harmful ways for people to earn their livelihoods. Ecotourism is one alternative economic activity that can use nature, if done carefully, with minimal harm. Ecotourism has spurred communities to protect rare ecological sites and has been modestly successful at generating currency for developing countries.

## PUBLIC OPINION ABOUT NATURAL RESOURCES

Every year the Gallup Organization conducts a comprehensive poll of Americans on environmental issues. The latest poll was conducted in March 2004. Poll participants were asked to express their level of

**TABLE 9.2**

### Public concern about loss of tropical rain forests, 2004

PLEASE TELL ME IF YOU PERSONALLY WORRY ABOUT THIS PROBLEM A GREAT DEAL, A FAIR AMOUNT, ONLY A LITTLE, OR NOT AT ALL. THE LOSS OF TROPICAL RAIN FORESTS?

|  | Great deal % | Fair amount % | Only a little % | Not at all % | No opinion % |
|---|---|---|---|---|---|
| 2004 Mar 8–11 | 35 | 26 | 23 | 15 | 1 |
| 2003 Mar 3–5 | 39 | 29 | 21 | 11 | * |
| 2002 Mar 4–7 | 38 | 27 | 21 | 12 | 2 |
| 2001 Mar 5–7 | 44 | 32 | 15 | 8 | 1 |
| 2000 Apr 3–9 | 51 | 25 | 14 | 9 | 1 |
| 1999 Apr 13–14 | 49 | 30 | 14 | 6 | 1 |
| 1991 Apr 11–14 | 42 | 25 | 21 | 10 | 2 |
| 1990 Apr 5–8 | 40 | 24 | 19 | 14 | 3 |
| 1989 May 4–7 | 42 | 25 | 18 | 12 | 3 |

SOURCE: "Please tell me if you personally worry about this problem a great deal, a fair amount, only a little, or not at all. The loss of tropical rain forests?," in *Poll Topics and Trends: Environment*, The Gallup Organization, Princeton, NJ, March 17, 2004 [Online] www.gallup.com [accessed March 30, 2004]

concern about two issues related to natural resources: loss of tropical rain forests and extinction of plant and animal species.

As shown in Table 9.2 only 35 percent of the respondents expressed a great deal of concern about the loss of tropical rain forests. This value is down from a high of 51 percent recorded in 2000. Concern about this issue has varied over the years, but made a noticeable drop in priority over the last three years. In 2002 just over a quarter of those asked (26 percent) expressed a fair amount of concern about the loss of tropical rain forests, while 23 percent felt a little concern, and 15 percent expressed no concern at all.

Poll participants were slightly more worried about the extinction of plant and animal species. (See Table 9.3.) In 2004 more than a third (36 percent) indicated they felt a great deal of worry about this issue, while 26 percent expressed a fair amount of worry, and 23 percent felt a little concerned. Only 15 percent of those asked expressed no worry at all. In 2000 the percentage of people expressing a great deal of concern about this issue was 45 percent.

In 2004 Gallup pollsters also asked people to express their approval or disapproval over opening up the Arctic National Wildlife Refuge in Alaska for oil exploration. As shown in Figure 9.10 a majority (55 percent) of people are opposed to the idea, while 41 percent support it. These values are very similar to the results obtained in Gallup polls conducted during 2001 and 2002.

TABLE 9.3

**Public concern about extinction of plant and animal species, 2004**

PLEASE TELL ME IF YOU PERSONALLY WORRY ABOUT THIS PROBLEM A GREAT DEAL, DEAL, A FAIR AMOUNT, ONLY A LITTLE, OR NOT AT ALL. EXTINCTION OF PLANT AND ANIMAL SPECIES?

|  | Great deal % | Fair amount % | Only a little % | Not at all % | No opinion % |
|---|---|---|---|---|---|
| 2004 Mar 8–11 | 36 | 26 | 23 | 15 | * |
| 2003 Mar 3–5 | 34 | 32 | 21 | 12 | 1 |
| 2002 Mar 4–7 | 35 | 30 | 22 | 12 | 1 |
| 2001 Mar 5–7 | 43 | 30 | 19 | 7 | 1 |
| 2000 Apr 3–9 | 45 | 33 | 14 | 8 | * |

SOURCE: "Please tell me if you personally worry about this problem a great deal, a fair amount, only a little, or not at all. Extinction of plant and animal species?," in *Poll Topics and Trends: Environment*, The Gallup Organization, Princeton, NJ, March 17, 2004 [Online] www.gallup.com [accessed March 30, 2004]

FIGURE 9.10

**Public opinion about opening up the Arctic National Wildlife Refuge for oil exploration, March 2003**

PLEASE SAY WHETHER YOU GENERALLY FAVOR OR OPPOSE OPENING UP THE ARCTIC NATIONAL WILDLIFE REFUGE IN ALASKA FOR OIL EXPLORATION

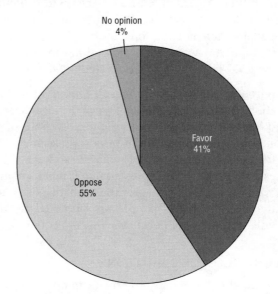

SOURCE: "Please say whether you generally favor or oppose opening up the Arctic National Wildlife Refuge in Alaska for oil exploration," in *Poll Topics and Trends: Environment*, The Gallup Organization, Princeton, NJ, March 17, 2004 [Online] www.gallup.com [accessed March 30, 2004]

# CHAPTER 10
# RENEWABLE ENERGY

## WHAT IS RENEWABLE ENERGY?

Imagine an energy source that uses no oil, produces no pollution, cannot be affected by political events and cartels, creates no radioactive waste, and yet is economical. Although that might sound impossible, some experts claim that technological advances could make a renewable energy-based economy achievable by the mid-twenty-first century.

Renewable energy is a term used to describe energy from sources that are naturally regenerated and are, therefore, virtually unlimited. These energy sources include the sun, wind, water, vegetation, and the heat of the earth.

Solar energy, wind energy, hydropower, and geothermal power are all renewable and clean sources of energy. Each of these alternative energy sources has advantages and disadvantages, and many observers hope that one or more of them may someday provide a substantially better energy source than conventional, fossil fuel burning methods. As the United States and the rest of the world continue to expand their energy needs, which puts a strain on the environment, alternative sources of energy continue to be explored in the hope that they might provide a higher percentage of America's (and the world's) future energy requirements.

## A HISTORICAL PERSPECTIVE

Before the nineteenth century most energy used came from renewable sources. People burned wood for heat, used sails to harness the wind and propel boats, and installed water wheels on streams to grind grain. The large-scale shift to nonrenewable energy sources began in the 1800s with the Industrial Revolution, a period marked by the rise of factories—first in Europe and then in North America. Coal was the most efficient fuel for the steam engine, which was perhaps the most important invention of the Industrial Revolution. By the turn of the twentieth century, coal had replaced wood as the main fuel source.

Coal's dominance, however, lasted only a few decades. In the early 1950s petroleum products such as gasoline and diesel fuel surpassed coal as the fuel of choice. (See Figure 10.1.) Until the early 1970s most Americans were unconcerned about the sources of the nation's energy. Supplies of coal and oil, which together provided more than 90 percent of U.S. energy, were believed to be plentiful. The decades preceding the 1970s were characterized by cheap gasoline and little public discussion of energy conservation.

That carefree approach to energy consumption ended in the 1970s. A fuel oil crisis brought on by political events in the Middle East made Americans more aware of the importance of developing alternative sources of energy to supplement and perhaps even replace fossil fuels. In major cities throughout the United States, gasoline rationing became commonplace, lower thermostat settings for offices and living quarters were encouraged, and people waited in line to fill their gas tanks. The crisis was over by the late 1970s and oil prices dropped, but the incident shocked America, where mobility and personal transportation are highly valued. As a result, President Jimmy Carter's administration encouraged federal funding for research into alternative energy sources.

In 1978 the U.S. Congress passed the Public Utilities Regulatory Policies Act (PURPA; PL 95-617), which was designed to help the struggling alternative energy industry. The act exempted small producers from state and federal utility regulations and required existing local utilities to buy electricity from the smaller producers. PURPA encouraged the growth of small-scale electric power plants, especially those fueled by renewable sources. The renewable industries responded by growing rapidly, gaining experience, improving technologies, and lowering costs. This act was a major factor in the development of the commercial renewable energy market.

FIGURE 10.1

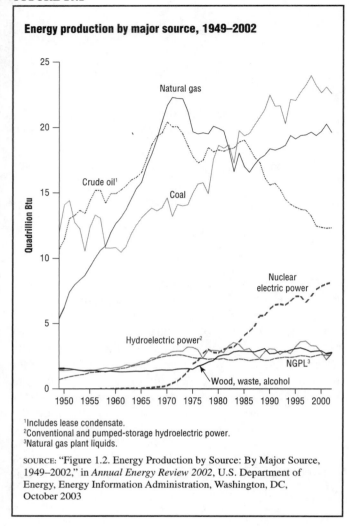

**Energy production by major source, 1949–2002**

¹Includes lease condensate.
²Conventional and pumped-storage hydroelectric power.
³Natural gas plant liquids.

SOURCE: "Figure 1.2. Energy Production by Source: By Major Source, 1949–2002," in *Annual Energy Review 2002*, U.S. Department of Energy, Energy Information Administration, Washington, DC, October 2003

FIGURE 10.2

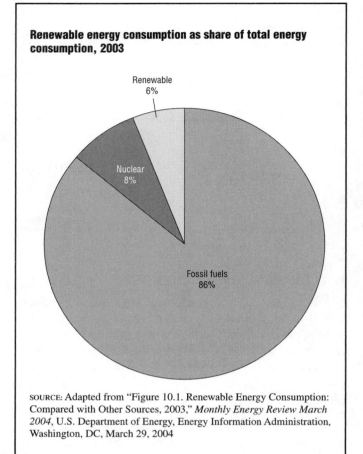

**Renewable energy consumption as share of total energy consumption, 2003**

SOURCE: Adapted from "Figure 10.1. Renewable Energy Consumption: Compared with Other Sources, 2003," *Monthly Energy Review March 2004*, U.S. Department of Energy, Energy Information Administration, Washington, DC, March 29, 2004

In the 1980s President Ronald Reagan decided that private-sector financing for the short-term development of alternative energy sources was best. As a result, he proposed the reduction or elimination of federal expenditures for alternative energy sources. Although funds were severely cut, the U.S. Department of Energy (DOE) continues to support some research and development to explore alternate sources of energy. Table 10.1 shows a detailed breakdown of the nation's energy sources between 1949 and 2002. Renewable energy comprised 9 percent of the total in 1949.

### How Much of Today's Energy Is Renewable?

In 2003 renewable energy accounted for only 6 percent of total U.S. energy consumption. (See Figure 10.2.) Of that hydroelectric power accounted for 44 percent, followed by wood (34 percent) and waste (10 percent). Geothermal energy sources, alcohol fuels, and wind and solar power were minor sources. (See Figure 10.3.)

Figure 10.4 shows the historical contributions of hydroelectric power, wood, and waste to renewable ener-gy resources since 1973. Consumption of hydroelectric and wood power has varied up and down over time, while waste shows a gradual upward trend.

Electric utility companies have historically been the biggest consumers of renewable energy because they use it to generate electricity. In 2003 electric power generation accounted for 59 percent of renewable energy consumption. It was followed by industrial uses with 30 percent of the total. Consumption by residential, commercial, and transportation sectors was minor. Industrial use has outpaced residential use since the late 1950s.

### BIOENERGY

The term bioenergy refers to energy that is generated using biomass—organic material such as wood, agricultural waste from plants and animals, seaweed and algae, and municipal solid waste (MSW) or garbage. These raw materials can be converted into liquid or gaseous biofuels or used directly to provide heat, electricity, or combined heat and power. Figure 10.5 shows common materials used to make bioenergy. The by-products of biomass conversion can be used for fertilizers and chemicals.

Bioenergy from wood, waste, and alcohol fuels accounted for an estimated 2.9 quadrillion British thermal

**TABLE 10.1**

**Energy production by source, 1949–2002**

(Quadrillion Btu)

| | Fossil fuels | | | | | | | Renewable energy[1] | | | | | | |
|---|---|---|---|---|---|---|---|---|---|---|---|---|---|---|
| Year | Coal | Natural gas (dry) | Crude oil[2] | Natural gas plant liquids | Total | Nuclear electric power | Hydroelectric pumped storage[3] | Conventional hydroelectric power | Wood, waste, alcohol[4] | Geothermal | Solar | Wind | Total | Total |
| 1949 | 11.974 | 5.377 | 10.683 | 0.714 | 28.748 | 0 | 5 | 1.425 | 1.549 | NA | NA | NA | 2.974 | 31.722 |
| 1950 | 14.060 | 6.233 | 11.447 | 0.823 | 32.563 | 0 | 5 | 1.415 | 1.562 | NA | NA | NA | 2.978 | 35.540 |
| 1951 | 14.419 | 7.416 | 13.037 | 0.920 | 35.792 | 0 | 5 | 1.424 | 1.535 | NA | NA | NA | 2.958 | 38.751 |
| 1952 | 12.734 | 7.964 | 13.281 | 0.998 | 34.977 | 0 | 5 | 1.466 | 1.474 | NA | NA | NA | 2.940 | 37.917 |
| 1953 | 12.278 | 8.339 | 13.671 | 1.062 | 35.349 | 0 | 5 | 1.413 | 1.419 | NA | NA | NA | 2.831 | 38.181 |
| 1954 | 10.542 | 8.682 | 13.427 | 1.113 | 33.764 | 0 | 5 | 1.360 | 1.394 | NA | NA | NA | 2.754 | 36.518 |
| 1955 | 12.370 | 9.345 | 14.410 | 1.240 | 37.364 | 0 | 5 | 1.360 | 1.424 | NA | NA | NA | 2.784 | 40.148 |
| 1956 | 13.306 | 10.002 | 15.180 | 1.283 | 39.771 | 0 | 5 | 1.435 | 1.416 | NA | NA | NA | 2.851 | 42.622 |
| 1957 | 13.061 | 10.605 | 15.178 | 1.289 | 40.133 | (s) | 5 | 1.516 | 1.334 | NA | NA | NA | 2.849 | 42.983 |
| 1958 | 10.783 | 10.942 | 14.204 | 1.287 | 37.216 | 0.002 | 5 | 1.592 | 1.323 | NA | NA | NA | 2.915 | 40.133 |
| 1959 | 10.778 | 11.952 | 14.933 | 1.383 | 39.045 | 0.002 | 5 | 1.548 | 1.353 | NA | NA | NA | 2.901 | 41.949 |
| 1960 | 10.817 | 12.656 | 14.935 | 1.461 | 39.869 | 0.006 | 5 | 1.608 | 1.320 | 0.001 | NA | NA | 2.929 | 42.804 |
| 1961 | 10.447 | 13.105 | 15.206 | 1.549 | 40.307 | 0.020 | 5 | 1.656 | 1.295 | 0.002 | NA | NA | 2.953 | 43.280 |
| 1962 | 10.901 | 13.717 | 15.522 | 1.593 | 41.732 | 0.026 | 5 | 1.816 | 1.300 | 0.002 | NA | NA | 3.119 | 44.877 |
| 1963 | 11.849 | 14.513 | 15.966 | 1.709 | 44.037 | 0.038 | 5 | 1.771 | 1.323 | 0.004 | NA | NA | 3.098 | 47.174 |
| 1964 | 12.524 | 15.298 | 16.164 | 1.803 | 45.789 | 0.040 | 5 | 1.886 | 1.337 | 0.005 | NA | NA | 3.228 | 49.056 |
| 1965 | 13.055 | 15.775 | 16.521 | 1.883 | 47.235 | 0.043 | 5 | 2.059 | 1.335 | 0.004 | NA | NA | 3.398 | 50.676 |
| 1966 | 13.468 | 17.011 | 17.561 | 1.996 | 50.035 | 0.064 | 5 | 2.062 | 1.369 | 0.004 | NA | NA | 3.435 | 53.534 |
| 1967 | 13.825 | 17.943 | 18.651 | 2.177 | 52.597 | 0.088 | 5 | 2.347 | 1.340 | 0.007 | NA | NA | 3.694 | 56.379 |
| 1968 | 13.609 | 19.068 | 19.308 | 2.321 | 54.306 | 0.142 | 5 | 2.349 | 1.419 | 0.009 | NA | NA | 3.778 | 58.225 |
| 1969 | 13.863 | 20.446 | 19.556 | 2.420 | 56.286 | 0.154 | 5 | 2.648 | 1.440 | 0.013 | NA | NA | 4.102 | 60.541 |
| 1970 | 14.607 | 21.666 | 20.401 | 2.512 | 59.186 | 0.239 | 5 | 2.634 | 1.431 | 0.011 | NA | NA | 4.076 | 63.501 |
| 1971 | 13.186 | 22.280 | 20.033 | 2.544 | 58.042 | 0.413 | 5 | 2.824 | 1.432 | 0.012 | NA | NA | 4.268 | 62.723 |
| 1972 | 14.092 | 22.208 | 20.041 | 2.598 | 58.938 | 0.584 | 5 | 2.864 | 1.503 | 0.031 | NA | NA | 4.398 | 63.920 |
| 1973 | 13.992 | 22.187 | 19.493 | 2.569 | 58.241 | 0.910 | 5 | 2.861 | 1.529 | 0.043 | NA | NA | 4.433 | 63.585 |
| 1974 | 14.074 | 21.210 | 18.575 | 2.471 | 56.331 | 1.272 | 5 | 3.177 | 1.540 | 0.053 | NA | NA | 4.769 | 62.372 |
| 1975 | 14.989 | 19.640 | 17.729 | 2.374 | 54.733 | 1.900 | 5 | 3.155 | 1.499 | 0.070 | NA | NA | 4.723 | 61.357 |
| 1976 | 15.654 | 19.480 | 17.262 | 2.327 | 54.723 | 2.111 | 5 | 2.976 | 1.713 | 0.078 | NA | NA | 4.768 | 61.602 |
| 1977 | 15.755 | 19.565 | 17.454 | 2.327 | 55.101 | 2.702 | 5 | 2.333 | 1.838 | 0.077 | NA | NA | 4.249 | 62.052 |
| 1978 | 14.910 | 19.485 | 18.434 | 2.245 | 55.074 | 3.024 | 5 | 2.937 | 2.038 | 0.064 | NA | NA | 5.039 | 63.137 |
| 1979 | 17.540 | 20.076 | 18.104 | 2.286 | 58.006 | 2.776 | 5 | 2.931 | 2.152 | 0.084 | NA | NA | 5.166 | 65.948 |
| 1980 | 18.598 | 19.908 | 18.249 | 2.254 | 59.008 | 2.739 | 5 | 2.900 | 2.485 | 0.110 | NA | NA | 5.494 | 67.241 |
| 1981 | 18.377 | 19.699 | 18.146 | 2.307 | 58.529 | 3.008 | 5 | 2.758 | 2.590 | 0.123 | NA | NA | 5.471 | 67.007 |
| 1982 | 18.639 | 18.319 | 18.309 | 2.191 | 57.458 | 3.131 | 5 | 3.266 | 2.615 | 0.105 | NA | NA | 5.985 | 66.574 |
| 1983 | 17.247 | 16.593 | 18.392 | 2.184 | 54.416 | 3.203 | 5 | 3.527 | 2.831 | 0.129 | NA | NA | 6.488 | 64.106 |
| 1984 | 19.719 | 18.008 | 18.848 | 2.274 | 58.849 | 3.553 | 5 | 3.386 | 2.880 | 0.165 | (s) | (s) | 6.431 | 68.832 |
| 1985 | 19.325 | 16.980 | 18.992 | 2.241 | 57.539 | 4.076 | 5 | 2.970 | 2.864 | 0.198 | (s) | (s) | 6.033 | 67.647 |
| 1986 | 19.509 | 16.541 | 18.376 | 2.149 | 56.575 | 4.380 | 5 | 3.071 | 2.841 | 0.219 | (s) | (s) | 6.132 | 67.087 |
| 1987 | 20.141 | 17.136 | 17.675 | 2.215 | 57.167 | 4.754 | 5 | 2.635 | 2.823 | 0.229 | (s) | (s) | 5.687 | 67.608 |
| 1988 | 20.738 | 17.599 | 17.279 | 2.260 | 57.875 | 5.587 | 5 | 2.334 | 2.937 | 0.217 | (s) | (s) | 5.489 | 68.951 |
| 1989 | 21.346 | 17.847 | 16.117 | 2.158 | 57.468 | 5.602 | 5 | R2.837 | 3.062 | R0.317 | 0.055 | R0.022 | R6.294 | R69.364 |
| 1990 | 22.456 | R18.326 | 15.571 | 2.175 | R58.529 | 6.104 | −0.036 | R3.046 | R2.662 | R0.336 | 0.060 | R0.029 | R6.133 | R70.729 |
| 1991 | 21.594 | 18.229 | 15.701 | 2.306 | 57.829 | 6.422 | −0.047 | R3.016 | 2.702 | R0.346 | 0.063 | R0.031 | R6.158 | R70.362 |
| 1992 | 21.629 | 18.375 | 15.223 | 2.363 | 57.590 | 6.479 | −0.043 | 2.617 | 2.847 | 0.349 | 0.064 | 0.030 | 5.907 | 69.933 |
| 1993 | 20.249 | 18.584 | 14.494 | 2.408 | 55.736 | 6.410 | −0.042 | 2.892 | 2.804 | 0.364 | 0.066 | 0.031 | 6.157 | 68.262 |

**TABLE 10.1**

**Energy production by source, 1949–2002** [CONTINUED]

(Quadrillion Btu)

| Year | Fossil fuels | | | | | Nuclear electric power | Hydroelectric pumped storage[3] | Renewable energy[1] | | | | | | Total |
|---|---|---|---|---|---|---|---|---|---|---|---|---|---|---|
| | Coal | Natural gas (dry) | Crude oil[2] | Natural gas plant liquids | Total | | | Conventional hydroelectric power | Wood, waste, alcohol[4] | Geothermal | Solar | Wind | Total | |
| 1994 | 22.111 | 19.348 | 14.103 | 2.391 | 57.952 | 6.694 | −0.035 | 2.683 | 2.939 | 0.338 | 0.069 | 0.036 | 6.065 | 70.676 |
| 1995 | 22.029 | R19.082 | 13.887 | 2.442 | R57.440 | 7.075 | −0.028 | 3.205 | 3.068 | 0.294 | 0.070 | 0.033 | 6.669 | R71.156 |
| 1996 | 22.684 | R19.344 | 13.723 | 2.530 | R58.281 | 7.087 | −0.032 | 3.590 | 3.127 | 0.316 | 0.071 | 0.033 | 7.137 | R72.472 |
| 1997 | 23.211 | 19.394 | 13.658 | 2.495 | 58.758 | 6.597 | −0.041 | 3.640 | 3.006 | 0.325 | 0.070 | 0.034 | 7.075 | 72.389 |
| 1998 | 23.935 | 19.613 | 13.235 | 2.420 | 59.204 | 7.068 | −0.046 | 3.297 | 2.835 | 0.328 | 0.070 | 0.031 | 6.561 | 72.787 |
| 1999 | 23.186 | 19.341 | 12.451 | 2.528 | 57.505 | 7.610 | −0.062 | 3.268 | R2.885 | 0.331 | 0.069 | 0.046 | R6.599 | R71.652 |
| 2000 | 22.623 | R19.662 | 12.358 | 2.611 | R57.254 | 7.862 | −0.057 | 2.811 | R2.907 | 0.317 | 0.066 | 0.057 | R6.158 | R71.218 |
| 2001 | R23.053 | R20.227 | 12.282 | R2.547 | R58.109 | 8.028 | −0.090 | R2.201 | R2.678 | R0.311 | R0.065 | R0.068 | R5.324 | R71.372 |
| 2002P | 22.554 | 19.561 | 12.314 | 2.561 | 56.990 | 8.145 | −0.089 | 2.668 | 2.756 | 0.304 | 0.064 | 0.106 | 5.899 | 70.946 |

[1]End-use consumption and electricity net generation.
[2]Includes lease condensate.
[3]Pumped storage facility production minus energy used for pumping.
[4]Alcohol is ethanol blended into motor gasoline.
[5]Included in "Conventional Hydroelectric Power."
R=Revised. P=Preliminary. NA=Not available. (s)=Less than 0.0005 quadrillion Btu.
Note: Totals may not equal sum of components due to independent rounding.

SOURCE: "Table 1.2. Energy Production by Source, 1949–2002," in *Annual Energy Review 2002*, U.S. Department of Energy, Energy Information Administration, Washington, DC, October 2003

FIGURE 10.3

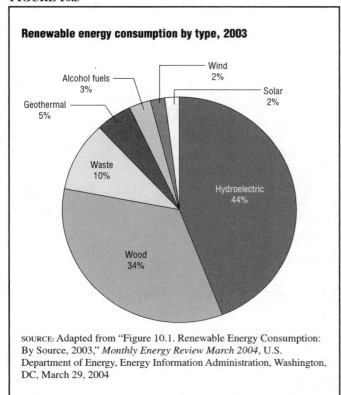

**Renewable energy consumption by type, 2003**

- Wind 2%
- Alcohol fuels 3%
- Solar 2%
- Geothermal 5%
- Waste 10%
- Hydroelectric 44%
- Wood 34%

SOURCE: Adapted from "Figure 10.1. Renewable Energy Consumption: By Source, 2003," *Monthly Energy Review March 2004*, U.S. Department of Energy, Energy Information Administration, Washington, DC, March 29, 2004

units (Btu) of energy in 2003, comprising 47 percent of renewable energy consumed. (See Figure 10.3.)

## Biomass Conversion

There are two types of biomass conversion processes: thermochemical conversion and biochemical conversion. Thermochemical conversion uses heat to produce chemical reactions in biomass. Direct combustion is the easiest and most commonly used method. Materials such as dry wood or agricultural wastes are chopped and burned to produce steam, electricity, or heat for industries, utilities, and homes. Wood burning in stoves and fireplaces is one example. The burning of agricultural wastes is also becoming more widespread. In Florida, for example, sugar cane producers use the residue from the cane to generate much of their energy.

Pyrolysis, also called gasification or carbonization, uses heat to break down biomass to yield liquid, gaseous, and solid fuels. Converting wood to charcoal is an example of this process.

The second type of conversion process, biochemical conversion, uses enzymes, fungi, or other microorganisms to convert high-moisture biomass into either liquid or gaseous fuels. Bacteria convert manure, agricultural wastes, paper, and algae into methane, which is used as fuel. Sewage treatment plants have used anaerobic (without oxygen) digestion for many years to generate methane gas. Small-scale digesters have been used on farms,

primarily in Europe and Asia, for hundreds of years. The DOE estimates that many thousands of biofuel plants are in use in Korea, and perhaps half a million plants operate in China.

Another type of biochemical conversion process, fermentation, uses yeast to decompose carbohydrates to yield ethyl alcohol (ethanol) and carbon dioxide. Sugar crops, grains (corn, in particular), potatoes, and other starchy crops are common feedstocks that supply the sugar for ethanol production.

## Wood

Wood energy was the first energy source in America's industrialization. Wood, the most commonly used biofuel, is still used to heat millions of homes every year. In 2003 it provided 2.1 quadrillion Btu, accounting for 34 percent of renewable energy consumption (See Figure 10.3.)

When wood is widely used as a fuel in an area, deforestation can occur, resulting in the possibility of soil erosion and mud slides. Burning wood, as with the burning of fossil fuels, also pollutes the environment.

## Ethanol and Methanol—Important Agricultural By-Products

Ethanol is a colorless, nearly odorless, flammable liquid derived from fermenting plant material that contains carbohydrates in the form of sugar. U.S. ethanol production has grown steadily since 1980, reaching 2.81 billion gallons in 2003, according to the U.S. Department of Energy. Production capacity is largely located in the farming regions of the upper Midwest. (See Figure 10.6.)

Most of the ethanol manufactured for use as fuel is derived from corn, wood, and sugar. A mixture of 10 percent ethanol and 90 percent gasoline is usable in any internal combustion engine without the need to modify the motor. Although the DOE claims that the demand for alcohol/gasoline blends is increasing because alcohol can substitute for lead as an octane booster, there is little question that the development of ethanol depends more on the continued support of farm state legislators than any economic benefit.

Ethanol is difficult and expensive to produce in bulk. Methanol-blend fuels have also been tested successfully. (Methanol is methyl alcohol.) Using methanol instead of diesel fuel virtually eliminates sulfur emissions and reduces other environmental pollutants usually emitted from trucks and buses. Burning biofuels in vehicle engines creates a "carbon cycle" in which the earth's vegetation can in turn make use of the products of combustion and, therefore, reduce net greenhouse gases. (See Figure 10.7.) Producing methanol from biofuels, however, is costly.

Some scientists believe ethanol made from wood, sawdust, corncobs, or rice hulls could liberate the alcohol

FIGURE 10.4

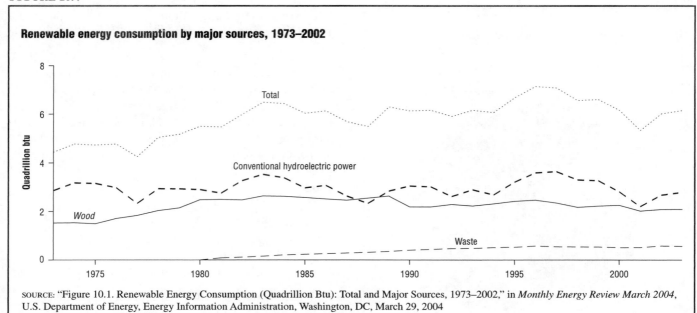

**Renewable energy consumption by major sources, 1973–2002**

SOURCE: "Figure 10.1. Renewable Energy Consumption (Quadrillion Btu): Total and Major Sources, 1973–2002," in *Monthly Energy Review March 2004*, U.S. Department of Energy, Energy Information Administration, Washington, DC, March 29, 2004

fuel industry from its dependence on food crops such as corn and sugar cane. Worldwide, there are enough corn-cobs and rice hulls left over from annual crop production to produce more than 40 billion gallons of ethanol.

Advocates of wood-derived ethanol believe that the eventual result of wood-to-ethanol conversion research could create a sustainable liquid fuel industry that does not rely on pollution-generating fossil fuels. For instance if new trees were planted to replace those that were cut for fuel, they would be available for later harvesting and, in the meantime, contribute to the prevention of global warming by continuing their carbon dioxide processing function. Other scientists warn that a huge demand for transportation fuels could create a demand for wood that might accelerate the destruction of old-growth forests and endanger ecosystems. Without care-ful attention to forestry practices, ethanol production might aggravate rather than solve the fuel problem.

**Municipal Solid Waste**

Each year millions of tons of municipal solid waste (MSW), or garbage as it is commonly called, are buried in landfills and city dumps. This method of disposal is not only costly but is becoming increasingly difficult as some land-fills across the nation are near capacity. Many communities have discovered that they can solve two problems at once by constructing waste-to-energy (WTE) plants. Not only is garbage burned and reduced in volume by 90 percent, ener-gy in the form of steam or electricity is generated in a cost-effective way. The potential energy benefit is significant.

WASTE-TO-ENERGY PLANTS. The two most common WTE plant designs are the mass burn (also called direct combustion) and the refuse derived fuel (RDF) systems.

Most WTE plants in the United States use the mass burn system. This system's advantage is that the waste does not have to be sorted or prepared before burning, except for removing obviously noncombustible, oversized objects. The mass burn eliminates expensive sorting, shredding, and transportation machinery that may be prone to break down.

Waste is carried to the plant in trash trucks and dropped into a storage pit. Large overhead cranes lift the garbage into a furnace feed hopper that controls the amount and rate of waste that is fed into the furnace. Next, the garbage is moved through a combustion zone so that it burns to the greatest extent possible. The burning garbage produces heat, and that heat is used to produce steam. The steam can be used directly for industrial needs or heat can be sent through a turbine to power a generator to produce electricity.

RDF systems process waste to remove noncombustible objects and to create homogeneous and uniformly sized fuel. Large items such as bedsprings, dangerous materials, and flammable liquids are removed by hand. The trash is then shredded and carried to a screen to remove glass, rocks, and other material that cannot be burned. The remaining material is usually sifted a second time with an air separator to yield fluff. The fluff is sent to storage bins before being burned, or it can be compressed into pellets or briquettes for long-term storage. This fuel can be used as an energy source by itself in a variety of systems, or it can be used with other fuels such as coal or wood.

The major obstacle to increasing the use of municipal WTE plants is their effect on the environment. Noise from trucks, fans, and processing equipment at RDF plants can

**FIGURE 10.5**

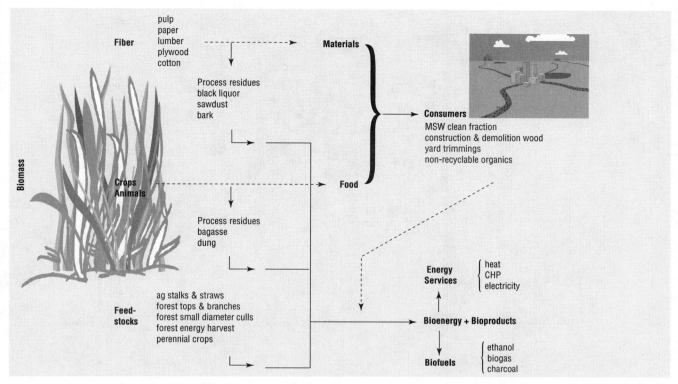

**Biomass to bioenergy**

SOURCE: "Biomass to Bioenergy," in *Bioenergy Image Gallery*, U.S. Department of Energy, Oak Ridge National Laboratory, Oak Ridge, TN, May 7, 2002 [Online] http://bioenergy.ornl.gov/gallery/index.html [accessed March 31, 2004]

be unpleasant for nearby residents. The emission of particles into the air is controlled by electrostatic precipitators, and most gases can be eliminated by proper combustion techniques. There is concern, however, about the amounts of dioxin (a very dangerous air pollutant) that are often emitted from these plants.

LANDFILL GAS RECOVERY. Landfills contain a large amount of biodegradable matter. Gas is created because of the lack of oxygen that helps the growth of methagens—types of bacteria that produce methane gas and carbon dioxide. In the past, as landfills aged, these gases built up and leaked out. This gas leakage prompted some communities to drill holes and burn off the flammable and dangerous methane.

The energy crisis of the 1970s made this methane gas an energy resource too valuable to waste, and efforts were made to find an inexpensive way to tap the gas. The first landfill gas-recovery site was finished in 1975 at the Palos Verdes Landfill in Rolling Hills Estates, California. Depending on the extraction rates, most existing sites can produce gas for about 20 years.

In a typical operation, garbage is allowed to decompose for several months. When a sufficient amount of

methane gas has developed, it is piped out to a generating plant where it is burned to produce electricity. In its purest form methane gas is equivalent to natural gas and can be used in exactly the same way.

The advantages of tapping gas from a landfill go beyond the energy provided by the methane. When internal pressure forces methane gas to seep into the air, it carries very unpleasant odors into the surrounding neighborhoods. The released methane can also be a danger because, if it accumulates and is accidentally ignited, it can explode. Extracting the methane gas for energy eliminates both of these problems.

## HYDROPOWER

Hydropower is the energy that comes from the natural flow of water. Usually, the power is harnessed by taking advantage of gravity when water falls from one level to another. The energy of falling water is converted into mechanical energy. In the past, water's energy was harnessed by waterwheels to grind grain or turn saws. Modern technology uses water's energy to turn turbines that create electricity. Hydropower is a renewable, nonpolluting, and reliable energy source.

FIGURE 10.6

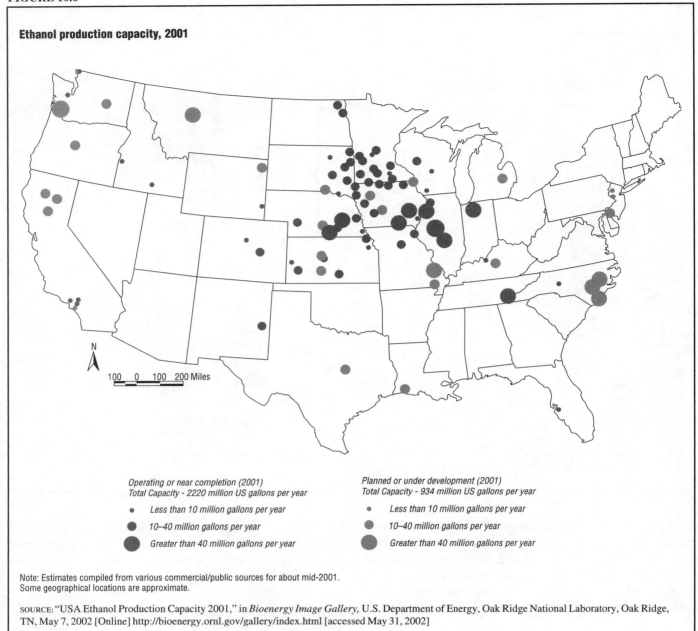

**Ethanol production capacity, 2001**

*Operating or near completion (2001)*
*Total Capacity - 2220 million US gallons per year*

- Less than 10 million gallons per year
- 10–40 million gallons per year
- Greater than 40 million gallons per year

*Planned or under development (2001)*
*Total Capacity - 934 million US gallons per year*

- Less than 10 million gallons per year
- 10–40 million gallons per year
- Greater than 40 million gallons per year

Note: Estimates compiled from various commercial/public sources for about mid-2001. Some geographical locations are approximate.

SOURCE: "USA Ethanol Production Capacity 2001," in *Bioenergy Image Gallery,* U.S. Department of Energy, Oak Ridge National Laboratory, Oak Ridge, TN, May 7, 2002 [Online] http://bioenergy.ornl.gov/gallery/index.html [accessed May 31, 2002]

Hydropower is still the only means of storing large quantities of electrical energy for almost instant use. This is done by holding water in a large reservoir behind a dam with a hydroelectric power plant below. The dam creates a height from which water flows. The fast-moving flow of water from the dam pushes the turbine blades that turn the rotor part of the electric generator. When the coils of wire on the rotor sweep past the generator's stationary coil, electricity is produced. Whenever power is needed at peak times, water valves are opened and, in a short amount of time, turbine generators produce extra power. The Hoover Dam, located on the Colorado River at the Nevada-Arizona border, is the site of one of the largest U.S. hydroelectric plants.

According to the Energy Information Administration's March 2004 *Monthly Energy Review,* in 2003 hydroelectric power generated about 2.8 quadrillion Btu of energy in the United States, down substantially from when it peaked at 3.64 billion quadrillion Btu in 1997. (See Table 10.1.) Hydropower accounted for 44 percent of the nation's renewable energy consumption in 2003.

**Advantages and Disadvantages of Hydropower**

Small hydropower plants in the United States are costly to build but quickly become cost-efficient because of their low operating costs. One of the disadvantages of small hydropower generators is their reliance on rain and melting snow to fill reservoirs, a problem especially dur-

**FIGURE 10.7**

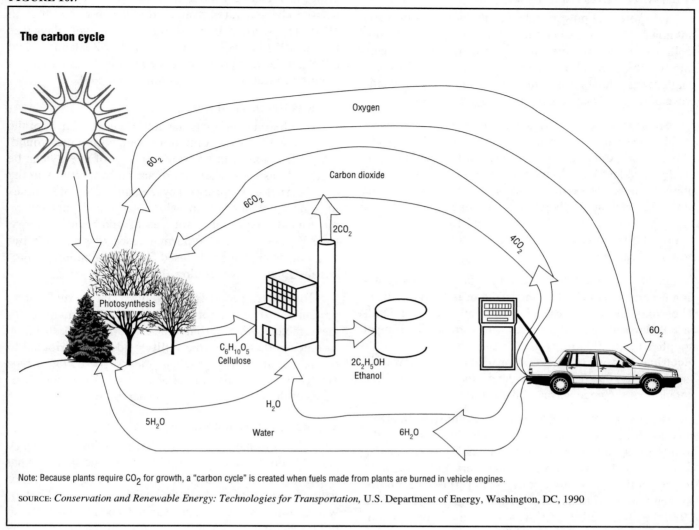

**The carbon cycle**

Oxygen

$6O_2$

Carbon dioxide

$6CO_2$

$2CO_2$

$4CO_2$

$6O_2$

Photosynthesis

$C_6H_{10}O_5$
Cellulose

$2C_2H_5OH$
Ethanol

$H_2O$

$5H_2O$

Water

$6H_2O$

Note: Because plants require $CO_2$ for growth, a "carbon cycle" is created when fuels made from plants are burned in vehicle engines.

SOURCE: *Conservation and Renewable Energy: Technologies for Transportation,* U.S. Department of Energy, Washington, DC, 1990

ing years with drought conditions. Other concerns include the difficult search for the proper terrain on which to build a hydroelectric power plant; the high cost of construction; and the ecological concern that dams could ruin streams, dry up waterfalls, and interfere with marine life habitats.

Large hydropower plants suffer from the same problems except that they rarely lack sufficient water since they can be built only on very large rivers. There is little potential to build new, large, hydropower plants in the United States because plants have already been built at all of the best sites.

**New Directions in Hydropower Energy**

Since almost all power sites have already been developed, hydropower's contribution to U.S. energy generation should remain relatively constant—although existing sites can become more efficient as new generators are added.

Most of the new development in hydropower is occurring in developing nations that see it as an effective method of supplying power to their growing populations. Major hydropower development programs are tremendous public works projects requiring huge amounts of money, most of it borrowed from the developed world. Leaders of developing countries believe that, in the long run, despite threats to the environment, the dams will pay for themselves by bringing cheap electric power to their people.

While developing nations have utilized only a small portion of their large-scale hydropower potential, the United States and Europe have developed a major proportion of their potential. In addition, dams are now less favored because of their harm to the environment. Large-scale hydropower development has virtually stopped in the United States, with not one new dam being approved for federal funding since the late 1980s.

Dams in the United States were usually constructed entirely with federal monies. Since 1986, however, any new dam proposed in the United States must receive half its funding from local governments. Any new major supplies of hydroelectric power for the United States will most likely come from Canada.

## Other Alternatives Using Water

The potential power locked in the world's oceans is unknown. However, since the ocean is not as easily controlled as a river or water that is directed through canals into turbines, unlocking that potential power is far more challenging. Three ideas being considered are tidal plants, wave power, and ocean thermal energy conversion.

TIDES AND WAVES. A tidal plant uses the power generated by the tidal flow of water as it ebbs (flows back out to sea). A minimum tidal range of three to five yards is generally considered necessary for an economically feasible plant. Canada, for example, has built a 20-megawatt unit at the Bay of Fundy, where the tidal range—15 yards—is the largest in the world. The largest existing tidal facility is the 240-megawatt plant at the La Rance estuary in northern France.

In 2000 the world's first commercial wave power station became operational on the island of Islay in Scotland. The small system is called a Limpet, short for land-installed marine powered energy transformer. The Limpet supplies up to 500 kilowatts of electricity to the island's electrical power grid. Several projects are underway in Japan and the Pacific region to determine a way to use the potential of the huge waves of the Pacific.

OCEAN THERMAL ENERGY CONVERSION. In 1881 Jacques Arsene d'Arsonval, a French physicist, was the first to propose tapping the thermal energy of the ocean. Not until 1974, however, was a laboratory and test facility for ocean thermal energy conversion (OTEC) technologies built at the National Energy Laboratory of Hawaii. In 1980 the DOE built OTEC-1, a test site on board a converted U.S. Navy tanker. In 1980 Congress enacted two laws to promote the commercial development of OTEC technology—the Ocean Thermal Energy Conversion Act (PL 96-320) later modified by PL 98-623, and the Ocean Thermal Energy Conversion, Research, and Development and Demonstration Act (PL 96-310).

OTEC uses the temperature difference between the ocean's warm surface water and the cooler water in its depths to produce heat energy that can power a heat engine to generate electricity. OTEC systems can be installed on ships, barges, or offshore platforms with underwater cables that transmit electricity to shore. In addition to providing power, OTEC systems can be used to desalinate water, provide air conditioning and refrigeration, and produce methanol, ammonia, hydrogen, aluminum, chlorine, and other chemicals.

## GEOTHERMAL ENERGY

Since ancient times humans have exploited the earth's natural hot water sources. Although bubbling hot springs became public baths in ancient Rome, using naturally occurring hot water and underground steams to produce power is a relatively modern development. The first electricity to be generated from natural steam was in Italy in 1904. The world's first natural steam power plant was built in 1958 in a volcanic region of New Zealand. A field of 28 geothermal power plants covering 30 square miles in northern California was completed in 1960.

### What Is Geothermal Energy?

Geothermal energy is the natural, internal heat of the earth trapped in rock formations deep within the ground. Only a fraction of this vast storehouse of energy can be extracted, usually where there are large fractures in the earth's crust. Hot springs, geysers, and fumaroles (holes in or near volcanoes from which vapor escapes) are the most easily exploitable sources of geothermal energy. (See Figure 10.8.) Geothermal reservoirs provide hot water or steam that can be used for heating buildings, processing food, and generating electricity.

To produce power from a geothermal energy source, pressurized steam or hot water is extracted from the earth and directed toward turbines. The electricity produced by turbines is then fed into a utility grid and distributed to residential and commercial customers. By the late 1990s electricity from this source accounted for almost two-thirds of the world's geothermal energy use.

### Types of Geothermal Energy

Like most natural energy sources, geothermal energy is usable only when it is concentrated in one spot, in what is called a "thermal reservoir." The four basic categories of thermal reservoirs are hydrothermal (dry steam and hot, or wet, steam), dry rock, geopressurized, and magma (rock so hot it has liquefied) reservoirs. Most of the known areas for geothermal power in the United States are located west of the Mississippi River. (See Figure 10.9.)

Hydrothermal reservoirs consist of a heat source covered by a permeable formation through which water circulates. Dry steam is produced when hot water boils underground and some of the steam escapes to the surface under pressure. Once at the surface impurities and tiny rock particles are removed and the steam is then piped directly to the electrical generating station. These systems are the cheapest and simplest form of geothermal energy.

Hot steam systems are created when underground water is heated to more than 700 degrees Fahrenheit by the surrounding hot rock or magma, but the water remains liquid because of intense pressure. When the water is brought to the surface and the pressure is reduced, a small amount of water becomes steam that is then separated and used to power an electrical generating plant.

Dry rock formations are the most common geothermal source, especially in the west. To tap this source of energy, water is injected into naturally hot rock formations to produce steam or water for collection.

FIGURE 10.8

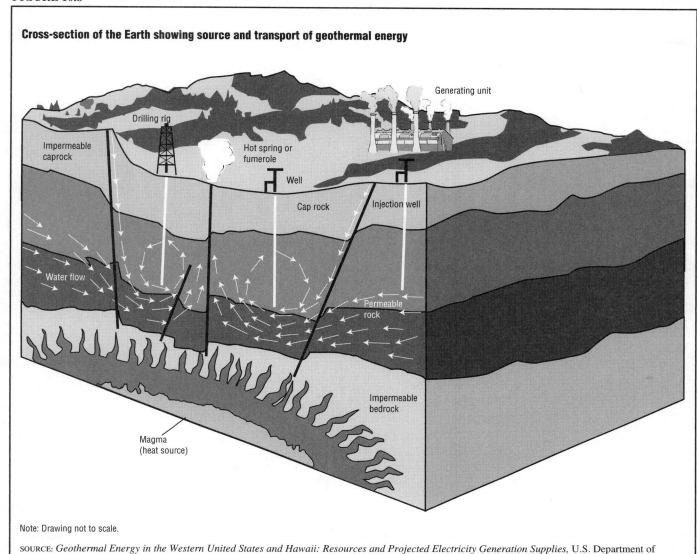

**Cross-section of the Earth showing source and transport of geothermal energy**

Note: Drawing not to scale.

SOURCE: *Geothermal Energy in the Western United States and Hawaii: Resources and Projected Electricity Generation Supplies,* U.S. Department of Energy, Energy Information Administration, Washington, DC, 1998

Geopressurized reservoirs are sedimentary formations containing hot water and methane gas. Supplies of geopressurized energy remain uncertain, however, and drilling is expensive. Scientists hope that advanced technology will eventually permit the commercial exploitation of the methane content in these reservoirs.

Magma resources are found where molten or partially liquefied rock is located from 10,000 to 33,000 feet below the earth's surface. Because magma is so hot, ranging from 1,650 degrees to 2,200 degrees Fahrenheit, it is a good geothermal resource. The process for extracting energy from magma is still in the experimental stages.

## Disadvantages of Geothermal Energy

Geothermal plants are expensive because they must be built near the source. Other drawbacks include low efficiency, bad odors from sulfur released in processing, noise, lack of access for most states, and potentially harmful pollutants (hydrogen sulfide, ammonia, and radon). Also, poisonous arsenic or boron are often found in geothermal waters. Serious environmental concerns have been raised over the release of chemical compounds, potential water contamination, the collapse of land surface around the area from which the water is being drained, and potential water shortages resulting from massive withdrawals of water.

## American Production of Geothermal Energy

Geothermal energy ranked fourth in U.S. renewable energy production in 2003 after hydroelectric, wood, and waste energy. Geothermal energy accounted for 0.3 quadrillion Btu in 2003, down from the early 1990s when it approached 0.4 quadrillion Btu. In 2003 it represented only 5 percent of renewable energy consumed in the United States.

**FIGURE 10.9**

**Geothermal resources map**

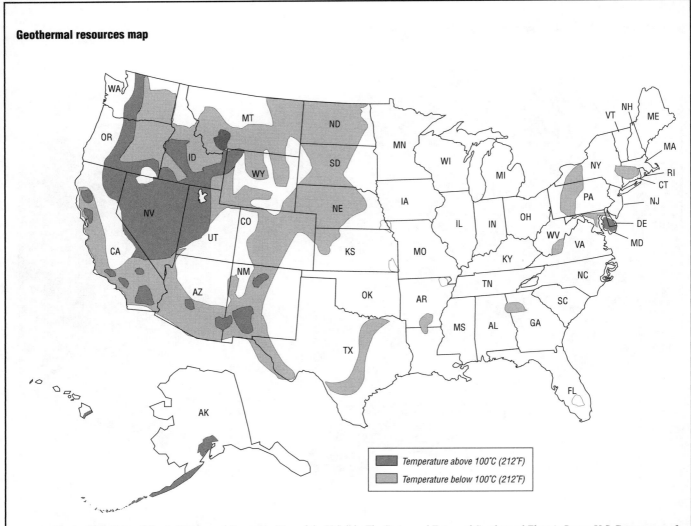

SOURCE: Charles F. Kutscher, "Fig. 1: Geothermal Resources Map of the U.S.," in *The Status and Future of Geothermal Electric Power,* U.S. Department of Energy, National Renewable Energy Laboratory, Golden, CO, August 2000

Public-sector involvement in the geothermal industry began with the passage of the Geothermal Steam Act of 1970 (PL 91-581), which authorized the U.S. Department of the Interior to lease geothermal resources on federal lands. Although the United States is the greatest producer of geothermal power with 28 percent of the world's capacity, the geothermal industry in the United States is losing steam. Oil prices have remained relatively low, and most of the easily exploited geothermal reserves have already been developed. In addition, utility companies and independent power producers argue over who should build additional generating capacity and what prices should be paid for the power. As a result, U.S. geothermal capacity has dropped. Growth in the American market depends on the regulatory environment, oil price trends, and the success of unproven technologies for economically exploiting some of the presently inaccessible geothermal reserves.

### World Production of Geothermal Energy

During the oil crisis of the 1970s, when energy was at the forefront of the international agenda, governments scrambled to find domestic alternatives to imported oil. As public interest grew, research dollars became available and a large number of geothermal energy plants were built. Although interest has since faded, geothermal power's commercial development worldwide has continued at a slow but steady pace.

Since 1979 worldwide geothermal electrical generating capacity has nearly tripled. According to the International Geothermal Association, there were nearly 8,000 megawatts of installed capacity in 2000. The U.S. accounted for 28 percent of the total, followed by the Philippines (24 percent), Italy (10 percent), Mexico (9 percent), and Indonesia and Japan (7 percent each). Together these countries accounted for 85 percent of total capacity.

Worldwide geothermal capacity is still less than that of a handful of average-sized coal-fired power plants. World geothermal reserves are immense but unevenly distributed. These reserves fall mostly in seismically active areas at the margins or borders of Earth's nine major tectonic plates. Exploited reserves represent only a small fraction of the overall potential—many countries are believed to have in excess of 100,000 megawatts of geothermal energy available.

A few nations in the developing world—El Salvador, Kenya, Bolivia, Costa Rica, Ethiopia, India, and Thailand—have considerable steam reserves available for power generation. Debt-ridden developing nations that have substantial unexploited geothermal reserves are especially eager to use them instead of relying on costly fossil fuel imports for their energy needs.

## WIND ENERGY

Wind energy is really a form of solar energy. Winds are created by the uneven heating of the atmosphere by the sun, the irregularities of the earth's surface, and the rotation of the earth. As a result, winds are strongly influenced by local terrain, water bodies, weather patterns, vegetation, and other factors. This wind flow, when "harvested" by wind turbines, can be used to generate electricity.

Wind machines have changed dramatically from those that were common in the 1800s. Early windmills produced mechanical energy to pump water and run sawmills. In the late 1890s Americans began experimenting with wind power to generate electricity. Their early efforts produced enough electricity to light one or two modern light bulbs.

Compared to the pinwheel-shaped farm windmills that can still be seen dotting the American rural landscape, state-of-the-art wind turbines look more like airplane propellers. Their sleek, high-tech aerodynamics and fiberglass design allow them to generate an abundance of electricity, while they also produce mechanical energy and heat. (See Figure 10.10.)

Unlike solar energy systems, wind systems produce renewable energy at night as well as during the day. During the 1990s industrial and developing countries alike began using wind power as an adaptable source of electricity to complement their existing power sources and to bring electricity to remote regions. Wind turbines cost less to install per unit of kilowatt capacity than either coal or nuclear facilities. After installing a windmill there are few additional costs, particularly as the fuel (wind) is free.

Wind speeds are generally highest and most consistent in mountain passes and along coastlines. Europe has the greatest coastal wind resources, and clusters of wind turbines, or wind farms, are being developed in much of

**FIGURE 10.10**

Horizontal axis wind turbines are used to create energy on a wind farm in Altamont Pass, California. *(Photograph by Keven Schafer. Corbis Corporation. Reproduced by permission.)*

Europe and Asia. Denmark, the Netherlands, China, and India are especially interested in fostering the development of domestic wind industries. In the United States it is estimated that sufficient wind energy is available to provide more than one trillion kilowatt-hours of electricity annually.

### Energy Production by Wind Turbines

Wind is the world's fastest growing energy source. According to the American Wind Energy Association, worldwide capacity at the end of 2003 was approximately 39,000 megawatts, up 26 percent from the year before. Europe generated almost three-quarters of the world's wind power. U.S. capacity at the end of 2003 was 6,374 megawatts, accounting for 16 percent of the total. Nearly 1,700 of those megawatts were added during 2003.

The wind industry in the United States began with research projects in California. In 1981 the state erected 144 relatively small turbines capable of generating a combined total of 7 megawatts of electricity. Within a year the number of turbines had increased ten times, and by 1986 had multiplied 100-fold. The 1980s saw an explosion of wind technology in California. By 1995 the state produced enough wind power to supply all of San Francisco's residents. Experts point out that California's dominance had less to do with wind availability than tax credits that were offered by the state until 1999. Approximately 32 percent of the total 6,374 megawatts of installed wind power capacity in the United States was located in California at the end of 2003. (See Figure 10.11.)

During the 1990s wind energy facilities began to appear in other states, particularly in the Midwest. Texas now supplies 20 percent of the nation's wind power capacity, thanks to legislative policies that require the state's utilities to provide renewable generating capacity. Together, California and Texas generate more than half of the country's wind power.

FIGURE 10.11

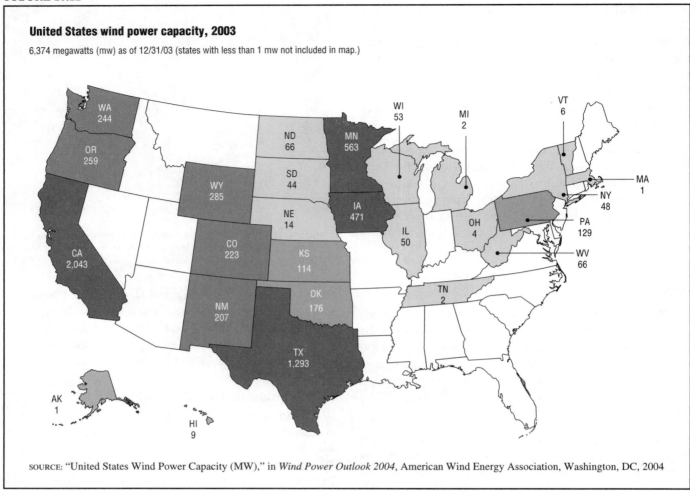

**United States wind power capacity, 2003**

6,374 megawatts (mw) as of 12/31/03 (states with less than 1 mw not included in map.)

WA 244
OR 259
CA 2,043
AK 1
HI 9
WY 285
CO 223
NM 207
ND 66
SD 44
NE 14
KS 114
OK 176
TX 1,293
MN 563
IA 471
WI 53
IL 50
MI 2
OH 4
TN 2
WV 66
PA 129
NY 48
VT 6
MA 1

SOURCE: "United States Wind Power Capacity (MW)," in *Wind Power Outlook 2004*, American Wind Energy Association, Washington, DC, 2004

Studies show that several states, especially the plains states, have wind speeds sufficient to supply electricity. Twelve states—North Dakota, South Dakota, Texas, Kansas, Montana, Nebraska, Wyoming, Oklahoma, Minnesota, Iowa, Colorado, and New Mexico—contain 90 percent of the U.S. wind energy potential. Refinements in wind turbine technology may enable a substantial portion of the nation's electricity to be produced by wind energy.

## Problems in the U.S. Wind Industry

Wind energy capacity grew very slowly until the late 1990s, adding less than 300 megawatts of capacity between 1990 and 1998. Crude oil prices fell during this time, making oil and gas the lowest-cost fuel sources. Concern about reducing the federal budget resulted in a change in federal policy toward renewable energies. Furthermore, some people were concerned about the uncertainty involving electric utility deregulation. U.S. producers received around $.03 to $.04 per kilowatt-hour in price guarantees while wind producers in Germany, Denmark, and India were guaranteed more than twice that amount. As a result investors often found investing in wind energy in the United States too risky. However, the tide began to turn in the late 1990s. Between 1999 and 2003 the country's wind generating capacity expanded at an average rate of 28 percent annually.

## Development of Wind Energy throughout the World

During the decade following the 1973 oil crisis, more than 10,000 wind machines were installed worldwide, ranging in size from portable units to multi-megawatt turbines. In developing villages small wind turbines recharge batteries and provide essential services. In China small wind turbines allow people to watch their favorite television shows, a major reason for the increased demand for turbines in China.

Although wind power supplies less than 0.1 percent of the world's electricity, it is one of the fastest growing energy sources. The most ambitious wind energy program is planned for India, which expects to provide enough electrical power to serve 5 million customers. India is expected to be the most rapidly growing market for wind turbines and, if the planned program is successful, wind may supply more energy for India than the country's nuclear program.

Interest in wind energy has been driven in part by the declining cost of capturing wind energy—from more than

$.25 per kilowatt-hour in 1980 to $.05 per kilowatt-hour for new turbines in the late 1990s. This makes wind power nearly competitive with gas- and coal-powered plants, even before considering wind's environmental advantages.

Encouraged by improved technology, falling costs, and government incentives like tax credits and guaranteed prices, wind power is booming across Europe. One reason for the growth in the industry is that the European Union wants to diversify its energy sources while clamping down on pollution. Almost no country supports the expansion of nuclear power and, in many areas, wind power is becoming economically viable.

Wind energy is produced under entirely different circumstances in the United States than in Europe. American entrepreneurs, seeing wind energy as a potentially profitable business, built large wind farms with huge numbers of turbines. When oil prices fell and tax credits were cut, growth stalled. In Denmark, Germany, Sweden, and the Netherlands, wind energy began as a grassroots movement, with small groups of politically motivated investors installing one or a few machines at a time. In Spain, Britain, and Greece, the clusters were larger because money was provided by local governments and utilities. The latest trend in Europe is to build wind farms offshore where there is more wind and fewer complaints that they clutter the landscape.

## Advantages and Disadvantages of Using Wind Energy

The main problem with wind energy is that the wind does not always blow. Some people object to the whirring noise of wind turbines or do not like to see wind turbines clustered in mountain passes and along shorelines because they interfere with scenic views. Some environmentalists have charged that the wind turbines are responsible for the loss of some species of endangered birds that fly into the blades. Finally, as with all types of renewable energy, wind power is more expensive to produce than energy generated through conventional means, at least when the price of oil remains low.

On the other hand, generating electricity with wind offers many environmental advantages. Wind farms do not emit climate-altering carbon dioxide, acid rain–forming pollutants, respiratory irritants, or nuclear waste. Because wind farms do not require water to operate, they are especially well suited to semiarid and arid regions. Wind farming also offers the added benefit of reducing soil loss on land prone to wind erosion because turbines capture the wind and decrease its potential for downwind destruction.

## SOLAR ENERGY

Ancient Greek and Chinese civilizations used glass and mirrors to direct the sun's rays to start fires. Solar energy (energy from the sun) is a renewable, widely avail-

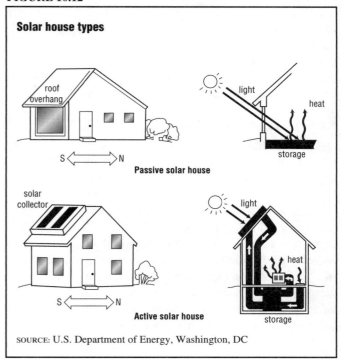

**FIGURE 10.12**

**Solar house types**

**Passive solar house**

roof overhang

S ← → N

light

heat

storage

**Active solar house**

solar collector

S ← → N

light

heat

storage

SOURCE: U.S. Department of Energy, Washington, DC

able energy source that generates neither pollution nor hazardous waste. Solar-powered cars have already competed in long-distance races, and solar energy has been used routinely for many years to power spacecraft. Although many people consider solar energy a product of the space age, the Massachusetts Institute of Technology built the first solar-powered house in 1939.

Solar radiation is nearly constant outside Earth's atmosphere, but the amount of solar energy, or insolation, reaching any point on Earth varies with changing atmospheric conditions such as clouds and dust, and the changing position of Earth relative to the sun. In the United States insolation is greatest in the West and Southwest. Nevertheless, almost all U.S. regions could use solar energy.

## Passive and Active Solar Energy Collection Systems

Passive solar systems such as greenhouses, or windows with a southern exposure, use heat flow, evaporation, or other natural processes to collect and transfer heat. They are considered the least costly and least difficult systems to implement. (See Figure 10.12.)

Active solar systems use mechanical methods to control the energy process. They require collectors and storage devices as well as motors, pumps, and valves to operate the systems that transfer heat. Collectors consist of an absorbing plate (solar panel or collector) that transfers the sun's heat to a working fluid (liquid or gas), a translucent cover plate that prevents the heat from radiating back into the atmosphere, and insulation on the back of the collector panel to further reduce heat loss. (See Figure 10.12.) Excess solar energy is transferred to a storage facility so it may be used to provide power on cloudy days. In both active and passive systems the conversion of solar energy into a form of power is made at the site where it is used. The most common and least expensive active solar systems are used for heating water.

## Solar Thermal Energy Systems

A solar thermal energy system uses intensified sunlight to heat water or other fluids to more than 750 degrees Fahrenheit. Mirrors or lenses constantly track the sun's position and focus its rays onto solar receivers that contain fluid. Solar heat (energy) is transferred to the water that, in turn, powers a steam-driven electric generator. In a distributed solar thermal system, the collected energy powers irrigation pumps, provides electricity for small communities, or captures normally wasted heat from the sun in industrial areas. In a central solar thermal system, the energy is collected at a central location and used by utility networks for a large number of customers.

Other solar thermal energy systems include solar ponds and trough systems. Solar ponds are lined ponds filled with water and salt. Because salt water is denser than fresh water, the salt water migrates to the bottom and absorbs the heat, while the fresh water on top keeps the salt water contained and traps the heat. Trough systems use U-shaped mirrors to concentrate the sunshine on water or oil-filled tubes.

## Photovoltaic Conversion Systems

The photovoltaic (PV) cell solar energy system converts sunlight directly into electricity without the use of mechanical generators. PV cells have no moving parts, are easy to install, require little maintenance, do not pollute the air, and usually last up to 20 years. PV cells are commonly used to power small devices such as watches or calculators. They are also used on a larger scale to provide electricity for rural households, recreational vehicles, and businesses. Solar panels using PV cells have generated electricity for space stations and satellites for many years. Solar panels have also provided electricity for a few major buildings in the United States.

Since PV systems produce electricity only when the sun is shining, a backup energy supply is needed. PV cells produce the most power around noon when sunlight is the most intense. A PV system typically includes storage batteries that provide electricity during cloudy days and at night.

The use of PV technology is expanding both in the United States and abroad. Although PV systems have a higher initial cost than conventional power plants, they have a much lower operating cost.

## Using Solar Energy

Because it is difficult to measure solar energy directly, other measures are often used as an indicator. According to the DOE there were 84 low-temperature collector manufacturers in 1979. That number dropped to ten by 2001. (See Figure 10.13.) Total shipments of solar thermal collectors peaked in 1981 at more than 21 million square feet and declined sharply after that. According to the DOE's *Annual Energy Review 2002* (Washington, DC, 2003), since 1991 there has been a steady increase in shipments, including a sharp spike in 2001 up to 11 million square feet.

By 1999 the market for solar thermal collectors for space heating had virtually disappeared. However, solar energy became increasingly popular as a means for heating swimming pools and water used in the home. Most of the solar thermal collector market is driven by this residential demand (primarily in the Sunbelt states), with only a small proportion for commercial purposes. Some state and municipal power companies have added solar systems as adjuncts to their regular power sources during peak hours.

## Advantages and Disadvantages of Solar Energy

The primary advantage of solar energy is its inexhaustible supply, while its primary disadvantage is its reliance on a consistently sunny climate to provide continuous electrical power. Very few areas of the United States have enough constant sunshine to make this system an efficient replacement for conventional methods. In addition, a large amount of land area is necessary for the most efficient collection of solar energy for solar thermal units.

PV cell solar energy systems are probably the most attractive form of solar energy production. A PV cell system is nonpolluting, silent, and can be operated by computer. In addition, it is less expensive to operate because there are no turbines or other moving parts, which makes maintenance minimal. Above all the fuel source (sunshine) is free and plentiful. The disadvantage of a PV cell energy system is the initial cost. As shown in Figure 10.14 total shipments by PV cells increased dramatically over the last decade, reaching 98 thousand peak kilowatts during 2001.

## Future Development Trends

Interest in PV cell solar energy systems is particularly high in rural and remote areas where it is impractical to extend traditional electrical power lines. In some remote areas PV cells are used as independent power sources for communications or for the operation of water pumps or refrigerators. This use will most likely increase where the traditional use of an electrical cord is a problem.

Although solar power still costs more than fossil fuel energy, utilities could turn to solar energy to provide extra power on extremely hot or cold days. Some people believe

FIGURE 10.13

## Number of manufacturers of solar thermal collectors, 1974–1984 and 1986–2001

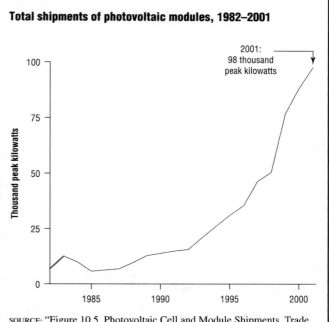

[1]Collectors that generally operate in the temperature range of 140 degrees Fahrenheit to 180 degrees Fahrenheit but can also operate at temperatures as low as 110 degrees Fahrenheit.
[2]Collectors that generally operate at temperatures below 110 degrees Fahrenheit. Note: Data were not collected for 1985. Medium-temperature collectors include special collectors.

SOURCE: "Figure 10.3. Solar Thermal Collector Shipments by Type, Price and Trade: Number of U.S. Manufacturers, 1974–1984 and 1986–2001," in *Annual Energy Review 2002*, U.S. Department of Energy, Energy Information Administration, Washington, DC, October 2003

FIGURE 10.14

## Total shipments of photovoltaic modules, 1982–2001

2001: 98 thousand peak kilowatts

SOURCE: "Figure 10.5. Photovoltaic Cell and Module Shipments, Trade, and Prices: Total Shipments, 1982–2001," in *Annual Energy Review 2002*, U.S. Department of Energy, Energy Information Administration, Washington, DC, October 2003

that building solar energy systems to provide peak power capacity would be cheaper in the long run than building new and expensive diesel fuel generators, utility regulators may decide that the price of fossil fuel power must include the hidden cost of fossil fuel damage to the environment caused by acid rain and the greenhouse effect.

## HYDROGEN—A FUEL OF THE FUTURE?

Hydrogen is the lightest and most abundant chemical element. It contains only one proton and one electron. From an environmental point of view it is also the ideal fuel. Its combustion as a fuel or conversion to electricity produces only water vapor—it is entirely carbon-free. The National Aeronautics and Space Administration (NASA) has used liquid hydrogen for decades to power rockets into space.

Three-quarters of the mass of the universe consists of hydrogen; however, it is combined with other elements, as in water. The elemental, combustible, form of hydrogen is not found in the atmosphere. Hydrogen-containing compounds cannot be converted into pure gas hydrogen without the expenditure of energy. Most elemental hydrogen is currently obtained by separating it from hydrocarbons, such as natural gas, gasoline, or methanol. This process is called "reforming." However, reforming requires energy in the form of heat. Therefore, with today's technology little or nothing can be gained from an overall energy point of view.

Hydrogen has considerable potential as a clean fuel and, because it is a gas, can be distributed with essentially the same technology as natural gas. Scientists are researching ways to economically produce hydrogen gas. The possibility of a transition to hydrogen has been considered for more than a century, and many see hydrogen as the logical "third-wave" fuel—hydrogen gas following oil, just as oil replaced coal decades earlier.

The most promising application of hydrogen is in fuel cells. A fuel cell works like a battery, except it never runs down as long as hydrogen is supplied. It contains two electrodes—a negative electrode or anode and a positive electrode or cathode—surrounding an electrolyte. An electrolyte is a substance that can conduct electricity. Hydrogen is fed to the anode, and the atoms are separated into protons and electrons that travel along different paths to the cathode. The electrons travel along an external circuit creating an electricity flow. The protons travel through the electrolyte to the cathode, where oxygen is fed in. The protons mix with the oxygen atoms and electrons, forming water and generating heat.

Research in the United States is focused on the use of traditional hydrocarbon fuels (like gasoline) as the hydrogen source for vehicles containing a fuel "reformer" and a fuel cell. This takes advantage of the existing infrastructure of gasoline stations.

In March 2004 the DOE introduced its "Hydrogen Posture Plan," which outlines a plan to create a hydrogen-based transportation energy system in the United States. President Bush requested that $227 million be spent on research in 2005 to support this hydrogen fuel initiative.

FIGURE 10.15

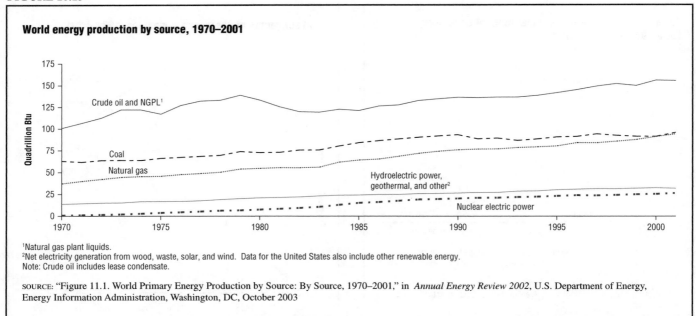

**World energy production by source, 1970–2001**

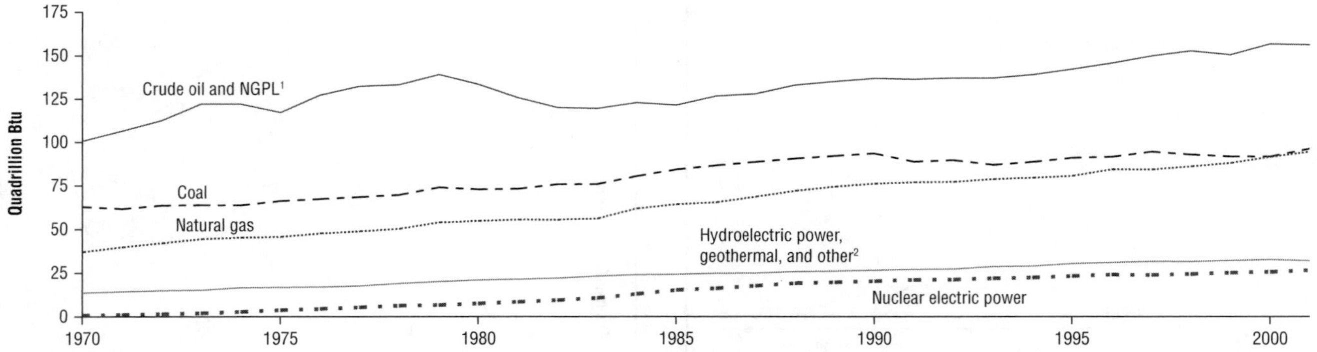

Quadrillion Btu
175
150
125
100
75
50
25
0
1970
1975
1980
1985
1990
1995
2000
Crude oil and NGPL[1]
Coal
Natural gas
Hydroelectric power, geothermal, and other[2]
Nuclear electric power


[1]Natural gas plant liquids.
[2]Net electricity generation from wood, waste, solar, and wind. Data for the United States also include other renewable energy.
Note: Crude oil includes lease condensate.

SOURCE: "Figure 11.1. World Primary Energy Production by Source: By Source, 1970–2001," in *Annual Energy Review 2002*, U.S. Department of Energy, Energy Information Administration, Washington, DC, October 2003

## THE FUTURE OF RENEWABLE ENERGY

Renewable energy accounted for 8 percent of the total world energy production in 2001. (See Figure 10.15.) Global energy consumption is expected to increase by 60 percent through the next two decades, reaching 600 quadrillion Btu by 2020. Most of the growth will be in developing countries, particularly in Asia.

In 2003 renewable energy contributed only 6 percent of the total energy consumed by the United States. The nation's fleet of automobiles has become less energy efficient, and consumers are less interested in energy-saving technology. Many economists believe that only a return of higher energy prices will cause Americans to once again reduce their energy use and consider renewable energy sources. Despite their environmental advantage over fossil fuels, renewable energies have never attracted enough financial support from the government, the public, or energy companies to the point that they can be cost competitive with fossil fuel power.

# IMPORTANT NAMES AND ADDRESSES

**American Geophysical Union (AGU)**
2000 Florida Ave. NW
Washington, DC 20009-1277
(202) 462-6900
Toll-free: 1-800-966-2481
FAX: (202) 328-0566
E-mail: service@agu.org
URL: http://www.agu.org/

**American Lung Association**
61 Broadway, 6th Floor
New York, NY 10006
(212) 315-8700
Toll-free: 1-800-LUNGUSA
(1-800-586-4872)
E-mail: info@lungusa.org
URL: http://www.lungusa.org

**U.S. Bureau of Reclamation**
1849 C St. NW
Washington, DC 20240-0001
(202) 513-0501
FAX: (202) 513-0315
URL: http://www.usbr.gov/

**Centers for Disease Control
and Prevention**
1600 Clifton Rd.
Atlanta, GA 30333
(404) 639-3311
Toll-free: 1-800-311-3435
URL: http://www.cdc.gov

**Clean Water Fund**
4455 Connecticut Ave. NW, Suite A300-16
Washington, DC 20008-2328
(202) 895-0432
FAX: (202) 895-0438
E-mail: cwf@cleanwater.org
URL: http://www.cleanwaterfund.org

**U.S. Climate Change Science Program**
1717 Pennsylvania Ave. NW, Suite 250
Washington, DC 20006
(202) 223-6262
FAX: (202) 223-3065
E-mail: information@climatescience.gov
URL: http://www.climatescience.gov

**Council on Environmental Quality**
722 Jackson Pl. NW
Washington, DC 20503
(202) 395-5750
FAX: (202) 456-6546
URL: http://www.whitehouse.gov/ceq

**U.S. Department of Energy (DOE)**
1000 Independence Ave. SW
Washington, DC 20585
(202) 586-5575
Toll-free: 1-800-342-5363
FAX: (202) 586-4403
E-mail: the.secretary@hq.doe.gov
URL: http://www.energy.gov

**Environmental Business
International Inc.**
4452 Park Blvd., Suite 306
San Diego, CA 92116
(619) 295-7685
FAX: (619) 295-5743
E-mail: ebi@ebiusa.com
URL: http://www.ebiusa.com

**Environmental Defense Fund**
257 Park Ave. South
New York, NY 10010
(212) 505-2100
FAX: (212) 505-2375
E-mail: members@environmental
defense.org
URL: http://www.edf.org

**Environmental Industry
Associations (EIA)**
4301 Connecticut Ave., #300
Washington, DC 20008
(202) 244-4700
Toll-free: 1-800-424-2869
FAX: (202) 966-4824
E-mail: jleca@envasns.org
URL: http://www.envasns.org

**Environmental Protection Agency (EPA)**
Ariel Rios Building
1200 Pennsylvania Ave. NW

Washington, DC 20460
(202) 272-0167
URL: http://www.epa.gov/

**U.S. Fish and Wildlife Service**
1849 C St. NW, Room 3359
Washington, DC 20240-0001
(202) 208-7407
Toll-free: 1-800-344-9453
E-mail: contact@fws.gov
URL: http://www.fws.gov

**USDA Forest Service**
1400 Independence Ave. SW
Washington, DC 20250-0003
(202) 205-8333
E-mail: webmaster@fs.fed.us
URL: http://www.fs.fed.us/

**Freshwater Society**
2500 Shadywood Rd.
Excelsior, MN 55331
(952) 471-9773
FAX: (952) 471-7685
E-mail: freshwater@freshwater.org
URL: http://www.freshwater.org

**Friends of the Earth**
1717 Massachusetts Ave. NW, Suite 600
Washington, DC 20036-2002
Toll-free: 1-877-843-8687
FAX: (202) 783-0444
E-mail: foe@foe.org
URL: http://www.foe.org

**General Accounting Office**
441 G St. NW
Washington, DC 20548
(202) 512-4800
E-mail: webmaster@gao.gov
URL: http://www.gao.gov/

**U.S. Geological Survey**
12201 Sunrise Valley Dr.
Reston, VA 20192
(703) 648-4000
Toll-free: 1-888-275-8747

FAX: (703) 648-4888
URL: http://www.usgs.gov

**Greenpeace**
702 H St. NW
Washington, DC 20001
(202) 462-1177
Toll-free: 1-800-326-0959
E-mail: greenpeace.usa@wdc.
greenpeace.org
URL: http://www.greenpeaceusa.org

**Idaho National Engineering and
Environmental Laboratory**
2525 North Fremont Ave.
P.O. Box 1625
Idaho Falls, ID 83415
Toll-free: 1-800-708-2680
E-mail: info@inel.gov
URL: http://www.inel.gov

**Izaak Walton League of America**
707 Conservation Ln.
Gaithersburg, MD 20878
(301) 548-0150
Toll-free: 1-800-453-5463
FAX: (301) 548-0146
E-mail: general@iwla.org
URL: http://www.iwla.org

**National Acid Precipitation
Assessment Program**
NOAA, Mail Code R/SAB
1315 East-West Highway
Silver Spring, MD 20910
(301) 713-0460, ext. 202
FAX: (301) 713-3515
E-mail: napap@noaa.gov
URL: http://www.oar.noaa.gov

**National Aeronautics and Space
Administration (NASA)**
Goddard Space Flight Center
Code 130, Office of Public Affairs
Greenbelt, MD 20771
(301) 286-8955
FAX: (301) 286-1707
E-mail: gsfcpao@pop100.gsfc.nasa.gov
URL: http://www.gsfc.nasa.gov/

**National Audubon Society**
700 Broadway
New York, NY 10003
(212) 979-3000
FAX: (212) 979-3188
E-mail: jbianchi@audubon.org
URL: http://www.audubon.org

**National Coalition against the Misuse
of Pesticides**
701 E St. SE, Suite 200
Washington, DC 20003
(202) 543-5450
FAX: (202) 543-4791
E-mail: info@beyondpesticides.org
URL: http://www.beyondpesticides.org

**National Mining Association (NMA)**
101 Constitution Ave. NW, Suite 500 East

Washington, DC 20001-2133
(202) 463-2600
FAX: (202) 463-2666
E-mail: craulston@nma.org
URL: http://www.nma.org

**National Oceanic and Atmospheric
Administration (NOAA)**
14th St. and Constitution Ave. NW,
Room 6217
Washington, DC 20230
(202) 482-6090
FAX: (202) 482-3154
E-mail: answers@noaa.gov
URL: http://www.noaa.gov

**National Parks Conservation Association**
1300 19th St. NW, Suite 300
Washington, DC 20036
Toll-free: 1-800-628-7275
FAX: (202) 659-0650
E-mail: npca@npca.org
URL: http://www.npca.org

**National Safety Council**
1121 Spring Lake Dr.
Itasca, IL 60143-3201
(630) 285-1121
FAX: (630) 285-1315
E-mail: info@nsc.org
URL: http://www.nsc.org

**National Weather Service Climate
Prediction Center**
5200 Auth Rd.
Camp Springs, MD 20746
(301) 763-8000 ext. 7163
E-mail: Carmeyia.Gillis@noaa.gov
URL: http://www.cpc.ncep.noaa.gov

**National Wildlife Federation**
11100 Wildlife Center Dr.
Reston, VA 20190-5362
Toll-free: 1-800-822-9919
URL: http://www.nwf.org

**Natural Resources Defense Council
(NRDC)**
40 West 20th St.
New York, NY 10011
(212) 727-2700
FAX: (212) 727-1773
E-mail: nrdcinfo@nrdc.org
URL: http://www.nrdc.org

**The Nature Conservancy**
4245 North Fairfax Dr., Suite 100
Arlington, VA 22203-1606
(703) 841-5300
Toll-free: 1-800-628-6860
E-mail: comment@tnc.org
URL: http://nature.org

**U.S. Nuclear Regulatory Commission**
Office of Public Affairs (OPA)
Washington, DC 20555
(301) 415-8200
Toll-free: 1-800-368-5642

E-mail: opa@nrc.gov
URL: http://www.nrc.gov

**Pesticide Action Network, North America**
49 Powell St., Suite 500
San Francisco, CA 94102
(415) 981-1771
FAX: (415) 981-1991
E-mail: panna@panna.org
URL: http://www.panna.org

**Public Citizen**
1600 20th St. NW
Washington, DC 20009
(202) 588-1000
E-mail: member@citizen.org
URL: http://www.citizen.org

**Rachel Carson Council, Inc.**
P.O. Box 10779
Silver Spring, MD 20914
(301) 593-7507
FAX: (301) 593-6251
E-mail: rccouncil@aol.com
URL: http://members.aol.com/rccouncil/
ourpage

**Sierra Club**
National Headquarters
85 Second St., 2nd Floor
San Francisco, CA 94105-3441
(415) 977-5500
FAX: (415) 977-5799
E-mail: information@sierraclub.org
URL: http://www.sierraclub.org

**Union of Concerned Scientists**
2 Brattle Sq.
Cambridge, MA 02238-9105
(617) 547-5552
FAX: (617) 864-9405
E-mail: ucs@ucsusa.org
URL: http://www.ucsusa.org

**The Wilderness Society**
1615 M St. NW
Washington, DC 20036
Toll-free: 1-800-843-9453
E-mail: member@tws.org
URL: http://www.wilderness.org

**World Wildlife Fund**
1250 24th St. NW
Washington, DC 20037
(202) 293-4800
E-mail: PIResponse@wwfus.org
URL: http://www.worldwildlife.org

**Worldwatch Institute**
1776 Massachusetts Ave. NW
Washington, DC 20036-1904
(202) 452-1999
FAX: (202) 296-7365
E-mail: worldwatch@worldwatch.org
URL: http://www.worldwatch.org

# RESOURCES

The Environmental Protection Agency (EPA) monitors the status of the nation's environment and publishes a wide variety of materials on environmental issues. Publications consulted for this book include *EPA's Draft Report on the Environment 2003* (2003), *2000 National Water Quality Inventory* (2002), *Water on Tap: What You Need to Know,* (2003), *Factoids: Drinking Water and Ground Water Statistics for 2003* (2004), *Protecting and Restoring America's Watersheds* (2001), *Maps of Lands Vulnerable to Sea Level Rise* (2001), *EPA Acid Rain Program 2002 Progress Report* (2003), *Hypoxia and Wetland Restoration* (2002), *What Are Wetlands?* (2003), and *The Plain English Guide to the Clean Air Act* (1993).

Also useful from the EPA were *National Biennial RCRA Hazardous Waste Report: Based on 2001 Data* (2003), *Managing Your Hazardous Waste: A Guide for Small Businesses* (2001), *Guide for Industrial Waste Management* (1999), *Latest Findings on National Air Quality: 2002 Status and Trends* (2003), *Municipal Solid Waste in the United States: 2001 Facts and Figures* (2003), *National Source Reduction Characterization Report* (1999), *Report on the Supply and Demand of CFC-12 in the United States 1999* (1999), *Light-Duty Automotive Technology and Fuel Economy Trends 1975 through 2004* (2003), *Drinking Water: Past, Present, and Future* (2000), and *Inventory of U.S. Greenhouse Gas Emissions and Sinks: 1990–2001* (2003). The EPA's Endocrine Disrupter Screening and Testing Advisory Committee (EDSTAC) published *EDSTAC Final Report* (1998) on endocrine disrupters. The EPA also provided *Update: National Listing of Fish and Wildlife Advisories* (2003) and the *Toxic Release Inventory Public Data Release* (2003).

The Energy Information Administration (EIA) of the U.S. Department of Energy's (DOE) *Emissions of Greenhouse Gases in the United States 2002* (2003) was a source of data on global warming. The EIA also published *International Energy Outlook 2003* (2003), *International Energy Annual 2001* (2003), *Future U.S. Highway Energy Use: A Fifty Year Perspective* (2001), *Annual Energy Review 2002* (2003), *Monthly Energy Review* (March 2004), *Alternatives to Traditional Transportation Fuels 1998* (1999), and *The Waste Isolation Plant* (1999). *Yucca Mountain Studies* (1990) and bulletins of the Office of Civilian Radioactive Waste Management of the U.S. Department of Energy were also helpful. Oak Ridge National Laboratory produced *Transportation Energy Data Book: Edition 23* (2003). The Idaho National Engineering and Environmental Laboratory of the DOE published *National Spent Nuclear Fuel Program* (2004).

Data from the Centers for Disease Control and Prevention's (CDC) *Morbidity and Mortality Weekly Report, Surveillance Summaries, Healthy People 2000,* and *Health, United States, 2003* were invaluable. The National Institute of Environmental Health Sciences (NIEHS) of the National Institutes of Health prepared the *NIEHS Report on Health Effects from Exposure to Power-Line Frequency Electric and Magnetic Fields* (1999) and *Questions and Answers about Electric and Magnetic Fields Associated with the Use of Electric Power* (1995).

The National Aeronautics and Space Administration (NASA—Goddard Space Flight Center) publishes a variety of materials on environmental and space issues. Useful in this book were *NASA Facts* (2004), *Understanding Our Changing Climate* (1997), and *Looking at the Earth from Space* (1994). The U.S. Department of Transportation produced *Critter Crossings* (2000), a study of the conflicts between humans and wildlife along U.S. roads and highways. The U.S. Government Accounting Office has published many useful reports on environmental issues, including energy, air and water quality, mobile phone health issues, wetlands, nuclear waste, landfills, and pollution. The U.S. Climate Change Science Program and Subcommittee on Global Change Research published *The U.S. Climate Change Science Program Vision for the*

*Program and Highlights of the Scientific Strategic Plan* (2003). The *Climate Action Report* (2002), prepared by the UN Framework Convention on Climate Change, was useful in explaining the scientific community's determinations about global warming.

Numerous other U.S. government publications were used in the preparation of this book. They included *Policy Implications of Greenhouse Warming* (1990), *National Report on Sustainable Forests—2003,* and *America's Forests: 2003 Health Update* (2003) from the U.S. Forest Service, an agency under the U.S. Department of Agriculture (USDA). The USDA also provided *Statistical Highlights of United States Agriculture, 2002/2003* (2003). The U.S. Geological Survey (USGS), Denver, CO, documents the use of the nation's waters every five years. *Estimated Use of Water in the United States in 2000* (2004) was used in the preparation of this book. The USGS also produced *Sustainability of Ground Water Resources* (1999), *Fertilizers—Sustaining Global Food Supplies* (1999), *Materials Flow and Sustainability* (1998), and *Water of the World* (undated).

The National Conference of State Legislatures publishes reports on a wide variety of environmental topics. Especially useful were the *Legislators' Guide to Alternative Fuel Policies and Programs* (1997), *Legisbrief—Lead Hazard Disclosures in Real Estate Transactions* (1997), and *Two Decades of Clean Air: EPA Assesses Cost and Benefits* (1998). The American Wind Energy Association produced *Wind Power Outlook 2004* (2004).

*Environmental Market Outlook to 2010, Briefing for EPA-NACEPT* (2002), prepared by Environmental Business International, San Diego, CA, an environmental research and consulting group, was the source of data on the status of the environmental industry.

Material from the Gallup Organization's public opinion surveys was extremely useful, as was information from the Endangered Species Coalition on plant species used in medications. Also helpful was *Garbage Then and Now* (undated) by the National Solid Waste Management Association. The American Lung Association published *State of the Air: 2004* (2004). The National Safety Council produced *Reporting on Climate Change: Understanding the Science* (2000), *A Guide to the U.S. Department of Energy's Low-Level Radioactive Waste* (2002), and *A Reporter's Guide to Yucca Mountain* (2001).

The J. G. Press provided its important biennial study "The State of Garbage in America" from *Biocycle Magazine.* Resources for the Future provided information on the Superfund in *Superfund's Future: What Will It Cost?* (2001). The Environmental Defense Fund, New York, NY, discussed the leading environmental concerns of Americans and also published *Evaluation of Erosion Hazards,* prepared by the H. John Heinz III Center for Science, Economics, and the Environment for the Federal Emergency Management Agency (April 2000) on the erosion of U.S. coastlines.

# INDEX